FROM LUMINOUS HOT STARS TO STARBURST GALAXIES

Luminous hot stars represent the extreme upper mass end of normal stellar evolution. Before exploding as supernovae, they live out their lives of only a few million years with prodigious outputs of radiation and stellar winds which dramatically affect both their evolution and environments.

A detailed introduction to the topic, this book connects the astrophysics of massive stars with the extremes of galaxy evolution represented by starburst phenomena. A thorough discussion of the physical and wind parameters of massive stars is presented, together with considerations of their birth, evolution, and death. Hll galaxies, their connection to starburst galaxies, and the contribution of starburst phenomena to galaxy evolution through superwinds, are explored. The book concludes with the wider cosmological implications, including Population III stars, Lyman break galaxies, and gamma-ray bursts, for each of which massive stars are believed to play a crucial role.

This book is ideal for graduate students and researchers in astrophysics who are interested in massive stars and their role in the evolution of galaxies.

PETER S. CONTI is an Emeritus Professor at the Joint Institute for Laboratory Astrophysics (JILA) and the Astrophysics and Planetary Sciences Department at the University of Colorado.

PAUL A. CROWTHER is a Professor of Astrophysics in the Department of Physics and Astronomy at the University of Sheffield.

CLAUS LEITHERER is an Astronomer with the Space Telescope Science Institute, Baltimore.

Cambridge Astrophysics Series

Series editors:

Andrew King, Douglas Lin, Stephen Maran, Jim Pringle and Martin Ward

Titles available in the series

FROM LUMINOUS HOT STARS TO STARBURST GALAXIES

PETER S. CONTI
University of Colorado, Boulder

PAUL A. CROWTHER
University of Sheffield

CLAUS LEITHERER
Space Telescope Science Institute, Baltimore

CAMBRIDGE
UNIVERSITY PRESS

CAMBRIDGE UNIVERSITY PRESS
Cambridge, New York, Melbourne, Madrid, Cape Town, Singapore, São Paulo, Delhi

Cambridge University Press
The Edinburgh Building, Cambridge CB2 8RU, UK

Published in the United States of America by Cambridge University Press, New York

www.cambridge.org
Information on this title: www.cambridge.org/9780521791342

First published 2008   **1005511750**

Printed in the United Kingdom at the University Press, Cambridge

A catalog record for this publication is available from the British Library

Library of Congress Cataloging in Publication data
Conti, P. S. (Peter S.)
From luminous hot stars to starburst galaxies / Peter S. Conti, Paul A. Crowther, Claus Leitherer.
   p.   cm. – (Cambridge astrophysics series ; 45)
Includes bibliographical references and index.
ISBN 978-0-521-79134-2
1. Starbursts.   2. Gamma ray bursts.   3. Stars – Evolution.
4. Stellar winds.   I. Crowther, Paul A.   II. Leitherer, Claus.   III. Title.
QB806.5.C66   2008
523.1′125–dc22

                                                        2008031554

ISBN 978-0-521-79134-2 hardback

Dedicated to our supportive wives: Carolyn, Suzie, and Irina.

Contents

Preface

This monograph had its origins about a decade ago when it became apparent that the field of luminous hot stars was rapidly expanding in extent and depth and connections to extragalactic astrophysics, in particular starburst galaxies, were first recognized. At that time a decade had passed since the 1988 *O stars and Wolf–Rayet stars* NASA monograph of Conti and Underhill. Since then, there have been far reaching advances in the astrophysics of these stars and of star-forming galaxies, both locally and at high-redshift, together with the way each affect their surroundings. On the observational side, progress interpreting the spectra of luminous hot stars allows their physical parameters to be derived with unprecedented accuracy. Studies of their ubiquitous stellar winds plus their dependence upon the element abundances has provided the impetus for revised stellar evolutionary model calculations. For the first time, additional physics, such as rotational mixing and magnetic fields, is being considered.

The advent of the Hubble Space Telescope in 1990 provided us with a UV wavelength range of greatly increased sensitivity just where most of the energy of hot stars is emitted. The use of newly commissioned 8-m ground-based telescopes has opened the door to studies of more distant hot stars in external galaxies, with different initial abundances and star formation histories compared to the Milky Way. In addition, techniques for the identification of star-forming galaxies at an epoch when the Universe was as little as one tenth of its present age have been developed, and exploited with these instruments. More recently, the Spitzer Space Telescope has provided a window on the mid-IR wavelengths of these same stars, so they may be studied in their birthplaces. Hot luminous stars form ionized hydrogen regions (HII) which can readily be analyzed in the Milky Way and other galaxies and provide an accurate, albeit indirect, method of determining the properties of their exciting stars.

Astronomers speak of starburst galaxies as those with substantial numbers of hot luminous stars, as evidenced by their extensive and very luminous (giant) HII regions. Typically such star formation activity comes about due to interactions between gas-rich galaxies, or gas collisions within the cores of dwarf irregular galaxies, where the material from each component is mixed and shocked, rapidly forming new stars. Such a mode of massive star formation stretches throughout the history of the Universe, being considerably more important in the earliest times when the Universe was smaller; possibly this mode of star birth dominates the formation of the first stars. The detection of this very first stellar generation is expected to be feasible with the James Webb Space Telescope in the next decade.

It is appropriate to now take stock and to review and summarize all the advances in these intimately related fields. In the past, massive stars were perhaps considered to be a peripheral topic in contemporary astrophysics, due to their rarity. However, developments in a number of fields now make them central to, for example, studies of the first (Population III) stars, which were biased towards high mass stars, and gamma-ray bursts, some of which originated from massive star progenitors. This is, we believe, the first monograph that connects the properties and evolution of massive stars with starburst galaxies. In another decade, these two subdisciplines will each become much too large to include in a single volume.

Acknowledgements

We are indebted to those who read part of the first draft of the manuscript: Bob Blum, You-Hua Chu, Ed Churchwell, Chris Fryer, Jay Gallagher, Henny Lamers, Andre Maeder, Phil Massey, Paolo Mazzali, Max Pettini, Jo Puls, Daniel Schaerer, and Stan Woosley. Their comments have substantially added to our presentation of the material. Thanks too to James Furness for proof reading chapters, and CUP for letting us piece the manuscript together at our own pace. PSC appreciates nearly continuous support of his research by the National Science Foundation for over three decades. PAC would like to thank the Royal Society for their support through their University Research Fellowship scheme.

1

Introduction

1.1 Motivation

The aim of this book is to describe the connection between the physics and evolution of relatively nearby luminous hot stars and more distant starburst phenomena occurring all the way to cosmological distances. There have been recent significant advances in our knowledge concerning hot stars and their contribution to highly energetic star formation episodes. The Hubble Space Telescope (HST) and Spitzer in particular have provided many new insights into these areas. Recent observations of the near infrared regions have also greatly aided our view of star formation processes. This book is aimed at those extragalactic astronomers who would be interested in hot star astrophysics and those stellar astrophysicists concerned with galaxy evolution. It would be of use in graduate astrophysics courses at a level suitable for advanced students. This monograph will be one of the first of its kind in spanning the connection from the astrophysics of hot stars to that of newly forming galaxies.

1.2 Observed properties

The Hertzsprung–Russell diagram

What are stars? They are fully gaseous, ionized, gravitationally bound entities, emitting large amounts of radiation over many wavelengths. In normal stars, the gas is in hydrostatic pressure equilibrium under the ideal gas law equation of state. It is held in balance against the inward forces of gravity by the outward pressure of radiation generated from nuclear reactions in the stellar interior (or from gravitational contraction adjustments). By using the phrase "luminous hot stars" we mean to describe those massive ones that inhabit the upper left-hand portion of the Hertzsprung–Russell (H-R) diagram. This fundamental empirical relationship, known for nearly one hundred years, plots stellar absolute magnitude along the ordinate. The brightness is plotted increasing *upward*, thus the value of the magnitude becomes smaller and even negative in those peculiar units so familiar to astronomers. The H-R diagram has as its horizontal axis the stellar color (or spectral type) such that the value along the abscissa increases going from hotter to cooler stars, bluer to redder objects, "earlier" to "later" spectral types, from left to right. Most stars are found in a relatively narrow band reaching from the lower right (faint and cool) to the upper left (bright and hot). This is called the main sequence (MS) and stars along it are often referred to as dwarfs, with an appropriate color prefix such as red, yellow, or blue. Stars above and to the right of the MS are called giants or supergiants, dependent on their distance from it, also with color prefixes.

Stars near to but below the MS are called sub dwarfs. Finally, on an H-R diagram of nearby stars a few are found well below the MS; these are termed white dwarfs (even though they are blue) and they represent the final state of evolution of low and intermediate mass stars.

In more precise physical terminology, the H-R diagram is a plot of stellar luminosity L_* (emitted energy) *upwards versus* effective ("surface") temperature T_{eff} increasing *right to left*. The T_{eff} is *defined* by the equation

$$L_* = 4\pi R_*^2 \sigma T_{eff}^4, \qquad (1.1)$$

where R_* is the radius of the star from which radiation is emitted and σ is the Stefan–Boltzmann constant. L_* and T_{eff} are parameters which readily lend themselves to estimation.[1] The important properties of luminous hot stars, that is, their luminosities, T_{eff}, masses, and chemical compositions will be considered in Chapter 3. Chapter 2 will discuss photometry and spectroscopy of early-type stars. Chapter 4 will describe the observations of stellar winds which are ubiquitous amongst massive stars, plus the physics behind them, which involves radiation pressure. According to a fundamental theorem of stellar structure, the luminosity and temperature of a star are unique properties of its mass and composition. Stellar rotation and magnetic fields introduce perturbations to this relationship, but these factors are usually small compared to the effects of mass, M_*,[2] and composition[3] on the structure.

MS stars utilize nuclear reactions that fuse ("burn") hydrogen to helium in their central cores, with exothermic energy release. The more massive the star, the more luminous it will be on the MS, with $L_* \approx M_*^{4.5}$ for solar-type stars, decreasing to $L_* \approx M_*^{2.5}$ for high mass stars. Most stars near the Sun have a similar composition (or abundance, referring to the elements *other than* hydrogen and helium); thus the MS is relatively narrow. Stars with an initial abundance below solar values will lie on a different MS (the sub dwarfs) and evolve in a quantitatively different manner.

Stellar spectroscopy

The gaseous matter that makes up a star has a wide run in properties from the innermost core to the outer regions. Most stars are spherical in shape to a good approximation so there is a simple radial dependence of variables such as the gas density, temperature, and pressure. These decrease outwards monotonically following the perfect gas law such that the star is in hydrostatic equilibrium at every point. While the stellar cores are typically at a few 10^7 K, the surface temperatures are a few 10^3 to 10^4 K. The gas in the interior of a star is highly ionized and quite opaque to radiation. The energy created in the core diffuses slowly (in the Sun $\approx 10^6$ years) outward to the surface. The character of the radiation keeps in thermal equilibrium with its surroundings such that the γ rays generated in the core become visible wavelength radiation at the stellar surface.

But what is the "surface" of a gaseous object? This can be thought of as that region of the star from which the emergent radiation escapes (roughly where the continuum opacity is near unity). This "photosphere" is overlain by a slightly lower density and temperature

---

[1] Typically in units of solar luminosity, $L_\odot$, where $L_\odot$ is $\approx 4 \times 10^{33}$ erg s^{-1}.
[2] Typically in units of solar masses, $M_\odot$, where $M_\odot$ is $\approx 2 \times 10^{33}$ g.
[3] Normal stars, including the Sun, are composed mostly of hydrogen ($X \approx 70\%$, by mass) and helium ($Y \approx 28\%$). There is a small admixture ($Z \approx 2\%$) of all other elements, often called "metals" by astronomers.

regime where the line opacity for various relatively abundant elements and ions has dropped to near unity. This cooler "atmosphere" is the source of the absorption line spectra of normal stars. In most stars the geometric depth of the photosphere and atmosphere has some overlap and is *small* with respect to the stellar radius (hence can be treated as "plane-parallel"). This is the reason that the Sun's limb appears sharp. Gaseous material exists above a stellar atmosphere but normally its density is so low that it is normally not easily detectable directly (but, e.g., total solar eclipses reveal this material in the Sun, commonly known as a corona).

The characteristics of a stellar spectrum are a function of the properties of the stellar photosphere and atmosphere, which are closely coupled to L and T_{eff} in normal stars. The specific spectral features that will be found depend upon the absolute numbers of ions present (related to the abundance of the element and the ionization state) and the line transition probabilities. As the most common element, lines from hydrogen are visible in most stars and they are prominent in hot stars. Helium lines are readily seen only in the hottest stars, given the relatively high energy needed to excite the lowest electronic level. Generally speaking, only high excitation lines of ions of other elements are observed in hot stars in the optical regions. Ground state lines of hydrogen and helium, along with highly ionized common elements, are found primarily at far UV wavelengths. These outer boundaries of stars provide nearly all the direct observations of their properties. Until the latest or final stages of evolution are reached, the surface composition of stars represents their initial values and *not* that of the nuclear reactions in the core which otherwise remain hidden from direct view.[4] (Neutrinos from nuclear reactions in the interior of the Sun have been directly detected.)

Classification systems have utilized the appearance and strength of absorption lines and estimation of line ratios to determine their character, or spectral type. The spectral types for normal stars are labeled with the letters O, B, A, F, G, K, and M, in order from the hottest ("earliest") to the coolest ("latest"). Numerical subtypes distinguish differences within the spectral letters and go from number 0 to 9 (but with a few 0.5 divisions). Astronomers also utilize line widths and other line ratios to determine the atmospheric pressures (related to gas densities, local gravity, hence luminosity). These labels utilize roman numerals I–V, with the latter corresponding to the faintest luminosity class. Line widths are generally narrowest in the most luminous stars while the Balmer (and HeI) lines have Stark-broadened wings in the fainter ones. While five luminosity classes are easily distinguished among cool K and M stars with their wide dispersion in M_V, only I, III, and V are adopted for O type stars which have a more limited M_V range, as discussed in Chapter 2.

Three stars in Cygnus with broad emission lines of highly excited elements were first identified by Wolf & Rayet (1867) using a stellar spectroscope. These stars are spectacular in appearance, having strong broad emission lines rather than narrow dark absorption features as nearly all other stars, and the Sun, show. What a sight for the eye: brilliant bright colorful lines superimposed upon a rainbow continuum. These stars later became known as Wolf–Rayet (W-R) stars. It later became apparent that their spectra came in two "flavors", those with strong lines of helium and nitrogen (WN) and those with strong helium, carbon, and oxygen (WC). Subsequently it was found that some WN stars show hydrogen lines but this element has never been found in WC types.

---

[4] Exceptions are found for massive stars – see Chapter 5.

Distribution of stars in our Galaxy

Our Milky Way Galaxy is an extended stellar system containing $\approx 2 \times 10^{11}$ stars. The overall spatial morphology has been compared to that of a "pair of back-to-back fried eggs". The slightly flattened central region, or "bulge", out to a radius of ≈ 1 kpc, contains mostly old stars. A highly extended "disk", with radius out to about 15 kpc, has a spiral arm structure in its plane and extensive regions of relatively new star formation. The Sun is found at a galacto-centric distance of about 8 kpc. The entire Galaxy is rotating, differentially, with a period at the solar distance of some 200 Myr. Surrounding this flattened system, the source of most of the optical radiation of our Galaxy, is a spherical halo sparsely populated by old stars, with ages $\approx 12 \times 10^9$ years (12 Gyr). Recent star formation is found in and near the galactic center, and old stars are also dispersed throughout the disk. See also Chapter 7.

Extragalactic luminous hot stars

Hot stars range in luminosity up to a few $10^6 L_\odot$ and are individually detectable on 4-m class telescopes with distances up to 1 Mpc, which includes the extent of the Local Group of galaxies. Classification and spectroscopic analysis of normal stars with their absorption lines is difficult but selected surveys have been published, particularly for the Magellanic Clouds (MCs). W-R stars, with their strong emission line spectra, can easily be detected with narrow emission band imaging in the Local Group. Beyond these distances identification and study of luminous hot stars is challenging but the advent of 8–10 meter ground based telescopes has eased the observing time required. (Individual W-R stars are now being identified in galaxies with distances up to 5 Mpc.) The MC stars sample a range of initial abundances, from solar values to a factor of five smaller. Local Group galaxies have a range of abundances, from somewhat metal-rich values to well below solar metallicity.

It is well known that the details of stellar evolution, particularly for high mass stars having strong winds, depend on the initial composition. Astrophysicists need to continually examine the predictions of stellar evolution models of stars to compare their properties with those of real objects in the sky. Analysis of stellar properties over a wide range of metallicity will enable astronomers to perform stronger tests of stellar evolution theory.

For example, it has been found that there is an observed upper limit to the luminosity of red supergiants (RSGs) using the H-R diagrams of galaxies of the Local Group. This empirical result is now referred to as the Humphreys–Davidson (H-D) limit, following Humphreys & Davidson (1979). The red H-D limit is substantially below that of the hottest main sequence stars. Since massive stars evolve at roughly constant luminosity, the absence of stars in the uppermost right hand corner of the H-R diagram was initially a puzzle. Where are the helium burning stages of the most luminous stars if they are not found in the RSG region? The answer lies in assigning highly luminous *hot* stars with unique spectral properties, namely W-R stars. Gamov (1943) first suggested that the anomalous composition of W-R stars was the result of nuclear processed material being visible on their surfaces. Evolutionary issues are more thoroughly discussed for single stars in Chapter 5 and close binaries in Chapter 6.

Extensive classification surveys of individual stars, aside from W-R stars, and perhaps the brightest blue supergiants, will be difficult beyond the Local Group until even larger telescopes are available. Spectroscopic analyses will be tedious and direct tests of stellar evolution models more problematic. It is, however, possible to check certain aspects of stellar evolution theory using indirect methods, by investigating the collective properties of luminous hot stars.

1.3   Stellar atmospheres

Astrophysical models

Astrophysicists construct mathematical and physical models that utilize often incomplete (and possibly incorrect) theory to make predictions about observable phenomena. The astronomer, in turn, makes and analyzes observations to confront the theory so as to help improve it or show what must be discarded. Models can be used to interpret nearly all the aspects of the positions of stars on the H-R diagram. One class of models deals with the physics of the stellar photosphere and atmosphere (treated together). For luminous hot stars, as we shall see in Chapter 4, one must also consider the stellar wind as an integral aspect of that region.

Generally, radiative processes dominate over collisional processes within hot stars, such that the standard approach of local thermodynamic equilibrium (LTE) within a plane-parallel geometry no longer holds, requiring a more elaborate, computationally demanding approach, known as non-LTE, supplemented with the need to incorporate atomic line opacities. Conversely, for high mass cool supergiants, LTE is valid, but extensive molecular opacities are needed instead. Details are covered in Chapter 3.

Composition

The solar photosphere is dominated by hydrogen and helium, with 1–2% by mass in the form of metals, and represents the initial composition of the solar nebula 4.5 Gyr ago. In general, the surface composition of other stars share this characteristic, *not* that of the nuclear reactions in the core which otherwise remain hidden from our direct view, at least until the latest or final stages of evolution are reached. In the solar case, one exception is that neutrinos from nuclear reactions in the interior of the Sun have been directly detected.

In the case of OB stars, despite their youth, their surfaces may expose partially processed nuclear material, due to internal mixing. This is especially true for red supergiants, where nuclear processed material is dredged up in their convective envelopes. In the extreme cases of W-R stars, a combination of mixing and previous stellar winds reveal fully processed material from H or even He-burning. The evolution of massive stars is considered in Chapter 5.

1.4   Stellar winds

OB winds

During their sojourn on the main sequence, the extremes of luminosity and temperature for O and early B stars result in strong stellar winds, sufficient to carry away substantial amounts of material from the star. The existence of winds was first proposed by Beals (1929), with direct signatures established since the first rocket UV missions in the 1960s, which may be parameterized by a velocity and mass-loss rate. The former may be directly measured from observations, whilst the latter relies upon theoretical interpretation. A theoretical framework for winds from hot stars was devised, and developed soon after their detection, involving the driving of outflows via millions of UV spectral lines.

The critical aspect which distinguishes OB stars from solar type stars is their proximity to the Eddington limit, where radiation pressure balances gravity, plus their high temperature photospheres, where metal species are highly ionized. Mass-loss affects the stellar structure and a substantial modification of the evolution of massive stars occurs, in which the outer hydrogen-rich layers are removed, akin to peeling away the skin of an onion. Once the hydrogen fraction at the surface begins to fall, the star will turn back to hotter effective temperatures on the H-R diagram to become a W-R star. Details are covered in Chapter 5.

Wolf–Rayet winds

Specifically, WN stars have enhanced helium and nitrogen from hydrogen burning (during the CNO H-burning cycle) and WC stars show the products of the *triple-α* process. In other words, the stellar cores of these stars are partially revealed due to mass-loss, or mixing, in previous and current evolutionary stages. Given their composition anomalies, it is pretty obvious that W-R stars must be highly evolved. W-R stars are intrinsically luminous and found in close association with very massive stars. They are generally over-luminous for their masses, a property shared by other post-MS stars. It is thus natural to associate W-R stars with the late hydrogen and helium burning phases of the most luminous stars since no evolved counterparts appear in the RSG region of the H-R diagram.

The spectra of W-R stars are primarily that of broad emission lines superimposed upon a blue continuum. These come about because W-R stars have very strong stellar winds, the origin of which is tied up in their extremes of luminosity, temperature, and composition. The wind is sufficiently dense that optical depth of unity in the visual continuum arises in the outflowing material. The spectral features are formed even further out and are found primarily in emission. The line and continuum formation regions are, furthermore, geometrically extended compared to the stellar radius and their physical depth in the star is highly wavelength dependent. These topics are presented in more detail in Chapter 4. According to theory, winds from hot stars are intrinsically unstable, such that their outflows are highly structured. Indeed, observations of W-R stars, and most recently OB stars, confirm a clumpy nature. Since the majority of mass-loss diagnostics scale sensitively with density, it is difficult to confidently extract robust values. Structures within their inner envelopes cannot be directly imaged – hot, massive stars are simply too far away.

LBV and RSG outflows

There are also episodes of instability observed in which extensive ejection of matter from a hot star takes place during Luminous Blue Variable (LBV) phases, of which η Carinae is the most famous example. LBVs are apparently intermediate between the blue supergiant and W-R phases in the evolution of the most massive stars. Furthermore, there may be mixing of material from the helium rich core regions to the hydrogen rich outer parts of the stars, thus affecting the run of chemical composition.

RSGs also possess stellar winds, albeit denser and slower than other evolutionary phases of high mass stars, although there is no robust theoretical framework in their case. Radial pulsations are believed to provide the mechanism for transporting material to sufficient distance that dust grains can form, which permit winds to be driven outward via the intense continuum opacity from such high luminosity stars.

1.5   Evolution of single stars

Main sequence evolution models

Stellar evolution from the point at which hydrogen burning is first initiated can be sketched from stellar structure models – describing the behavior of the star from the core outwards to the photosphere – which are relatively mature. Energy transport from the center to the surface in normal stars is typically convective or radiative, depending on the temperature gradient. In high mass stars, the inner *core* is convective and the outer *envelope* is radiative; this situation is reversed in the Sun and low mass stars. While there is energy transport between convective and radiative regions there is no mixing of material between them in classical models. Matter within a convective region is, of course, well mixed.

As a star gradually converts hydrogen to helium in its core the composition changes and the stellar structure readjusts to keep equilibrium. The location of the star in the H-R diagram migrates "slowly" to slightly cooler temperatures (for massive stars) or to higher luminosities (for lower mass stars). Only the core mass participates in nuclear reactions unless there is mixing of hydrogen rich material downward from the outer zones of the star. For most stars the core mass fraction is about 10%; for higher mass stars, this is closer to 30%. Once the hydrogen in the core is nearly exhausted, the star very rapidly approaches post-MS evolution and leaves the MS (Chapter 5).

The evolution time for stars on the MS varies *inversely* as roughly the 3.5th power of the mass (since the fuel supply is proportional to M_* and the rate of using it up proportional to L_*). Stars of high stellar mass go through their entire evolution cycle before low mass stars have evolved much at all. Clusters of stars are known in which hundreds to thousands of stars are born at more or less the same time. Cluster evolution on the H-R diagram can readily be understood in terms of this stellar evolution timescale with its inverse dependence on mass. Initially, a cluster will be relatively blue and bright, as the hot stars on the MS dominate the luminosity. The most massive stars will be the first to "peel off" the MS in turn, in order of their mass. The position of the "turn-off" point of a cluster on the H-R diagram, the uppermost part of the MS, is a direct measure of its age. As the cluster evolves, the turn-off point moves to fainter and redder magnitudes along the MS as stars of lower and lower mass evolve away. Clusters are more fully considered in Chapter 7.

Post-main sequence evolution

Once a star begins to run out of hydrogen fuel in its core, the stellar structure changes dramatically to maintain hydrostatic equilibrium and all stars begin to migrate "rapidly" towards the upper right hand portion of the H-R diagram. This is the red supergiant (RSG) region for most high mass stars and the red giant region for low and intermediate mass stars. Although the surface temperatures are becoming cooler, the core temperatures and central densities are inexorably rising. At some point, a zone of material surrounding the now hydrogen-depleted core becomes hot (and dense) enough to support hydrogen fusion, here termed *shell* hydrogen burning. Eventually the temperature and density in the core are high enough that helium can begin to fuse into carbon (the *triple-α* process, or helium burning) under controlled conditions (for high and intermediate mass stars) or sporadically (for lower mass objects). The post-MS lifetime is typically about 10% of the MS lifetime for each mass, aside from the most massive stars. While clusters still young enough to contain mostly massive stars will appear relatively blue, as time goes on the color will be dominated by red supergiants and giants.

Lower mass stars are unable to process nuclear fuel beyond helium burning. This phase occurs as a series of "helium flashes" which not only mix processed material from the core to the surface but also result in rapid movements on the H-R diagram. Intermediate mass stars go through helium burning in a stable manner during the red giant stage, and massive stars do so during a RSG or W-R phase. High mass stars reach core temperatures high enough for further exothermic nuclear reactions to occur, and elements further up the periodic chart are produced in their interiors. All of these post helium burning reactions occur on timescales that are even more rapid than what has gone before. These stages end for low and intermediate mass stars with the ejection of the outer envelope, to be identified as a planetary nebula, and the uncovering of the white dwarf (WD) remnant. High mass stars collapse as their fuel is exhausted and a supernova (SN) results. Details are covered in Chapter 5.

It is expected that for single stars there ought to be a mass *above which* all stars will evolve to the W-R stage as sufficient mass-loss and/or mixing during blue supergiant phases will have occurred. There is some empirical evidence that this occurs at around 25–30 $M_\odot$ for solar composition. The initial stellar mass corresponding to the H-D limit for RSG lies close to this value, according to stellar models with mass-loss and rotational mixing (e.g., Meynet & Maeder 2003). This suggests that within a limited mass range some single W-R stars might be post-RSG stars. For close binaries, the critical mass for production of a W-R star might be lower. Binary star evolution will be discussed in Chapter 6.

Star death

Eventually all stars will run out of fuel or will no longer have the right conditions for further nuclear reactions to occur. In the case of low and intermediate mass stars the stellar cores have already reached a regime in which the ideal gas law no longer applies. Instead, their temperatures and densities are such that electron degeneracy governs the equation of state. These cores, when uncovered, are called white dwarfs, with typically 0.6 $M_\odot$. There is an upper limit to the mass of a white dwarf, the Chandrasekhar limit, of about 1.4 $M_\odot$. Type Ia supernovae occur when a white dwarf exceeds the Chandrasekhar limit, as a result of accretion from a binary or the merger of two sufficiently massive white dwarfs. As such, Type Ia SN may be used as standard candles, from which observations of distant SN indicate the presence of a cosmological constant (Perlmutter *et al.* 1999).

When high mass stars run out of fuel they collapse under the influence of gravity; electron degeneracy in the core is not stable. This collapse triggers an SN explosion which ends in a stable configuration such as a neutron star or a black hole. These kinds of supernovae are now assigned labels of type II, Ib or Ic. A small fraction of the rare type Ic supernovae whose cores produce black holes also experience one flavor of gamma ray burst (GRB) explosion. Intermediate and low mass stars will end in the white dwarf configuration. The dividing main-sequence mass for these separate end points according to the models is at about 8 $M_\odot$. We will take this as a convenient definition for luminous stars: those with initial masses *larger* than 8 $M_\odot$, irrespective of their subsequent evolution. On the main sequence, this mass corresponds (following, e.g., Schaller *et al.* 1992) to a few 10^3 $L_\odot$. The (effective) temperature for this mass and luminosity is around $T_{\rm eff} \sim 19\,000$ K on the main sequence. Stars of 8 $M_\odot$ live about 5×10^7 years (50 Myr) during their hydrogen burning stages. The largest mass stars ($\approx 100\ M_\odot$) live a few Myr years (their core mass fractions are near 30%, rather than 10%).

The end points of stellar evolution, the white dwarf, neutron star, and black hole state, are permanent and their timescales are not counted when one is considering the lifetime of the stellar object. A small fraction of extremely massive stars at very low metallicities may end their lives with a direct collapse to a black hole, without an accompanying supernova explosion, or leave no remnant behind via a hypothetical pair-instability SN. Additional details concerning star death for massive stars will be considered in Chapter 5.

1.6 Binaries

Binary fraction

Most, but probably not all stars are formed in binary systems. Binary detection is primarily from systematic spectroscopic observations taken over periods of days to years in which periodic Doppler shifts can be identified. This method, which is distance independent, has even been successfully used to search for duplicity of stars in nearby galaxies. The shorter the period, the larger the velocity range and the easier it is to find double stars. Binaries with periods longer than a decade or so, however, are difficult to identify with this method. A few binaries have been first detected by eclipses in which the light (from the combined system) diminishes in a periodic manner. This photometric method is also distance independent, but requires that the observers are aligned very close the orbital plane of the binary system.

Relatively nearby binaries can be identified if proper motion measures indicate a periodic motion in the sky. This led to the discovery of the binary nature of Sirius, with a period of about 30 years. Eventually the secondary was observed in the glare of the primary and found to be an under-luminous star, now realized to be a white dwarf. If star separations are sufficient, nearby binaries can be detected from their double nature on the plane of the sky. Periods in these cases are typically tens of years, or more. This type of detection is also highly distance dependent and is only suitable for relatively nearby stars.

A plot of the frequency of binaries as a function of period reveals a broad maximum at about a week, for massive stars, with a long tail towards longer periods. The shape and length of the tail would tell us whether or not all massive stars are binaries, but the detection methods described above are inadequate to fully describe the long period distribution function.

Massive binaries

The study of massive binaries provides the most robust method for measuring stellar mass, for which instances of up to $80 M_\odot$ have been derived, although the upper limit to the mass of stars may be a factor of two higher. Accurate values for high mass stars are very difficult to establish in general. Close massive binaries, each possessing stellar winds, lead to a wind–wind interaction, details of which are studied by complex hydrodynamics.

For sufficiently close massive binaries there may be substantial mass-loss via a "Roche lobe overflow" process in which the initially more massive primary loses material to the less massive secondary star, resulting in evolution quite different from the single star case. Mass transfer may force a white dwarf into a neutron star state and a neutron star can become a black hole. Various exotic products may result, including high mass X-ray binaries, binary pulsars or OB runaways. A second, intrinsically fainter flavor of GRBs appears to be the result of merging neutron stars or a neutron star and black hole merger.

1.7   Birth of massive stars and star clusters

Pre-main sequence evolution

Most massive stars in our Galaxy are born within clusters formed during the collapse of giant molecular clouds (GMCs), containing substantial amounts of gas and dust. Within the Milky Way, these are closely confined to the galactic plane. Some massive stars form from lower mass molecular clouds that lead to small clusters, which may or may not be gravitationally bound, such as the Orion nebula (Trapezium) cluster. The dust obscuration at optical wavelengths has made direct observations of massive star birth difficult in many cases, but IR observations are beginning to pierce the veil and reveal the underlying stars. The interaction of the luminous star with the gas and dust of the birth environment can be observed at IR, mm, and cm wavelengths. For luminous hot stars the intensity and hardness of the radiation and the presence of stellar winds complicate the formation processes. The stellar radiation heats and dissipates the dust, dissociates the molecular gas, and ionizes the local hydrogen. The stellar wind blows material outwards, contributing to shocks and jets in the surroundings.

Star birth is not yet well understood in detail and is one of the outstanding unsolved problems in stellar astrophysics. The basic framework appears to be as follows (e.g., Shu *et al.* 1987): a molecular cloud collapses under the force of self-gravity, or from an external pressure force (i.e., a nearby supernova). The cloud fragments into smaller units of stellar mass, which can be identified as protostellar objects. Each object will undergo a core collapse on a rapid dynamical timescale. Given that the angular momentum of the parent cloud will be conserved, the centrifugal forces will steadily increase as each protostar continues to shrink. At some point these forces will begin to balance gravity in equatorial regions and the stellar object takes on the geometry of a spherical core and a disk-like envelope. Some of the disk material will continue to accrete onto the stellar core and some will be lost to the system. Material can also be ejected perpendicular to the disk in the form of jets. The central star is now described as a Young Stellar Object (YSO), which can be undergoing further contraction.

The energy source during the collapse and contraction phases is from gravitational potential energy. Contraction continues and the core temperature continues to heat and grow denser. Eventually, the light elements in the core such as deuterium and lithium are able to undergo nuclear reactions with hydrogen, providing a new energy source and slowing the gravitational contraction. Light element burning does not last very long as these elements are low in abundance. In any event, the core temperature and density continue upwards until hydrogen burning commences and the star begins its MS lifetime.[5]

The buried star with an accretion disk supporting the YSO phase has a timescale that is independent of the nuclear turn-on timescales and could well be longer. For massive stars, the YSO phase could well exist after the star arrives on the ZAMS and for some portion of the MS lifetime. The first observational evidence of massive star formation represents the hot core phase, followed by the detection of Lyman continuum photon production at radio wavelengths during the ultracompact (or hypercompact) HII region phase. Only later will the star become optically visible, as it clears its natal environment. The birth of massive objects will be the topic of Chapter 7.

---

[5] This is called the "zero age main sequence", or ZAMS.

Star clusters

In our Galaxy, massive stars are found in clusters ranging downwards from those containing hundreds to those with only a few, or perhaps just one O-type star. In other galaxies, very luminous star formation episodes are on-going, resulting in the formation of even more massive clusters. These episodes are termed "starbursts" and contain many thousands of O-type stars. Such massive clusters, often labeled "super star clusters" were also formed early in the history of our Galaxy, leaving behind what are now called globular clusters containing only old and highly evolved stars. The term "mini-starburst" has been used to describe events (in other galaxies) in which the number of O-type stars born is large, but there is no commonly agreed-upon distinction between "mini" and normal starbursts as to luminosities or total masses. A useful dividing line might be at 1000 O-type stars. Massive clusters in our Galaxy have been typically identified from their radio wavelength properties as the O stars ionize the hydrogen gas in their vicinity. The most luminous of these HII regions are commonly referred to as Giant HII regions, which contain at least 10 O-type stars.

The initial mass function

An important physical concept connected to the process of star formation is the initial mass function (IMF). This describes the fraction of stars as a function of the mass in the form

$$\phi(M_*) \propto M_*^{-\alpha}, \tag{1.2}$$

where this α is the *slope* of the IMF, with an empirically inferred value near 2.3 following Salpeter (1955). Thus there are always many more low mass stars compared to those of high mass. When astronomers talk about "determining the IMF" they are speaking about finding the value of the slope. An important question, not yet settled, is whether or not the IMF slope is universal (e.g., independent of the composition and location). We will return to the IMF problems in Chapter 7 for star clusters and Chapter 9 for larger systems.

1.8   The interstellar environment

Interstellar gas

Our Galaxy contains not only stars but also interstellar material in the form of gas and dust. Depending on the ionization state of hydrogen, four separate temperature/density regimes can be identified, each wanting to be in pressure equilibrium with its surroundings. The *cold* regions contain molecular hydrogen having temperatures of 10–20 K and number densities of $\approx 100 \, \text{cm}^{-3}$, or substantially larger (molecular clouds). *Cool* regions are those where the hydrogen is predominantly neutral, with temperatures of about 100 K, and densities between 10 and 1000 cm^{-3}. Lower densities presumably exist but are not currently detectable; the higher densities are considered to be diffuse interstellar clouds. In *warm* regions, the hydrogen is ionized, the temperature is around 10^4 K. In the warm intercloud medium, the density ranges between 0.1 and 10 cm^{-3}; higher density regions (10 to 10^4 cm^{-3}) ionized from stars within are called HII regions. Finally, in *hot* regions, the temperatures are typically 10^6 K, the hydrogen is fully ionized and helium, carbon, nitrogen, and oxygen are partially ionized. Here the temperatures are typically about 10^6 K and the densities are $\approx 0.01 \, \text{cm}^{-3}$.

The hot material in our Galaxy is found in the spherical halo, and at the interfaces between both hot stars and SN with their winds and with the surrounding warm gas. The cold, molecular gas is for the most part closely confined to the central plane of the Galaxy and highly clumped. The cool and the warm material is also primarily in the disk of the Milky Way but with a larger vertical scale height than the molecular gas.

The properties of the interstellar gas may be studied by analyzing its spectral features. In the cold gas, molecular emission and absorption lines are seen, primarily due to CO. In the cool gas, absorption and emission of the 21 cm hyperfine line of HI are found, along with narrow absorption due to CII, SiII, NaI, CaII, etc. In warm regions, nebular recombination emission lines of H and HeI, and forbidden lines of ionized CNO and other common elements dominate. The hot gas contains emission features of highly ionized lines of common elements, plus soft X-ray emission.

Interstellar dust

In addition to the interstellar gas, our Galaxy contains dust "grains", made up of irregularly shaped "cores" of silicon-rich or carbon-rich material, with an overlay of "ices" (e.g., CO_2, CO, H_2O, etc.). The distribution of dust particles follow a λ^{-4} law dependence, where λ is the wavelength, while the observed extinction is roughly proportional to λ^{-1}, indicating a range of grain sizes, from (at least) 0.1 to 1 μm, comparable to smoke, or soot. UV and optical radiation is very effectively absorbed by dust, whilst IR and radio radiation is effectively transparent.

Wind blown bubbles

Chapter 8 contains a more detailed description of interstellar environments, with a particular emphasis on the immediate surroundings of massive and luminous stars, either ionized HII regions formed at the beginning of their main sequence lives – due to their high surface temperatures – for which the Trapezium cluster serves as a familiar prototype, or wind blown bubbles produced by their continuous stellar outflows. In extreme situations, nebulae may be produced due to violent eruptions, such as that experienced by the infamous LBV η Car during the nineteenth century, when perhaps $10M_\odot$ was ejected over a couple of decades, producing the Homunculus reflection nebula.

1.9   From GHII regions to starburst galaxies

HII regions

HII regions are those *warm* regions in which the local interstellar hydrogen is ionized from the hot star or stars within. These resultant emission nebulae show permitted emission lines of neutral hydrogen and ionized and neutral helium. Forbidden emission lines of other elements in various stages of ionization such as carbon, nitrogen, and oxygen are also seen. The ionization of the surrounding hydrogen environment arises from stellar radiation emitted below the ionization (Lyman) limit at 912Å. Roughly speaking, single luminous stars with T_{eff} of 30 000 K or higher will have HII regions of sufficient size (≈ 1 pc) to be directly observable. It is convenient to describe the luminosity of an HII region in terms of the number of Lyman continuum photons $N(\text{LyC})$ emitted per second as this is a readily measurable quantity, derived from hydrogen recombination spectra or free–free radio continua. Anticipating a result to be discussed later, we will adopt $\log N(\text{LyC}) = 49$ photon s^{-1} as that coming from a "typical" O star (which is also roughly that of the Orion nebula cluster/Trapezium), to be a convenient scaling parameter. Analyses of the forbidden emission lines of an HII region using nebular models may be used to derive the electron temperature of the nebula, its density and its composition. These models are relatively mature and give realistic predictions of nebular properties if certain simplifying assumptions (e.g., spherical symmetry, homogeneous nebular density, etc.) are reasonable approximations to reality. Chapter 8 discusses this topic in greater depth.

One would also like to infer the T_{eff} of the exciting star(s), but this procedure requires modeling the emergent stellar ionizing radiation from the stellar wind/photosphere(s), in addition to doing the nebular analysis. The stellar models, while highly sophisticated, are not yet as complete or as accurate as the nebular calculations as the physics is much more complicated. However, for distant luminous hot stars that cannot be studied individually, this is the only procedure that can be followed.

Giant HII regions

The sizes and luminosities of HII regions are a very strong function of the numbers of their exciting stars and T_{eff}. The phrase giant HII (GHII) has been applied to the brightest HII regions. Although there is no "official" definition as to where the label changes, a convenient luminosity could be one that is 10 times the Orion nebula cluster (e.g., $\log N(\text{LyC}) = 50\,\text{s}^{-1}$). This is roughly a level corresponding to that from the hottest single luminous star on the main sequence and consistent with labeling found in the literature.

The brightest individual GHII regions, such as the Tarantula nebula (30 Doradus) in the LMC, are up to 1000 times more luminous ($\log N(\text{LyC}) = 52\,\text{s}^{-1}$) than the Orion nebula cluster. They are prominent in the spiral arms of galaxies and in other regions of recent star formation. The lifetimes of GHII regions are of the order of the hottest and most luminous stars, about 10 Myr. The emission line spectra of the brightest of these can be identified and analyzed to very large distances, say 100 Mpc. Thus we can, in principle, obtain information on the composition of the nebulae and numbers of exciting stars of GHII regions out to these distances.

The stellar content of the handful of GHII regions in the Local Group galaxies can be counted and analyzed separately from the nebular properties. The stellar statistics can then be compared with that inferred from spectroscopic analyses of the nebulae themselves. This is a fundamental calibration procedure that ties together nearby and local star forming regions with those more distant. This procedure will be considered in detail in Chapter 9.

Galaxies

The optical appearance of galaxies represents the collective emitted radiation of millions to billions of stars. Their integrated properties define the overall luminosity of the galaxy and its spectral energy distribution (SED). Spectroscopy of a galaxy will reveal the superimposed spectra of its stars, suitably weighted by their fractional numbers and luminosities. Galaxy spectra are therefore composed primarily of stellar absorption features, but emission lines are found under some circumstances. Our Galaxy is termed a spiral due to a predominant disk-like appearance, similar to all other flattened galaxies which have spiral arms. The other main categories of galaxies are labelled ellipticals and irregulars based upon their spatial morphology.

Historically, one of the first set of galaxies with prominent emission line spectra was described by Seyfert (1943), who found low and high excitation nebular emission lines in a sample of relatively nearby galaxies. These lines were found in addition to the normal galaxy stellar absorption features. The emission line activity was found to be in the galactic nuclei, the luminosity of which being a substantial part of the entire system. "Seyfert" galaxies are now understood to be representative of what are called "active galactic nuclei" (AGN). Subsequent analyses of their emission lines showed the gas excitation/ionization is *non-thermal*. It is produced by hot disk material surrounding and being accreted into a massive black hole.

Sargent & Searle (1970) called attention to two dwarf galaxies with nebular emission lines having a character similar to those seen in GHII regions. This emission is superimposed upon the normal galaxy stellar absorption features. Analyses of their emission lines and those of similar appearing galaxies were shown to follow *thermal* excitation and ionization, with an origin from the large numbers of luminous hot stars contained within the gas. While such objects were at first labeled "HII" galaxies, they might better be described as examples of the "starburst" phenomena. This is described more fully in Chapter 9.

1.10   Starburst phenomena

What is a starburst? Essentially, it is a localized or widespread region of *extraordinary* star formation. While starbursts are currently occurring in only a small fraction of already existing galaxies, this mode may have been significant early in the history of the Universe, possibly even among the "first" generation of stars. So, what do we mean by extraordinary? It is this aspect which will now require some elaboration.

A star*burst* is imagined to form relatively rapidly, that is, on a timescale short compared with MS lifetimes. The expectation is that an entire range of masses of stars will be produced, from an *upper mass limit* M_{up} down to some *lower mass limit* M_{low}. While it is convenient to model this event such that it occurs instantaneously, it is recognized that the actual time frame might extend over a few Myr. The evolution of a starburst can be reasonably well modeled and will be discussed in greater detail in Chapter 10. Here we need to note one important timescale. For the first 10 Myr, the presence of luminous hot stars ionizes the gaseous material surrounding the burst and one sees nebular emission lines. As with GHII regions, this excited gaseous material dominates the appearance of the star-forming region in the optical. The phase where nebular emission lines are seen is merely the first 10 Myr of the starburst episode. Afterwards most of the stars formed in the burst are still present but the luminosity of the burst gradually fades over time. This has important implications for starburst detection.

In perusing the very extensive literature there are many examples of phenomena which are labeled starbursts. Their identification has often come from various sources of anomalous phenomena. One frequently used criterion has been to label a galaxy as a starburst if it contains an inordinately bright nucleus and/or a nearby bright region or regions. Typically, the luminosity of the regions is comparable (within a couple of magnitudes) to that of the entire galaxy. An independent spectroscopic analysis is needed (in some cases) to distinguish thermally excited gas from a starburst from the non-thermal AGN using forbidden emission line ratios. In most cases there is a clear spectroscopic distinction between thermal and non-thermal excitation (i.e., between starbursts and AGN). Starbursts and AGN may co-exist in some cases.

Another search criterion has been to select galaxies with strong UV continua using objective prism surveys. A strong blue continuum can come about where large numbers of recently born relatively hot stars are present. While many objects with strong UV continua turn out to be AGN, quite a few harbor starbursts.

An alternative approach to identifying a starburst is to estimate the mass involved in the starburst episode from the number of stars produced and compare it with the total (molecular) gas mass of the galaxy. If the rate of using up the gas (which can produce stars) is so fast that it would exhaust the material in a time that is short (say, $\approx 10\%$) compared to the Hubble Time (the inverse of the Hubble Constant, H_0, related to the age of the Universe), then one has a starburst. One difficulty with this calculation is that one does not always have a good estimate of the molecular gas mass, let alone the total starburst mass. With this caveat in mind, this definition has an appeal which makes physical sense. Details of starbursts and their evolution are considered in Chapter 10.

1.11   Cosmological implications

The focus of this book emphasises our current knowledge about massive stars in the nearby Universe. However, studies of massive stars and starbursts locally impact upon several, rapidly developing topics with cosmological implications. Observations suggest that matter contributes just 30% of the energy density of the Universe (the remainder is so-called dark energy), of which normal baryonic matter comprises just 4% of the total energy density.

Population III stars

When the Universe was younger than a few hundred thousand years, the temperature was high enough that all of the hydrogen was ionized, that is the electrons were free and separate from the protons. Once the Universe cooled to about 3000 K, the electrons and protons combined to form hydrogen atoms, from which photons emitted may be observed as the 2.7 K Cosmic Microwave Background (CMB), first detected by Penzias & Wilson in 1965. Thus began the era of recombination, or so-called "dark ages" when the intergalactic medium (IGM) became mostly neutral.

This period ended when the first stars formed, apparently within dark matter mini-halos of mass $\sim 10^6 M_\odot$ at redshift $z \approx 20$–30. Within the currently accepted cold, dark matter model for the formation of structure within the Universe, galaxies formed as baryonic gas cooled at the centers of such halos. Mergers of halos and galaxies led to the hierarchical build-up of galaxy mass.

According to numerical simulations of the collapse of primordial gas clouds, this first generation of stars was quite unlike anything seen in the present-day Milky Way. In the absence of metal coolants, these first, so-called Population III stars, were massive and extremely hot, greatly enhancing their ionizing photon production, although it is not yet clear whether characteristic masses were 10 or $100 M_\odot$. Supernovae from Population III stars provided the first metal enrichment within the IGM, in some cases via a hypothetical catastrophic pair-instability process, leaving behind no compact remnant.

Lyman-break galaxies

The timescale over which galaxies assembled is unclear, particularly the bulges and disks which are the main components of present-day galaxies. Further, massive galaxies appear to be more common earlier than is expected from simulations. It is only within the last decade that normal galaxies have been detected in large numbers above redshift $z > 1$, when the Universe was approximately half its current age. Since most present-day galaxies are relatively old, it follows that they formed during a period of intense star formation, similar to today's starburst galaxies. During the 1990s, techniques were developed that exploited the strong discontinuity shortward of the Lyman series in most star forming galaxies, such that it became feasible to identify large numbers of high redshift Lyman-break galaxies (LBGs), which are the precursors of present-day elliptical and spiral galaxies. Standard nebular and interstellar techniques may be applied to such redshift galaxies and quasars, allowing chemical abundances to be studied in the early Universe.

Gamma ray bursts

Finally, Lyman-break galaxies and high redshift quasars are not the only probes of the early Universe. Since the cosmological nature of GRBs was established during the 1990s, the routine observations of high-z GRB afterglows discovered by the Swift satellite also permits their use as probes of the circumstellar and interstellar environment within their host star-forming galaxy. Indeed, a connection between the deaths of massive stars and the more common long-duration GRBs was established around the turn of the millennium when several nearby GRBs also showed the unmistakable signature of a massive star undergoing core-collapse, which had earlier been predicted in the "collapsar" scenario for GRBs. Apparently massive stars play a central role in a broad range of astrophysical situations!

2

Observed properties

The first measured parameter of an individual star is its brightness in the night sky, referred to by astronomers as an (apparent) magnitude m, for which the number gets larger as the star becomes fainter.[1] The next most important measurements are a color, that is a difference between magnitudes at two (or more) wavelengths, and/or a spectral type. For many years, photography was the detector of choice for astrophysical purposes, followed chronologically by photoelectric cells for photometry. It is only relatively recently that these have been nearly entirely supplanted by modern detectors such as charged-coupled devices (CCDs). Since photographic plates (and films) were most sensitive to visible radiation, spectral classification traditionally concentrated on these wavelengths. Brightness, color, and spectral types can lead to the bolometric luminosity and T_{eff} of the star, the fundamental parameters required to place it on an H-R diagram. Stellar structure models (Chapter 5) begin with specifying the stellar mass and composition and predict the L and T_{eff} at every point in the subsequent evolution. There is a continual examination and comparison of the L and T_{eff} being produced by models and those observed in real stars, with the idea to improve our understanding of stellar evolution. This confrontation will be a recurrent theme throughout this book

2.1   Apparent and absolute magnitudes

Apparent magnitudes m are related to the brightness by the general relationship

$$m = -2.5 \log(\text{apparent brightness}) + \text{constant}, \tag{2.1}$$

so magnitudes are on a logarithmic[2] scale and the values get smaller as the stars appear brighter. The m of a stellar object that will eventually be used to derive a luminosity is typically measured through a filter, commonly the V defined in the Johnson UBV system (ultra violet, blue, visual) for which Vega defines the zero-point at *all* wavelengths. Alternatively, the spectrophotometric (AB) system may be defined,

$$m_{\mathrm{AB}} = -2.5 \log f_\nu - 48.60, \tag{2.2}$$

---

[1] This peculiar arrangement is due to ancient history: The Greek and Arabic astronomers labeled the brightest stars in the sky as magnitude 1, and the faintest 5, thus the inverse relationship to brightness.

[2] The eye is a "logarithmic" detector. The magnitude scale was first quantitatively defined by Pogson (1858).

where the flux at frequency ν is f_ν with units $\mathrm{erg\,s^{-1}\,cm^{-2}\,Hz^{-1}}$. For the AB system the constant is selected such that Vega defines zero magnitude *only* at 5556Å. In general, we shall follow the Johnson system. The m_V is related to the absolute magnitude M_V by the relationship

$$M_\mathrm{V} = m_\mathrm{V} + 5 - 5 \, \log \, d - A_\mathrm{V}, \tag{2.3}$$

where d is the distance in parsecs (pc) and A_V is the visual extinction in magnitudes. M_V is thus defined as being the extinction-corrected m_V at a (standard) distance of 10 pc, and the distance modulus $\mathrm{DM} = m_\mathrm{V} - A_\mathrm{V} - M_\mathrm{V}$. The Sun has an absolute visual magnitude of $M_\mathrm{V} = +4.82$ mag, whilst typical luminous hot stars have $M_\mathrm{V} \sim -5$ mag, so they are of order 10 000 times brighter in the visual. m_V is measured though a broad filter, but the bolometric luminosity is the emergent radiation over all wavelength bands. Comparing early-type O stars with the Sun over all wavelengths adds a further factor of 10–100, so the most luminous stars exceed the Sun by a factor of a million. In magnitudes, one can write

$$M_\mathrm{Bol} = M_\mathrm{V} + \mathrm{BC}, \tag{2.4}$$

where M_Bol is the *bolometric* magnitude, equivalent to the bolometric luminosity,

$$M_\mathrm{Bol} = -2.5 \log L_*/L_\odot + 4.74 \tag{2.5}$$

and BC the bolometric *correction*, i.e. this corrects for the radiation emitted outside of the V-band filter. The BC is always negative, by definition, and is strongly dependent on the T_eff, i.e., it is appreciable for both hot (blue) stars and cool (red) sources, where most of the stellar radiation appears either shortward, or longward of the V-band.

For nearby stars, the effective temperature may be directly obtained from measurements of its angular diameter together with the total absolute flux, as measured at the Earth. The prime source remains the seminal work by Code *et al.* (1976) who combined angular diameters measured by the Narrabri stellar intensity interferometer by Hanbury Brown, Davis, & Allen (1974) – typically of order a few milli-arcsec (mas) – with UV, visual, and IR observations of O, B, A, and F stars, after correction for interstellar extinction. While this method is termed direct, the allowance for fluxes in the unobservable Lyman continuum are taken from model calculations, and so becomes increasingly reliant on suitable models for very hot stars, notably γ Vel (WC8+O), ζ Pup (O4 If) from their sample. Underhill *et al.* (1979) expanded greatly upon this approach for 160 OB stars. Since a significant fraction of the total energy in O and B is emitted shortward of 912Å, more sophisticated techniques need to be employed, as we shall see in Chapter 3.

In general, useful relationships between luminosity, radius and effective temperature are (e.g., Cox 1999)

$$M_\mathrm{Bol} = 42.36 - 5 \log R_*/R_\odot - 10 \log T_\mathrm{eff} \tag{2.6}$$

and

$$\log L_* = -3.147 + 2 \log R_*/R_\odot + 4 \log T_\mathrm{eff}, \tag{2.7}$$

where 4.74 is the M_Bol of the Sun ($L_\odot = 3.845 \times 10^{33}$ erg s^{-1}) and 5777 K has been adopted for $T_\mathrm{eff\odot}$. In principle, then, one obtains the L of an individual star by measurement of its m_V, obtaining a distance, adopting a BC, and utilizing the above equations. A T_eff will first need to be obtained so that the BC can be estimated. Various calibrations relating BC to $T_{\rm eff}$ exist in the literature, the most recent of which relating to hot stars is

$$\text{BC} = 27.58 - 6.80 \log T_{\rm eff} \tag{2.8}$$

according to Martins, Schaerer, & Hillier (2005) for massive O stars and

$$\text{BC} = 20.15 - 5.13 \log T_{\rm eff} \tag{2.9}$$

for early B supergiants according to Crowther, Lennon, & Walborn (2006).

The rapid development of stellar atmosphere models over the past couple of decades has resulted in major revisions to the bolometric corrections of O stars. For example, Code *et al.* (1976) originally derived an effective temperature of $T_{\rm eff} \sim 32\,500$ K for ζ Pup (O4 If) based upon a bolometric correction of –3.18 mag, whilst a modern spectroscopic determination of its temperature by Repolust, Puls, & Herrero (2004) indicates $T_{\rm eff} \sim 39\,000$ K, i.e. BC = -3.64 mag. The dependence of BC upon temperature is clearly very steep, so small errors in $T_{\rm eff}$ may lead to large errors in luminosity. For example, the stellar luminosity for an O3 V star with absolute magnitude $M_{\rm V} = -5.0$ mag is $500\,000 L_\odot$ if $T_{\rm eff} = 50\,000$ K, but only $370\,000 L_\odot$ if $T_{\rm eff} = 45\,000$ K.

The above equations cannot be used to determine the bolometric correction for W-R stars or A supergiants, although the latter are close to zero. More general calibration exists for low and intermediate mass A and B dwarfs. No reliable calibration has yet been developed for W-R stars. Individual cases range between BC = -2.5 mag for late-type WN stars, to BC = -6.0 mag for some early-type WN or WO stars.

2.2 Distances

It is, of course, no trivial matter to determine the distance of an individual hot star, and this represents the greatest difficulty with deriving stellar luminosities for galactic OB stars. Due to their large distances, direct determination via stellar parallax methods is only possible for a dozen cases obtained by the Hipparcos satellite, which performed precision astrometry. Amongst the closest examples Hipparcos measured were ζ Oph (O9.5 Vn) and τ Sco (B0.2 IV) at ~ 135 pc, a hundred times more distant from the Sun than the closest star, α Centauri. Instead, one resorts to obtaining spectral type-$M_{\rm V}$ calibrations using OB stars in associations or open clusters. In order to obtain absolute visual magnitudes, one needs to correct for interstellar extinction.

To a "zero th" approximation, the radiation emitted from a spherical star follows a "blackbody" law and thus the Stefan–Boltzmann relationship. In real stars the overall SED is modified by the (photospheric) continuum opacity and the presence of absorption, or emission, features. There is a straightforward connection between a stellar color and its $T_{\rm eff}$ (hotter stars are bluer and cooler stars are redder) but the exact relationship at any wavelength needs ultimately to be modeled. According to Wien's law, the peak of the stellar radiation for a luminous hot star with $\approx 30\,000$ K is emitted in the far UV regions. At optical wavelengths one is observing way out on the long wavelength tail of the SED for hot stars. In the UBV system, the hottest star $B - V$ and $U - B$ intrinsic colors are thus nearly degenerate (e.g., differing only by a few hundredths of a magnitude). For slightly cooler stars these colors begin to be temperature dependent.

The highly wavelength dependent extinction of starlight by interstellar dust dominates the colors of luminous stars in the Milky Way (Chapter 8). The observed optical colors of hot

Table 2.1 *Absolute magnitude scale, M_V, and intrinsic colors, $(B-V)_0$, for OB subtypes updated from Lesh (1968), Conti et al. (1983), Walborn et al. (2002), and FitzGerald (1970).*

Lum class	V		III		Ia	
	M_V	$(B-V)_0$	M_V	$(B-V)_0$	M_V	$(B-V)_0$
			O-type			
2–3	−5.9	−0.32	−5.9		−6.5	
4	−5.8	−0.32			−6.5	
5	−5.2	−0.32	−6.1		−6.6	
6–6.5	−5.1	−0.32	−5.7		−6.6	
7–7.5	−4.9	−0.32	−5.6		−6.6	
8–8.5	−4.7	−0.31	−5.2	−0.31	−6.2	
9	−4.4	−0.31	−5.2	−0.31	−6.2	−0.28
9.5	−4.2	−0.30	−5.1	−0.30	−6.4	−0.27
			B-type			
0	−4.0	−0.30	−5.1	−0.30	−6.6	−0.24
0.5	−3.8	−0.28	−4.8	−0.28		−0.22
1	−3.4	−0.26	−4.0	−0.26	−6.9	−0.19
1.5	−2.9	−0.25	−3.7	−0.25		−0.18
2	−2.2	−0.24	−3.3	−0.24	−7.1	−0.17
2.5	−1.8	−0.22	−2.5	−0.22		−0.15
3	−1.3	−0.20	−2.2	−0.20	−7.1	−0.13
4	−1.1	−0.18	−1.9	−0.18		−0.11
5	−0.9	−0.16	−1.6	−0.16	−7.0	−0.09
6	−0.5	−0.14	−1.4	−0.14		−0.07
7	−0.5	−0.13	−1.0	−0.12	−6.8	−0.04
8	−0.3		− 0.6	−0.10	−6.7	−0.01

stars provide information on the intervening interstellar medium rather than their physical parameters. The average total-to-selective extinction R_V in the optical is given by

$$R_V = A_V/E(B-V) = 3.1, \tag{2.10}$$

where A_V is the absorption in the V band and $E(B-V)$ is the color excess along individual sight lines, R_V may deviate significantly from 3.1, due to differing grain sizes. $B-V$ is an observed color and $(B-V)_0$ is the intrinsic color of an object, such that

$$E(B-V) = (B-V) - (B-V)_0. \tag{2.11}$$

Intrinsic $(B-V)_0$ colors of OB stars lie close to −0.3 for O and early B stars, while late B-type approach 0.0 mag (Table 2.1). Conti, Garmany, & Massey (1986) obtained a uniform $(B-V)_0 = -0.28$ for all dwarf and giant O stars in their study of LMC O stars, plus $(B-V)_0 = -0.24$ for late O supergiants. These agree well with synthetic UBVJHK photometry provided by Martins & Plez (2006).

These calibrations, as listed in Table 2.1, may then be applied to individual stars in the field. In spite of their relative proximity, luminosities of galactic OB stars are often more imprecise than those of stars in the Magellanic Clouds or beyond. Fortunately, one finds reasonably good consistency between calibrations for the Milky Way and elsewhere. Nevertheless, stars within a given spectral type show a substantial spread in M_V, which is thought to be typically ± 0.5 mag.

2.3 Massive stars in Local Group galaxies

Massive stars within the Milky Way are not randomly distributed, but are concentrated within either compact open clusters, extended groups, or associations within the spiral arms. Perhaps half of all galactic O stars are concentrated within OB associations, of which Humphreys (1978) carried out a very extensive study. Otherwise, the majority of the known O stars have been identified via optical spectroscopy of bright (Henry Draper, HD) catalog stars. An electronic catalog of galactic O stars is provided by Maíz-Apellániz *et al.* (2004).[3]

For the Magellanic Clouds, the objective prism surveys of Rousseau *et al.* (1978) and Azzopardi & Vigneau (1982) remain the primary sources for early-type stars in the field, although updates were provided by Fitzpatrick, Garmany, Massey and colleagues. Studies of OB associations and open clusters have subsequently been undertaken by various groups, notably Massey, Parker, & Garmany (1989) for New General Catalog (NGC) 346 in the SMC, and Parker (1993) for 30 Doradus in the LMC. Further afield, photographic surveys of M33 (NGC 598) were undertaken by Humphreys & Sandage (1980) and Ivanov, Freedman, & Madore (1993). Due to the large area upon the sky of M31 (NGC 224), initial CCD surveys focused upon small regions, which have only recently been extended to the entire galaxy using wide field instruments at medium sized telescopes (notably Massey *et al.* 2006). Aside from M31 and M33, the other star forming galaxies within the Local Group are NGC 6822, IC 1613 and IC 10. Basic properties for all star forming Local Group galaxies are presented in Table 2.2 whilst a schematic is indicated in Fig. 2.1.

Metallicity gradients are commonly observed in spiral galaxies, as deduced from strong forbidden lines or weak recombination lines in HII regions (e.g. Pagel & Edmunds 1981) or from analysis of luminous blue stars. For the Milky Way the oxygen abundance gradient is fairly shallow, with Δ (O/H) ~ -0.04 dex kpc^{-1}, with a similar gradient in M31. In contrast, M33 possesses a steep metallicity gradient according to Vila-Costas & Edmunds (1992). Direct chemical abundance determinations require a measurement of the electron temperature for which auroral lines (e.g. [OIII] 4363) become increasingly weak at high metallicity, such that the majority of abundance gradient studies have been based upon calibrations of strong forbidden line methods (e.g. [OII] 3727, [OIII] 5007), which are insensitive at high metallicity. Consequently, nebular oxygen abundances for the metal-rich galaxies quoted in Table 2.2 need to be treated with caution.

Since distances to individual early-type stars within the Milky Way or external galaxies are so large, one needs to correct observed UBV photometry for reddening. This is generally carried out through reddening-free indices, with which candidate OB stars can be identified. Bolometric corrections for blue and red supergiants are large. These are rather uncertain unless spectroscopic classifications are known, due to their extreme dependence upon temperature, such that care must be taken when interpreting observations. Without spectral types, colors of O stars are effectively degenerate, such that great care needs to be taken when constructing

---

[3] nemesis.stsci.edu/ jmaiz/GOSmain.html

Table 2.2 *Nebular oxygen abundances, distance moduli (DM) and foreground reddening to star forming galaxies within the Local Group (after Massey 2003). Metallicities are shown for M33 in its nucleus, and at its Holmberg radius (ρ_0).*

Galaxy	log(O/H)+12	DM mag	$E(B-V)$ mag
Milky Way ($d < 3$ kpc)	8.70	–	–
LMC	8.37	18.50	0.13
SMC	8.13	18.85	0.09
Phoenix	–	22.99	0.02
NGC 6822	8.25	23.48	0.38
IC 10	8.25	24.10	0.80
IC 1613	7.85	24.3	0.09
Pegasus	7.93	24.4	0.02
M31	9.00	24.4	0.16
M33 ($\rho = 0$)	8.85	24.5	0.13
($\rho = \rho_0$)	8.12		
WLM	7.77	24.83	0.02

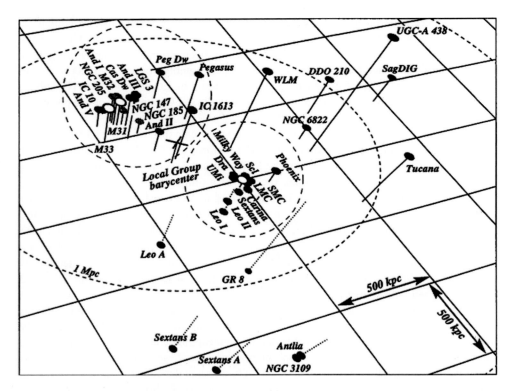

Fig. 2.1 3D representation of the Local Group in which the dashed ellipsoid marks a radius of 1 Mpc from the Local Group barycenter. The grid is parallel to the plane of the Milky Way, with galaxies above the plane indicated by solid lines, and those below indicated with dotted lines. Dashed circles enclose the M31/M33 and Milky Way subsystems. From Grebel (1999).

H-R diagrams from photometric data alone. Humphreys & McElroy (1984) and most recently Massey (2002) present H-R diagrams for large fields within the LMC and SMC, the latter of which are reproduced in Fig. 2.2. These support the following observational properties of massive stars.

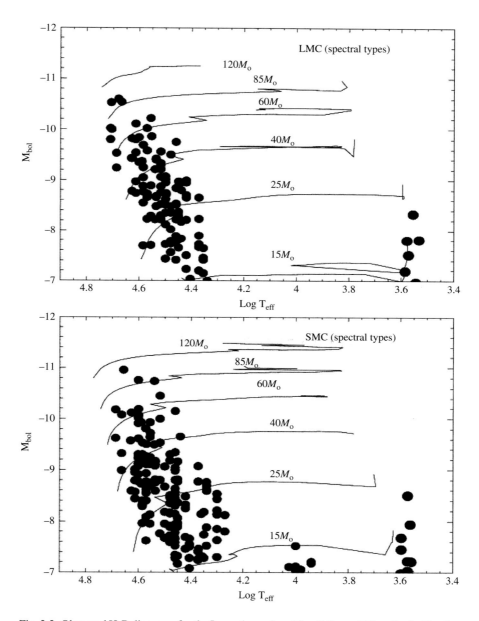

Fig. 2.2 Observed H-R diagrams for the Large (upper) and Small (lower) Magellanic Clouds, based upon spectroscopically derived effective temperatures, illustrating an absence of luminous RSGs (none brighter than $M_{Bol} = -9$ mag). Evolutionary tracks for 15–120 $M_\odot$ are presented for comparison. Reproduced from Massey (2002) by permission of the AAS.

1. The majority of massive stars inhabit a part of the H-R diagram close to the Main Sequence, as expected. Once spectral types are assigned, a cleaner Main Sequence is apparent.
2. A number of blue supergiants lie to the right of the Main Sequence, although there appears to be a deficit of luminous A supergiants above $M_{Bol} = -8.5$ mag. Since the bolometric correction of A-type stars is near zero, these stars are visually the brightest members of the Magellanic Clouds. Luminous blue variables occupy a part of the H-R diagram vertically above, although overlapping with, normal A and B supergiants.
3. A small number of RSG are observed, again with none observed above $M_{Bol} = -8.5$ to -9 mag. This supports the so-called Humphreys & Davidson (1979) limit, above which no red supergiants are observed, corresponding to an initial mass of $\sim 30 M_{\odot}$. The ratio of luminous blue (OBA) to red (MK) stars for the SMC and LMC is estimated to be $B/R \sim 4$, 10, respectively, significantly lower than $B/R \sim 30$ for the Solar Neighborhood (Langer & Maeder 1995).
4. There are no yellow (FG) supergiants or hypergiants within the Massey (2002) sample. Massive stars either occupy the red or blue part of the H-R diagram, such that the time spent within this phase is extremely brief.

The various massive stellar populations are now discussed in turn, with details of spectral classification deferred until later in this chapter.

Main sequence O and early B stars

Naturally, these are the most common massive stars observed in the star forming regions in the Milky Way and external galaxies. For main sequence stars, spectral types are a general guide to stellar mass, in the sense that an earlier subtype implies a more massive star, since surface gravities are uniformly $\log g \sim 4$. Main sequence O and early B stars tend to rotate rapidly, in some cases a significant fraction of break-up velocity. Since the most massive star within a young open cluster scales with the cluster mass (up to a probable stellar limit of $\sim 150 M_{\odot}$ according to Figer 2005), early O dwarfs are almost exclusively observed in young, massive star clusters such as R136 within the 30 Doradus complex (Massey & Hunter 1998) in the LMC or NGC 3603 in the Milky Way. Although these are the most massive stars known with high luminosities ($\log L/L_{\odot} = 4.5-6$), they are often difficult to observe since they are visually faint with respect to other massive stars due to their high effective temperatures, and corresponding large bolometric corrections. Physically, the closest cluster of O stars to the Sun is that powering the Orion nebula, M42, which has a distance of ~ 440 pc. Of these stars, θ^1 Orionis C = HD 37022 (O7 Vp) has the highest mass. From optical interferometry, Kraus *et al.* (2007) show that θ^1 C appears to be an eccentric ($e \sim 0.91$) system with a 10.9 year period involving the $\sim 34 M_{\odot}$ primary O star (C1), plus a secondary (C2) with $\sim 15 M_{\odot}$.

Blue supergiants

Collectively, O, B, and A supergiants are known as blue supergiants (BSGs), representing the evolutionary stage immediately following the Main Sequence. Visually BSGs are the brightest stars in external galaxies, with high luminosities ($\log L_*/L_{\odot} = 5-6.5$) and uniform visual colors, so candidates may easily be identified from broad-band imaging in external galaxies. Hα emission is a characteristic signature from follow-up spectroscopy, at least for the most luminous Ia supergiants, since BSGs possess extended atmospheres with

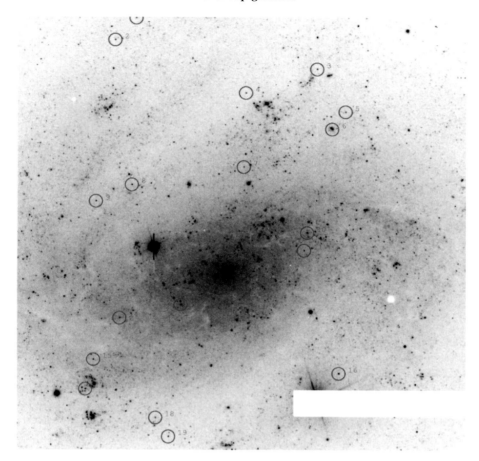

Fig. 2.3 VLT/FORS optical image of the southern spiral galaxy NGC 300 (6.8 × 6.8 arcmin) at a distance of 2 Mpc in which selected blue supergiants have been indicated. From Bresolin *et al.* (2002).

strong stellar winds. Within the Milky Way, Rigel (β Ori, B8 Iae), followed by Deneb (α Cyg, A2 Ia) are visually the brightest blue supergiants in the night sky. Early-type supergiants in the southern spiral galaxy NGC 300 are indicated in Fig. 2.3. From the figure, BSGs may have moved far from their birth cluster, so typically lie in the field, away from current star forming regions. Rotation rates are typically lower than for dwarfs of equivalent spectral type, although the absence of low $v_e \sin i$ supergiants suggests a "macroturbulence", ξ, which mimics the effect of rotation.

Luminous blue variables

The term LBV was first coined by Conti (1984) to describe a variable, luminous blue star. Examples are found in our Galaxy (e.g. AG Car, HR Car), the Magellanic Clouds (e.g. R127, S Dor), plus M31 and M33, the latter known historically as the Hubble–Sandage variables, after their discoverers E. Hubble and A. Sandage in the 1950s.

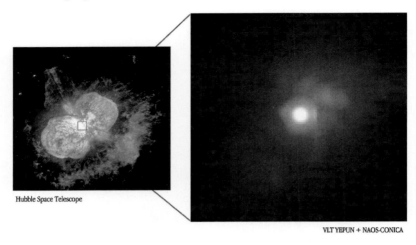

Hubble Space Telescope

VLT YEPUN + NAOS-CONICA

Fig. 2.4 Optical (left, HST/WFPC2) view of the Homunculus reflection nebula ejected from η Carinae during the nineteenth century. The lobes extend ~ 8.5 arcsec, equivalent to a physical radius of 0.1 pc; near-IR high spatial resolution image of the central object (right, ESO/NACO). ESO PR Photo 32a/03.

LBVs are believed to be represent a key stage in the evolution of the most massive stars, intermediate between the blue supergiant and Wolf–Rayet phases, since they are believed to eject many solar masses of material into the interstellar medium (ISM) over a relatively short time of 10^4 years. η Carinae, having ejected perhaps 10 $M_\odot$ over two decades during the nineteenth century, is the classic example of such behavior, as illustrated in Fig. 2.4. For this reason, the presence of an associated HII region, a potential remnant from such a recent mass-losing episode, may be a defining characteristic. The Pistol star in the Galactic Center Quintuplet cluster represents the best example of such an object which has not been observed to dramatically vary in its continuum flux, but is certainly a highly luminous blue supergiant with a characteristic ejecta nebula. Several theories for this intensive mass-loss have been proposed, but none has been reliably tested (Humphreys & Davidson 1994). Typically, mechanisms relate the importance of radiation pressure in their outer envelope, either via pulsational, dynamical or an Eddington-like instability. The latter might arise from an enhancement of opacity as the star moves to lower effective temperature or from the influence of rotation.

Several types of photometric variability are commonly discussed. LBVs exhibit micro-variations (~ 0.1 mag) over short timescales, and irregular variations of up to 1–2 mag. Classic examples of irregular variability over the past few decades include AG Carinae and HR Carinae in the Milky Way, and S Dor and R127 in the LMC. The visual light curve of AG Car over the past 50 years is presented in Fig. 2.5, obtained by dedicated amateur astronomers – notably Albert Jones and Frank Bateman – within the Variable Star Section of the Royal Astronomical Society of New Zealand (RASNZ). At visual maximum ($m_V \sim 6$ mag), the optical spectrum of AG Carinae is representative of an early A supergiant, whilst at visual minimum ($m_V \sim 7.5$ mag) it is more typical of an early B supergiant. At extreme minimum ($m_V \sim 8$ mag), AG Car has exhibited HeII emission lines, appropriate for O or WN stars. At this phase, it has been assigned composite Ofpe/WN9 or alternatively a very

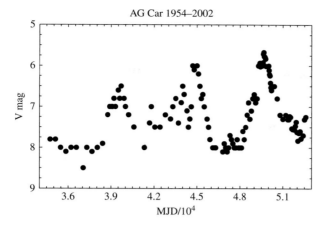

Fig. 2.5 Photometric light curve of AG Car over the past 50 years (MJD, Modified Julian Date), using datasets from the Variable Star Section of the RASNZ.

late-type WN spectral type (WN11). Since the difference in bolometric correction between an A and early B supergiant is ~ 1.5 to 2 mag, this irregular variability is thought to occur at more or less constant *bolometric* luminosity, as illustrated in Fig. 2.6. Some LBVs have even proceeded to spectral types as late as F or even G at visual maximum. Stellar winds of LBVs during their normal excursions across the H-R diagram are typical of normal OBA luminous supergiants, except that their wind velocities are atypically low.

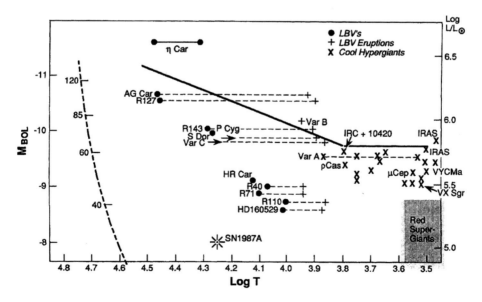

Fig. 2.6 Observed upper H-R diagram including location of luminous blue variables and cool hypergiants, plus Humphreys–Davidson limit (solid line) and ZAMS (dashed line). From Humphreys & Davidson (1994) with permission.

More remarkably, giant eruptions in LBVs have been observed, in which photometric variability exceeds 2 mag. Such outbursts are exceedingly rare, with only P Cygni and η Carinae known to have undergone such an eruption between the fifteenth and late nineteenth centuries. It is almost certain that other stars will have undergone such eruptions in recent history, but were not sufficiently bright to have been noticed by astronomers. P Cygni was discovered in 1600 by the Dutchman Bleau, when it suddenly appeared as a naked eye star (no star was apparent before). It remained bright for many years, before fading and re-appearing in 1655, and then fading again. At present it is around 5th magnitude, and so is at present the brightest LBV in the sky. P Cygni profiles, generally observed in UV resonance lines in hot star winds, are described as such because P Cygni displays these profiles in hydrogen and helium lines in its optical spectrum. Several cases of so-called "SN imposters" exist, including SN 1954J in NGC 2403, which are thought to be LBV giant eruptions instead of a genuine core-collapse SNe.

η Car, at the heart of the great Carina nebula, is known to have shed approximately $10 M_\odot$ during a couple of decades between 1837 and 1856 known as the "Great Eruption" (Davidson & Humphreys 1997). In 1843 it became the second brightest star in the sky with $m_V \sim -1$ mag, despite its distance of ~ 2500 pc, and so has one of the most remarkable photometric histories of any naked-eye star. The event remains unprecedented to this day, and corresponds to losing the equivalent mass of the Earth every 15 minutes over two decades. Such a large amount of material should be visible. Indeed it is, and has been named the Homunculus nebula, with a current size of ~ 0.1 pc, several hundred times larger than our Solar System. η Car underwent a second brightening between 1887 and 1895, known as the Lesser Eruption. At present, η Car along with all the other LBVs is too faint to be seen with the naked eye, although its bolometric output during this "quiescent" phase is $\sim 10^{6.7} L_\odot$ as deduced from its infrared dust luminosity. We shall return to this and other so-called ejecta nebulae in Chapter 8.

Because of the irregular nature of the LBVs, which only occasionally pass through the Humphreys–Davidson limit for red supergiants (Fig. 2.6), it is almost certain that many "dormant" LBVs await discovery, with all luminous blue supergiants potentially undergoing an LBV outburst at some point. One example is HDE 316285, an early B supergiant with a more extreme wind than P Cyg and spectroscopic similarities to η Car, yet only moderate spectroscopic/photometric variability has so far been observed.

B[e] supergiants

Rapidly rotating normal B stars, i.e. classical Be stars, are relatively common (see Section 2.5), with a typical B spectral appearance, plus narrow central emission at Hα. In contrast, B[e] stars, introduced by Swings (1976) additionally possess forbidden emission lines in their optical spectra, such as [Fe II] and [O I], in which the '[e]' notation follows that for forbidden lines. Walborn & Fitzpatrick (2000) present optical spectroscopy of B[e] supergiants. Another defining characteristic of B[e] stars is the presence of a near to mid-IR excess, due to hot (~ 1000 K) circumstellar dust – first detected in HD 45677 by J.-P. Swings – which is likely concentrated in a disk.

A number of Magellanic Cloud B[e] supergiants are known (e.g. Hen S134 in LMC, Hen S65 in SMC) with $\log (L_*/L_\odot) > 4.0$, that exhibit strong stellar winds, as indicated by P Cygni profiles of the Balmer lines, plus evidence for N-enhanced surface compositions. Their optical spectra typically display a hybrid appearance, i.e. narrow emission lines of low-excitation ions, plus broad emission lines from high-excitation ions, which have been interpreted as arising from a dense equatorial disk plus polar line-driven wind (Zickgraf *et al.* 1985). The formation mechanism for these disks remains unclear, although the answer is likely linked to rapid rotation. Lamers *et al.* (1998) provide a modern description of the various types of B[e] phenomena, and provide a catalog of Magellanic Cloud and Milky Way B[e] supergiants (e.g. GG Car), although uncertainties in distances to galactic B[e] stars hinders establishing their high luminosities.

B[e] supergiants have been contrasted with LBVs by Conti (1997). They occupy the same part of the H-R diagram – with each category accounting for 5–10 % of all B supergiants – but there is a general absence of outbursts and variability in B[e] supergiants. The photometric variability of B[e] supergiants is typically small (of order 0.1 mag), with the exception of R4 in the SMC, which has shown photometric variations of up to 0.5 mag. LBVs do not in general show a strong IR excess from a circumstellar disk, instead any IR excess observed likely originates from cool dust swept up in their nebulae. It is likely that B[e] supergiants represent the high rotational velocity tail of early-type stars prior to the RSG phase, since angular momentum would be lost during the RSG stage that would not be subsequently regained during the later blueward evolution.

Yellow hypergiants

Recalling Fig. 2.2, one may note a "yellow void" in which massive stars positively avoid the yellow part of the H-R diagram. Within the Milky Way, only a handful of yellow hypergiants are known. IRC+10420 is a useful prototype with an apparent A-F spectral type, has a high luminosity of $\log L_*/L_\odot = 5.7$ and is thought to be returning to the blue as a post-RSG since it possesses a complex circumstellar environment (Fig. 2.6). In their blueward evolution such stars are believed to enter a regime of increased dynamical instability, in which high mass-loss episodes occur. IRC+10420 has apparently undergone huge variability in mass-loss rate, 10^{-2} to $10^{-4} M_\odot$ yr^{-1}, via analysis of its circumstellar dust shell, which is much higher than for other Milky Way yellow hypergiants such as ρ Cas and HR 8752 (Blöcker *et al.* 1999).

Despite their rarity, three yellow hypergiants are observed in the young Milky Way open cluster Westerlund 1, which was discovered by B. Westerlund in 1961 and is presented in Fig. 2.7. This star cluster is thought to be the most massive young cluster in our Galaxy, with an approximate stellar mass of $M_{\rm clu} \sim 5 \times 10^4 M_\odot$, observed at an age of $\sim$4.5 Myr during which yellow hypergiants, red supergiants and Wolf–Rayet stars co-exist.

Red supergiants

Red supergiants, the immediate progenitors of most core-collapse supernovae, number amongst the brightest stars in the night sky – Betelgeuse (α Ori, M2 Iab) and Antares (α Sco, M1.5 Iab). Red supergiants possess highly extended atmospheres, with $\log g \sim 0$ and exceptionally large radii of up to $1500 R_\odot$ (7 AU). RSGs can easily be seen in external galaxies due to their high luminosity, most readily at near-IR wavelengths. RSGs are not observed to have absolute visual magnitudes exceeding $M_{\rm V} = -8.0$ mag, suggesting an upper luminosity (mass) limit. RSG in the Milky Way and LMC peak at M2 or M1 spectral type, for which bolometric corrections are –1 mag to –1.5 mag at most. Those within the SMC are known to exhibit earlier subtypes than Milky Way and LMC counterparts (Humphreys 1979),

Fig. 2.7 Composite 5×5 arcmin VRI image of Westerlund 1, a young massive Milky Way open cluster at a distance of 4–5 kpc. This cluster hosts red supergiants, yellow hypergiants and Wolf–Rayet stars (Clark *et al.* 2005). ESO PR Photo 09a/05.

which has been interpreted as being due to lower opacities in their atmospheres. The Humphreys–Davidson limit at $M_{\rm Bol} = -9.5$ mag ($\log L_*/L_\odot \sim 5.7$, Fig. 2.6) is considered to be the upper luminosity limit to RSGs, introduced by Humphreys & Davidson (1979), corresponding to initial masses of ~ 30–$35\ M_\odot$ based upon the current effective temperature scale for RSG and evolutionary models allowing for rotational mixing. The most massive stars apparently bypass the RSG phase. Within the Solar Neighborhood, dust production is dominated by intermediate-mass Asymptotic Giant Branch (AGB) stars, with RSGs playing only a minor role.

The fraction of RSG to BSG is known to increase at lower metallicity (Humphreys & McElroy 1984; Langer & Maeder 1995) from a comparison between the Milky Way and Magellanic Clouds. Regarding the possibility of RSG progenitors of W-R stars, remarkably few Milky Way open clusters host both types of star, suggesting a fairly narrow mass range in common (perhaps $M_{\rm init} \sim 25$–$30 M_\odot$), with the notable exception of Westerlund 1.

Wolf–Rayet stars

W-R stars within the Local Group have typically been discovered via techniques sensitive to their unusual broad emission line spectra, based on objective prism searches or interference filter imaging. Moffat & Shara (1983), and Massey & Conti (1983) independently developed narrow-band interference filter techniques that distinguished strong WR emission lines at He II λ4686 (WN stars) and C III λ4650 (WC stars) from the nearby continuum. Such techniques have been applied to regions of the Milky Way disk, Magellanic Clouds and other Local Group galaxies.

Standard UBV photometry does not permit W-R stars to be distinguished from normal hot stars. Broad-band measurements may overestimate the true continuum level by up to 1 magnitude in extreme cases, or more typically 0.5 mag for single early-type W-R stars due to their strong emission line spectra. Consequently, *narrow band ubv* filters were introduced by Smith (1968b) specifically designed for W-R stars to minimize the effect of emission lines (but their effect cannot be entirely eliminated). Most photometry of W-R stars has used the ubv filter system, to which the r filter was added by Massey (1984) The broad-band and narrow-band filter profiles are contrasted with a representative WC5 star in Fig. 2.8. The following relations relate the broad and narrow band optical indices for W-R stars

$$E(B - V) = 1.21 E_{\mathrm{b-v}} \tag{2.12}$$

$$A_{\mathrm{v}} = 4.1 E_{\mathrm{b-v}} = 1.11 A_{\mathrm{V}} \tag{2.13}$$

$$(B - V)_0 = 1.28 (b - v)_0 \tag{2.14}$$

$$M_{\mathrm{V}} = M_{\mathrm{v}} + 0.1. \tag{2.15}$$

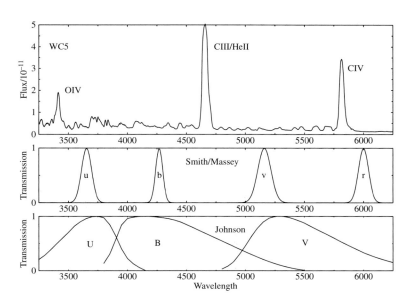

Fig. 2.8 Optical spectroscopy of HD 165763 (WC5) together with the Smith (1968b) ubv and Massey (1984) r narrow-band and Johnson UBV broad-band filters. Notice how the broader filters include the strongest two WC emission lines.

As with OB stars, observations in the ubv system are mostly used to determine the interstellar extinction, A_v, rather than to be able to say anything about the stellar properties of W-R stars.

A modern absolute magnitude–spectral type calibration for WN and WC stars has been provided by van der Hucht (2001), based on membership of galactic open clusters or OB associations, a revised version of which will be presented in the next chapter. Only γ Velorum, a WC binary system and the only naked eye W-R star visible (from the southern sky) was close enough for a reliable Hipparcos measurement, and even that remains controversial (Millour *et al.* 2007).

From membership of W-R stars in open clusters it is possible to investigate the initial masses of W-R stars empirically using turn-off masses for cluster member O stars obtained from isochrones of evolutionary models. Crowther *et al.* (2006d) provide a revised compilation of Milky Way clusters containing W-R stars. Distance estimates to field galactic W-R stars may be obtained using calibrations of near-IR absolute magnitudes for cluster member W-R stars, where interstellar extinction is much lower than visually (e.g. Hadfield *et al.* 2007).

Conti (1976) proposed that mass-loss could explain Wolf–Rayet stars, in which a massive O star loses a significant amount of mass via stellar winds, revealing first the CNO-burning products at its surface, and subsequently the He-burning products, which have been spectroscopically identified with the WN and WC phases. Such stars should be over-luminous for their mass, in accord with observations of W-R stars in binary systems. This general sequence has since become known as the "Conti scenario".

It is expected that there ought to be an initial mass, M_{init}, above which all single stars will evolve to the W-R stage, for which sufficient mass-loss and/or mixing will have occurred. Empirical evidence suggests this occurs at 25 and 40 $M_\odot$ for WN and WC stars at solar composition. The initial stellar mass corresponding to the Humphreys & Davidson (1979) limit for red supergiants (RSG) is $\sim 30 M_\odot$, according to stellar models with mass-loss (e.g., Levesque *et al.* 2005). Consequently, within a limited mass range some single WN stars are post-RSG stars, whilst evolution proceeds via an intermediate luminous blue variable (LBV) phase at higher mass. For close binaries, the critical mass for production of a W-R star has no such firm lower limit, since Roche lobe overflow or common envelope evolution could produce a W-R star in preference to an extended RSG phase.

From an evolutionary perspective, the absence of luminous RSGs and association of H-rich WN stars with young massive stellar populations (Langer *et al.* 1994; Crowther *et al.* 1995), suggests the following variation of the Conti scenario at solar metallicity. For the highest mass progenitors ($M_{init} \geq 75 M_\odot$)

$$O \rightarrow WN(\text{H-rich}) \rightarrow LBV \rightarrow WN(\text{H-poor}) \rightarrow WC \rightarrow SNIc,$$

whilst for stars with initial masses in the approximate range $M_{init} \sim 40 - 75 M_\odot$,

$$O \rightarrow BSG \rightarrow LBV \rightarrow WN(\text{H-poor}) \rightarrow WC \rightarrow SNIc,$$

and for lower mass $M_{init} \sim 25 - 40 M_\odot$,

$$O \rightarrow BSG \rightarrow RSG/LBV \rightarrow WN(\text{H-poor}) \rightarrow SNIb.$$

The role of the LBV phase is not yet settled – this stage may be circumvented in some high-mass cases, or alternatively dominate pre-WR mass-loss for the most massive stars (Smith & Owocki 2006). Conversely, the presence of dense, circumstellar shells around Type IIn SN indicates that some massive stars may even undergo core-collapse during the LBV phase. We will return to details of massive stellar evolution in Chapter 5.

From studies of Local Group galaxies, it has become clear that the absolute number and subtype distribution of W-R stars varies from galaxy to galaxy (Massey 2003). Essentially, the fraction of W-R to O stars is far smaller in metal-poor environments. For example, in the relatively metal-rich solar neighborhood, N(WR)/N(O) ∼0.15, while N(WR)/N(O) ∼ 0.01 in the metal-deficient SMC on the basis of only 12 W-R stars versus ∼1000 O stars. In addition, the subtype distribution of W-R stars favors nitrogen sequence WN stars at low metallicity. There are similar numbers of WN to WC stars known in the Milky Way, while WN stars exceed WC stars tenfold in the SMC. This is presented in Fig. 2.9, including recent determinations for galaxies beyond the Local Group up to 5 Mpc away, and illustrates a linear dependence of N(WC)/N(WN) with metallicity, with the notable exception of IC 10. IC 10 lies in the galactic plane so suffers from a high foreground extinction (recall Table 2.2). This contributes to the apparent deficiency of WN stars, which are more difficult to detect in external galaxies than WC stars due to weaker emission lines. The emission line equivalent widths of WN and WC stars are compared in Fig. 2.10, which illustrates how the strongest optical lines in the WN stars are far weaker than those of WC stars.

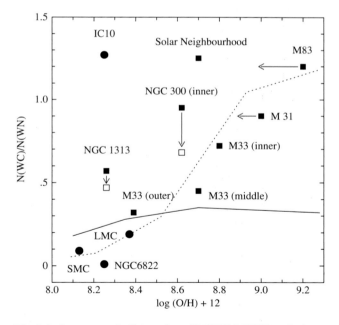

Fig. 2.9 Spectroscopically confirmed N(WC)/N(WN) ratio for nearby spiral (filled squares) and irregular (filled circles) galaxies. Open symbols for NGC 300 and NGC 1313 are corrected for photometric candidate WR sources. Predictions from evolutionary models from Meynet & Maeder (2005, solid lines) and Eldridge & Vink (2006, dotted lines). From Crowther *et al.* (2007).

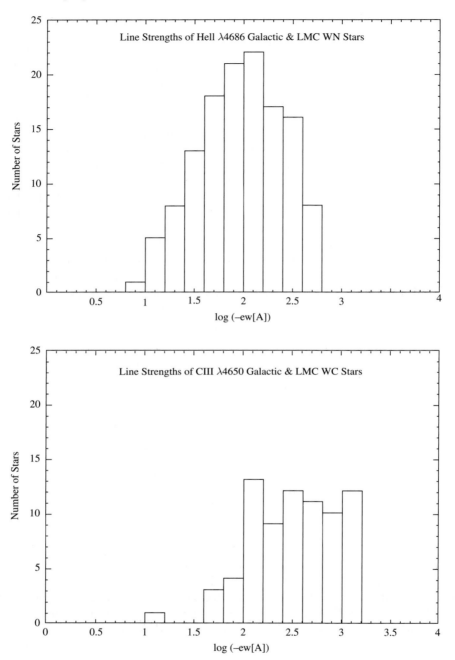

Fig. 2.10 Equivalent widths of the strongest optical lines in galactic and LMC WN stars (HeII 4686) and WC stars (CIII 4650). Reproduced from Massey & Johnson (1998) by permission of the AAS.

IC 10 aside, the reduced absolute W-R population and dominance of WN subtypes at low metallicity is due to the metallicity dependence of winds from precursor O stars, i.e. single stars for a given mass which are able to reach the WC phase at solar metallicity either end their life as an RSG or a WN star in low metallicity environments.

2.4   Spectral classification

Historically, both O and W-R subtypes were grouped together under the generic 'O' umbrella – signifying the presence of rare emission lines – within the original Henry Draper (HD) classes of Edward Pickering. In 1901, Annie Cannon sub-divided these into Oa through Oe, based upon the character of their emission lines, with strong, broad emission in a–c, weak Balmer emission in 'd', and weak He II and N III emission in 'e'. Those with merely absorption were called Oe5. In the 1920s, Harry Plaskett reorganized the O class into the more familiar modern notation, i.e. Wolf–Rayet for Cannon's classes Oa–c, with Od and Oe rearranged as Oe and Of, respectively, and the Oe5 group decimalized as O5 to O9.

Of course, the most common system of modern spectral classification builds upon that of Morgan, Keenan, & Kellman (1943), MK, who introduced a two-dimensional scheme involving letters and numbers to refer to spectral types and subtypes, plus roman numerals to refer to one of six luminosity classes, including Ia and Ib for luminous and less luminous supergiants, III for normal giants, and V for main sequence stars (dwarfs). Occasionally, class Ia-0 is used for the most luminous supergiants, although Ia^+ is now more commonly used for these hypergiants. Moderate-resolution blue spectroscopy is used to define natural groups of stars with similar spectral characteristics. *Standard stars* are typically selected for each spectral type and luminosity class. Classification of other stars is then carried out by obtaining their spectra and by using the spectral criteria and eye comparison to the standards. The division between massive and intermediate mass stars at 8–9 $M_\odot$ corresponds roughly to spectral type B1.5 on the main sequence. Thus O–early B spectral types are going to be the stars whose evolution we will follow in this book.

O–early B type stars

The MK system for these stars was modified and defined by Walborn (1971a). This was based upon spectra taken in the blue spectral region, on photographic *plates*, with a (reciprocal) dispersion of 63Å/mm. His classification system ranged from O4 through B2.5. Walborn (1971b) next identified a group of early O stars near η Carina, which extended his spectral subtype system in a natural way to O3. The classification for O–early B type stars utilizes HeI and HeII line ratios, along with SiII, SiIII and SiIV features. The leading lines are as follows: HeI: λ4471 and 4387 (all units here are Å); HeII: λ4542 and 4686; SiII: λ4128–30; SiIII: λ4552; SiIV: λ4089 and 4116. Basically, the appearance of the λ4542 HeII line distinguishes the hotter O from the cooler B spectral types. The O subtypes are defined by the ratio λ4471 HeI/λ4542 HeII which monotonically decreases towards earlier spectral types, such that the λ4471 line is very weak, or totally absent at O3. In luminosity V class the λ4686 HeII line behaves similarly to λ4542 although it is a bit stronger and it persists into the earliest B-type stars.

For early B spectral types, λ4542 HeII is absent and the HeI lines are strong. SiIV lines, which are seen throughout the O types, begin to fade at B0. SiIII lines appear and the SiIII/SiIV line ratio is useful between B0 and B1.5. At B2 and later subtypes, SiII appears as SiIII begins to fade and the SiII/SiIII ratio comes into use. In B and late-O type stars, the strength of these

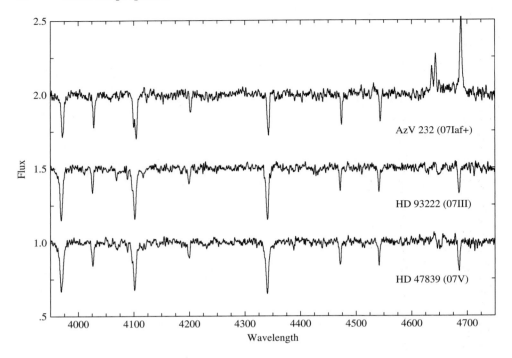

Fig. 2.11 Blue visual spectra of an O7 dwarf, giant, and supergiant, using spectroscopy from Walborn & Fitzpatrick (1990).

silicon features is luminosity dependent, being stronger in brighter stars. The helium lines, however, are less affected so that various silicon/helium line ratios are used as luminosity criteria. Among late-O–early-B stars, luminosity classes of Ia, Ib, III and V are recognized. Examples of an O7 dwarf, giant, and Ia supergiant are presented in Fig. 2.11.

In most O-type stars, though, a vastly different phenomenon begins to become apparent in the optical: the appearance of emission lines. Plaskett's Of subclass exhibited emission in λ4634–41 NIII triplet and λ4686 HeII. Walborn (1971a) pointed out that most O stars he observed showed the NIII lines in emission even though λ4686 remained in absorption or was not present. He labeled these as O((f)) and O(f) subtypes, respectively. We know now from detailed line modeling in O star atmospheres (e.g., Mihalas, Hummer, & Conti 1972) that NIII emission is a result of a fortuitous dielectronic recombination and implies nothing about the luminosity. On the other hand, the appearance of the λ4686 HeII line, whether in absorption, missing, or emission, *does* say something about the stellar wind structure and the luminosity. The status of λ4686 HeII is useful as a luminosity criterion. Following Walborn (1971a) the luminosity class of all but the latest type O stars is as follows: if the line is in absorption, the type is O V; if missing, O III; and if in emission, Of. The ((f)) and (f) nomenclature is (almost) redundant as nearly every O star, aside from the latest types, has λ4640 NIII in emission. A montage of blue optical spectra of O stars is presented in Fig. 2.12.

Concurrent with the developments of Walborn, an alternative classification scheme for O4–9.5 stars was put forward by Conti & Alschuler (1971), and later extended to O3 by Conti & Frost (1977). This had the advantage over Walborn's scheme in that it provided

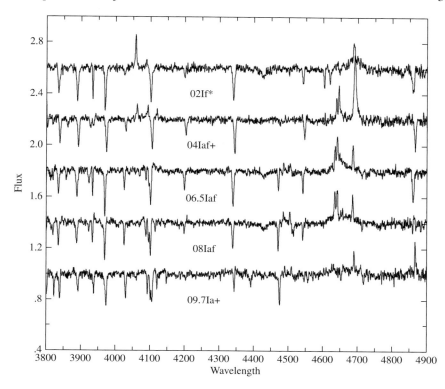

Fig. 2.12 A montage of blue optical spectra for O supergiants, using spectroscopy from Walborn & Fitzpatrick (1990) and Walborn *et al.* (2002).

a quantitative classification based upon the ratio of the equivalent widths of λ4471 HeI to λ4542 HeII. Overall, the two approaches are comparable, although subtle differences in nomenclature often result from the application of one scheme versus another. For example, 9 Sgr (HD 188001) has a Walborn spectral type of O7.5 Iaf versus O8 If according to Conti. The former scheme is in wider use, although not universally.

Recently, an O2 spectral type has been added by Walborn *et al.* (2002) such that O2–4 stars are defined by the ratio of *emission* lines at λ4058 NIV and λ4634–41 NIII, in contrast with the usual He *absorption* line criteria for OB stars. The principal motivation for this was that all O stars for which 4471 HeI was very weak or absent were previously grouped together as O3, yet exhibited a range of spectral morphologies, and effective temperatures from quantitative analysis. Main sequence stars with the highest ionization – NIV ≫ NIII emission – are re-defined as O2 stars. One famous example of an O2 supergiant is HD 93129A in the great Carina nebula which has exceptional properties, including a mass of ∼ 130$M_\odot$ (although this system has been identified as a binary). Whether O2 stars actually represent higher temperature, higher mass counterparts to O3 stars have still to be established. Examples of early O dwarfs are presented in Fig. 2.13.

The MK and Walborn classification systems depend on the spectra of stars in the vicinity of the Sun. These all have a composition similar to each other and to the Sun itself. How would one classify stars that had a substantially different abundance? For O stars this is not a significant problem since the spectral criteria involve hydrogen and helium lines for the most

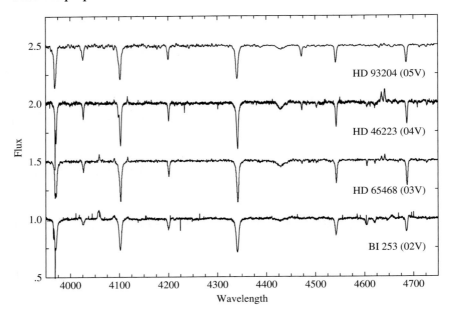

Fig. 2.13 Spectral classification of early O dwarfs, following Walborn *et al.* (2002).

part. Silicon ions do play a role in classification in the late O–early B stars and caution must be exercised in the cases of low metal abundance. For the LMC, with metal deficiencies of a factor of two, the effects on the classification are barely noticeable. For the SMC, with an abundance down by a factor of five, the silicon deficiency is obvious, but the silicon ion line ratios can be used with some confidence. Fortunately, the luminosity of these stars is known from their membership in the Magellanic Clouds and does not need to be established from their spectra. Recent revisions to the spectral classification of B supergiants in the SMC have been undertaken by Lennon (1997).

B stars with emission in the Balmer series (and sometimes HeI) have been known for a long time, and are referred to as Be stars. This is a very heterogeneous spectroscopic class, encompassing main sequence stars, supergiants, and very young stars (YSOs). The emission in the Main Sequence and YSO Be stars likely originates in a stellar disk. In the former objects, the disk is a result of ejection of material from a rapidly rotating star. In the latter objects, the disk arises during the stellar formation processes (see Chapter 7). Main Sequence Be stars are also called "classical Be stars". In Be supergiants, the emission comes from an extended wind, which may also have disk-like symmetry. A small set of O stars also have emission in the Balmer lines, labeled as Oe type, and can be thought of as analogs to the classical Be stars.

Walborn (1976) has reviewed and clarified the observations that certain O–early B stars had anomalously strong nitrogen or carbon lines, but otherwise normal features. These he had labeled as OBN or OBC stars. The OBC stars were also notable for having weak nitrogen lines, but the OBN stars have more or less normal appearing carbon features. OBN stars are found among both main sequence stars and supergiants, while the OBC stars are only in the luminous spectral classes. Other symbol or small letter suffixes have been introduced for anomalous OB spectra. These will be discussed in detail below.

Detailed atlases of λ1150–1800Å ultraviolet OB star spectra taken with the International Ultraviolet Explorer (IUE) and HST satellites have been compiled by Nolan Walborn and collaborators. Examples of P Cygni line profiles from NV 1240, SiIV 1400 and CIV 1550 in O supergiants and dwarfs are presented in Fig. 2.14. Copernicus, Hopkins Ultraviolet

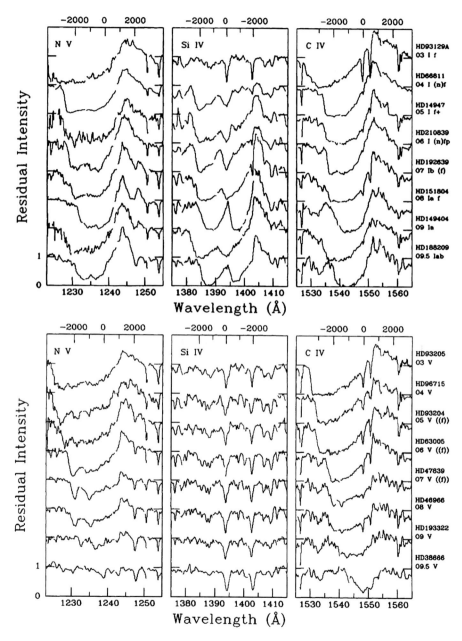

Fig. 2.14 Ultraviolet P Cygni line profiles in O supergiants (upper) and dwarfs (lower). Reproduced from Howarth & Prinja (1989) by permission of the AAS.

Telescope (HUT) and Far Ultraviolet Spectroscopic Explorer (FUSE) have extended studies to the far-UV λ912–1190Å, in which prominent P Cygni lines of numerous other ions are observed, notably CIII 977, NIII 989, OVI 1031, PV 1118 and CIII 1175.

In addition to optical spectral classifications, it is also possible to assign somewhat cruder ultraviolet or near-IR classifications. Most recently, it has become possible to obtain classification quality low-resolution spectroscopy of OB stars in the near-IR, specifically the K-band, as carried out in a pioneering study by M. Hanson. This has recently been extended to higher spectral resolution due to the advent of suitable instrumentation at large, 8–10 m ground-based telescopes, such as the Very Large Telescope (VLT) and Subaru. Examples of near-IR medium resolution spectroscopy of O dwarfs are presented in Fig. 2.15. As we shall see in Chapter 7, this has proved to be an extremely valuable resource by which we can study individual hot stars in regions that are inaccessible to optical or UV observations. At near-IR wavelengths, normal OB stars also show prominent absorption lines due to hydrogen and helium, with emission lines in leading members of the hydrogen series restricted to extreme OB supergiants, LBVs and W-R stars.

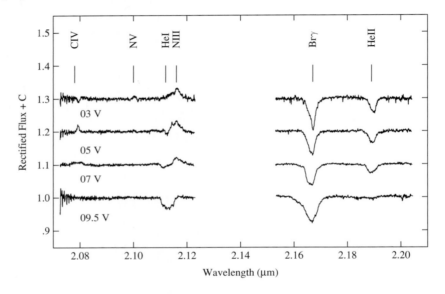

Fig. 2.15 K-band medium resolution spectroscopy of O dwarfs, drawn from Hanson *et al.* (2005a).

A-type and late-B supergiants

The standard MK classification criteria for late B supergiants relate the strength of SiII (λ4128, 4132Å) to HeI (4121Å) and MgII (4481Å) to HeI (4471Å) lines. Silicon and magnesium lines strengthen with respect to helium at later subtypes, as illustrated in the Walborn & Fitzpatrick (1990) digital atlas for Ia supergiants. B4 and B6 subtypes are rarely used. For A subtypes, the intensity of the Ca K line (3933Å) relative to the Hε+Ca H (3968Å) blend and/or Hδ is used. At A0, Ca K/(Hε+Ca H) ≤ 0.33, increasing to ∼ 0.9 at A7. AB supergiants may be discriminated by the strength of the Hγ line, which increases in absorption for later supergiants and from Ia to Ib luminosity classes (e.g. Azzopardi 1987). Lennon (1997) and

Evans & Howarth (2003) extended the classification of B and A supergiants to the low metallicity SMC.

Wolf–Rayet stars

Spectral classification of W-R stars follows Beals & Plaskett (1935) for nitrogen-rich (WN) and carbon-rich (WC) stars, in which decimal numbers are used for subtypes based upon ratios of emission lines. Two sequences were apparent in the first observations by Wolf & Rayet (1867) since their sample of Cygnus W-R stars included HD 191765 (WN), HD 192103 (WC), and HD 192641 (WC). The modern system for W-R classification is based upon that introduced by Smith (1968a).

WN spectral subtypes utilize a one-dimensional scheme, involving line ratios of NIII–V and HeII lines together with the appearance of HeI features, ranging from WN2 to WN9, extended to later subclasses by L. J. Smith and colleagues. At the earliest subtypes solely HeII and NV are present, with HeI and lower ionization stages of nitrogen absent, whilst at the latest subtypes, HeI is strong, HeII present, albeit weak, and NII-III present, with higher ionization stages absent.

Since the definition of some subtypes (e.g. WN8) involves the relative strength of nitrogen to helium emission lines, there may be an inherent metallicity dependence. Consequently, a scheme was devised that involved ratios of lines from either helium or nitrogen. Various multi-dimensional systems have been proposed, generally involving line strength or width, none of which have been generally adopted. The most recent was by Smith, Shara, & Moffat (1996). Representative W-R spectral standards are presented in Fig. 2.16.

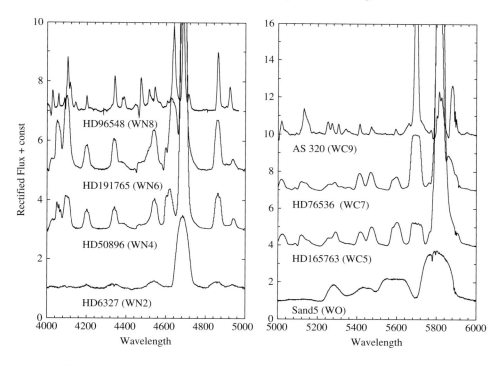

Fig. 2.16 A montage of optical spectroscopy of WN, WC, and WO stars.

However, complications arise for WN stars with intrinsically weak emission lines, such as HD 93131 in Carina OB1. From a standard spectroscopic viewpoint these stars formally possess mid-type WN spectral classifications. However, their spectral appearance is rather more reminiscent of Of stars than normal WN stars (Walborn & Fitzpatrick 2000), for which the 'ha' classifications was introduced by Smith *et al.* (1996) to indicate evidence for hydrogen and intrinsic absorption lines, as illustrated in Fig. 2.17. As we shall see in Chapter 5, such stars are widely believed to be massive O stars with relatively strong stellar winds at a rather early evolutionary stage, rather than the more mature, classical H-deficient WN stars. Indeed, Smith & Conti (2007) propose an alternative classification of WNH for such luminous H-rich stars.

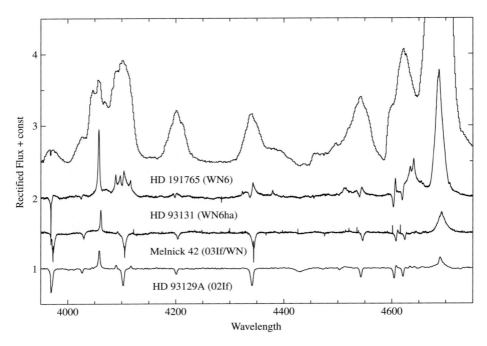

Fig. 2.17 Comparison between blue visual spectroscopy of the O2 If star HD 93129A, transition O3 If/WN6 star Melnick 42, weak-lined WN6ha star HD 93131, and the classical WN6 star HD 191765. This demonstrates a close morphological link between HD 93131, Melnick 42, and HD 93129A.

For a subset of WN stars, CIV λ5801–12 is unusually strong, leading to the classification of intermediate WN/C subtypes (Conti & Massey 1989). These stars are indeed considered to be at an intermediate evolutionary phase between the WN and WC stages, of relevance to mixing within Wolf–Rayet stars during their evolution.

WC spectral subtypes depend on the line ratios of CIII and CIV lines along with the appearance of OV, spanning WC4 to WC9 amongst massive stars.[4] Rare, oxygen-rich WO

---

[4] Some H-deficient central stars of planetary nebulae show carbon-sequence W-R spectral features that extend to later spectral types.

stars form an extension of the WC sequence to higher excitation, exhibiting strong OVI emission. The most recent scheme involves WO1 to WO4 depending on the relative strength of OV–VI and CIV emission lines. Representative examples of WC and WO stars are also presented in Fig. 2.16.

Line widths of early WC and WO stars are greater than late WC stars, although width alone is not a defining criterion for each spectral type. Not all W-R subtypes are observed in all environments. Early WN and WC subtypes are preferred in low metallicity galaxies, while the reverse is true at high metallicity. The subtype distribution of W-R stars in the Solar Neighbourhood, LMC and SMC is presented in Fig. 2.18. We shall address this aspect in Chapter 5.

UV and far-UV atlases of W-R stars have been presented elsewhere. Various near-IR atlases have also been produced in the J, H, and K-bands easily accessible from the ground, again showing a strong, broad emission line spectrum (see Crowther 2007 for details). Representative K-band spectra are presented in Fig. 2.19. As we will discuss in Chapter 3, some WC stars form hot dust, which can be seen as a strong excess in their near- to mid-IR spectral energy distributions. This excess emission strongly dilutes their characteristic wind emission-line spectrum in some cases.

Two recent infrared satellites, ISO and Spitzer, included mid-infrared spectrographs that allowed a number of nearby W-R stars to be spectroscopically observed in the thermal IR, where their emission line spectrum remains strong. This spectral range can be dominated by dust emission in some cases.

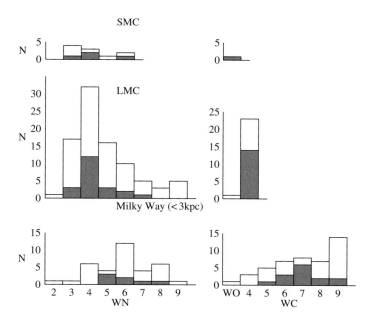

Fig. 2.18 Subtype distribution of Milky Way ($d < 3$ kpc), LMC and SMC W-R stars in which both close and visual WR binaries are shaded. Rare, intermediate WN/C stars are included in the WN sample. Adapted from Crowther (2007).

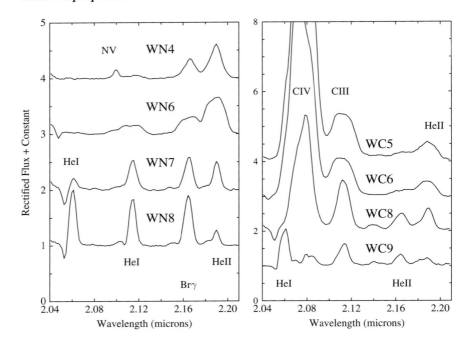

Fig. 2.19 K-band spectroscopy of WN and WC stars.

Transition objects and LBVs

As described above, a subset of O and B stars exhibit emission lines in their optical spectra. Although these are generally much weaker than in W-R stars, there are a number of stars with intermediate characteristics, which have been called transition objects between O-type and W-R.

There are two flavors of these so-called "slash" stars. N. Walborn labeled those very early O stars with anomalously strong HeII emission as O3If/WN of which several examples are now known in 30 Doradus in the LMC (e.g. Melnick 42). Whatever the nomenclature, very little distinguishes the most extreme O2–3 supergiants from the least extreme WN stars (recall Fig. 2.17). Indeed, from model studies the latter have been proposed to be H-burning objects, distinct from the main He-burning W-R sequence. In addition, a reconsideration of the spectral types of three galactic O6–8 Iafpe stars, including HD 152408, led to an alternative classification of WN9ha.

Walborn (1982) also identified a group of late O LMC emission line stars as Ofpe/WN9 subtypes, with apparently intermediate characteristics of Of and late WN stars Members of this group were re-classified as very late WN9–11 stars by Smith, Crowther, & Prinja (1994) on the basis that characteristic O star photospheric lines were absent, with additional examples identified in the Milky Way and M33. WN11 subtypes closely resemble extreme early-type B supergiants, except for the presence of HeII λ4686 emission. In the near-IR K-band these stars show strong Brγ and He I 2.058 μm emission lines, with He II 2.189 μm very weak. Such stars have also been detected in large numbers within the Galactic Center clusters, the brightest of which is known as the AF star, after its discovery by Allen, Hyland, & Hillier (1990) and Forrest *et al.* (1987).

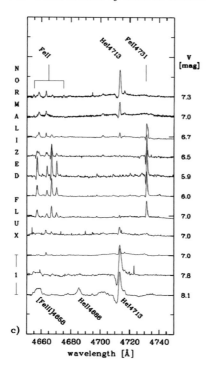

Fig. 2.20 Spectral changes in AG Car from 1989 (bottom) to 1999 (top), including an estimate of its visual brightness. Note the presence of He II 4686 in 1989. From Stahl *et al.* (2001).

Interest in this group of very late WN stars grew because of their high frequency of associated HII regions, plus an apparent connection with LBVs. One prototypical Ofpe/WN9 star (subsequently re-classified as WN11), R127 in the LMC, was later identified as an LBV, whilst a famous galactic LBV, AG Carinae, exhibited a WN11 spectrum at extreme visual minimum, as illustrated in Fig. 2.20. Spectroscopically, there are no distinguishing features unique to LBVs, since they share characteristics with A- or B-type hypergiants, i.e. emission in leading members of the hydrogen and helium series at optical or near-IR wavelengths, plus metal lines of Si II–III, Mg II, Na I, and Fe II. McGregor, Hyland, & Hillier (1988) undertook the first major near-IR surveys of early-type hypergiants in the Milky Way and Magellanic Clouds, revealing hot dust in some cases. Infrared photometric surveys such as the Two Micron All Sky Survey (2MASS) and GLIMPSE Spitzer legacy survey, only weakly affected by interstellar extinction, offer the possibility of identifying large numbers of extreme early-type supergiants through appropriate IR color criteria (e.g. Hadfield *et al.* 2007).

2.5 Observations of rotation and magnetic fields

During the formation process of all stars, including hot luminous OB stars, the conservation of angular momentum implies that they will commence their main sequence life rotating very fast, typically at 25 % of break-up velocity. Due to rotational broadening mechanisms, we are able to measure current OB (equatorial) rotational velocities v_e from optical or UV photospheric line profiles. Due to projection effects, $v_e \sin i$ is actually measured, where i is the inclination of the system on the sky as viewed by the observer, as originally set out by Otto Struve in 1929. The reduced contribution to the line profile from the stellar limb, where the projected rotational velocities are highest (due to limb darkening), was first incorporated by Carroll in 1933, whilst "gravity darkening", outlined by von Zeipel (1924), was used for the system of rotational standards introduced by Slettebak *et al.* (1975).

Gravity darkening relates the polar temperature (T_p) and equatorial temperature (T_e) to the effective temperature of the star. By definition,

$$T_{eff}^4 = \int T^4 dA / \int dA \tag{2.16}$$

where A is the surface area, such that for a given rotation velocity, $T_p \geq T_e$, i.e. the equatorial regions of rotating stars are cooler than their poles. In the extreme case of a centrally condensed star of polar radius R_p, rotating at critical or break-up, velocity, v_{crit},

$$v_{crit} = \sqrt{\frac{2GM}{3R_p}}, \tag{2.17}$$

the equatorial radius, $R_e = 1.5R_p$ and $T_{eff} = 0.8T_p$, rather than $R_e = R_p$, and $T_{eff} = T_e = T_p$ for the non-rotating case. For a typical main-sequence O3V star with a mass of $\sim 70 M_\odot$, $v_{crit} = 800 \, \text{km s}^{-1}$, whilst for a B2V star with $10 M_\odot$, $v_{crit} = 475 \, \text{km s}^{-1}$.

For OB stars, Conti & Ebbets (1977) measured rotational velocities from the observed line profile widths at half depth, calibrated against rotational standards. These standards were established by comparison with numerical models accounting for Roche geometry and von Zeipel gravity darkening, although a few rapid rotators served as calibrators. Subsequent studies (e.g. Howarth *et al.* 1997) compared UV observations of early type stars with "spun up" versions of narrow-line star templates, via a convolution with a rotational-broadening function. Amongst early-type stars, τ Sco (B0.2 IV) is the most famous example of a very slow rotator with $v_e \sin i < 5 \, \text{km s}^{-1}$. This approach neglects limb and gravity darkening, such that model line widths will universally exceed $v \sin i$ since the fast-rotating, gravity darkened equatorial regions will make a smaller contribution to the line profile than is assumed in the convolution. Note that line profile models typically assume uniform rotation, i.e. latitude independent angular velocities at the stellar surface.

The observed distribution of $v_e \sin i$ for early-type stars is shown in Fig. 2.21. There is an absence of narrow-lined spectra amongst galactic O stars for all luminosity classes in general, and amongst early-type supergiants in particular. From a sample of over 400 OB stars, Howarth *et al.* (1997) determined a median value of $v_e \sin i = 90 \, \text{km s}^{-1}$, which deprojects to a value 10 % higher.

From Fig. 2.21, the O main sequence distribution is much broader than that of supergiants, with only three supergiants amongst 33 stars with $v_e \sin i > 200 \, \text{km s}^{-1}$. This can be understood in that as a star evolves from the ZAMS, its rotational velocity will decrease due to conservation of angular momentum. The absence of slow rotators amongst O supergiants suggests another broadening mechanism beyond rotation, which has been generically labeled "macroturbulence". To date, the physical origin of macroturbulence remains unknown.

Most normal O and B stars are rotating at highly sub-critical velocities, with $v_e \sin i / v_{crit} = 0.1-0.5$. From Fig. 2.21, Be stars ("e" for emission) have higher rotational velocities than

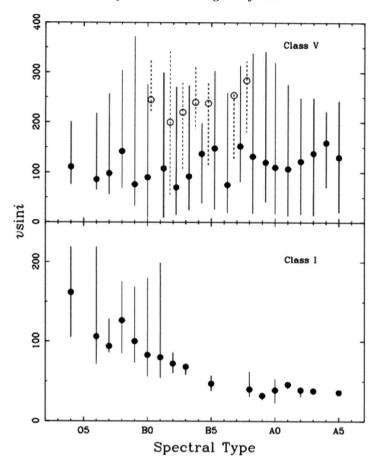

Fig. 2.21 Observed distribution of $v \sin i$ for dwarf and supergiant OBA stars. Circles show median values (Be stars are open symbols) whilst vertical bars show the total ranges with extreme values clipped. From Howarth (2004).

normal B main sequence stars, with $v_e \sin i / v_{\mathrm{crit}} \geq 0.5$. However, since the techniques typically used to measure rotational velocities neglect gravity darkening, the effective equatorial gravity is closer to zero than has been historically adopted. The only way to determine reliable rotation rates for rapidly rotating Be stars is to model line profiles allowing for von Zeipel darkening and using Roche models.

Rotation is much more difficult to measure in W-R stars, since photospheric features are absent. Attempts have been made, but generally without much success. Spectropolarimetry does permit information to be obtained regarding geometry, and suggests that $\sim 85\%$ of galactic W-R stars do not deviate significantly from spherical geometry, with a few notable exceptions (Harries, Hillier, & Howarth 1998). LBVs, meanwhile, do seem to possess globally distorted winds, in at least 50% of cases studied to date.

Magnetic fields of O stars impact upon their stellar interiors and atmospheres and hence on their evolution (Spruit 2002). Magnetic fields either originate from the star formation process or are produced in their interiors by dynamo processes. Unfortunately, there are very few direct observational constraints for the strength of B-fields. In several attempts at characterizing the B-fields that could be hosted at the surfaces of bright O stars, such as ζ Pup or ζ Ori, no detections have yet been made since Zeeman signatures are very weak in O stars, of order 0.1 percent peak to peak for a non-rotating star with a 1 kG dipole field, due to the fact that the intrinsic profiles of these objects are broad. The Zeeman signatures have been detected within photospheric lines in the most massive Orion nebula cluster star, θ^1 Ori C (O7 Vp), by Donati *et al.* (2002), with a strong dipole field observed (~ 1 kG), apparently inclined with respect to the apparent rotation axis.

As we will show in Chapter 5, typical neutron stars possess fields of strength $10^{12\pm1}$ G. Highly magnetic neutron stars (magnetars) possess fields of 10^{15} G. When scaled up to their anticipated Wolf–Rayet progenitor star, magnetic fields of ~ 100–1000 G would be inferred. Thus far, an upper limit to the magnetic field from just one Wolf–Rayet star, HD 50896 (WN4) has been obtained, revealing < 25 G, at the extreme low end of neutron star field strengths (St-Louis *et al.* 2007).

3

Stellar atmospheres

A detailed discussion of stellar atmospheres is beyond the scope of this book. Nevertheless, our means of studying the properties of hot massive stars relies upon our ability to properly interpret the stellar continuum and line information typically formed in the thin boundary layer between the unseen interior and effectively vacuum interstellar medium. An excellent monograph on the topic of stellar photospheres is provided by Gray (2005), whilst more advanced techniques are introduced by Mihalas (1978).

With respect to normal stars, our interpretation of hot, luminous stars is hindered by two effects. Firstly, the routine assumption of LTE breaks down for high-temperature stars, and particularly for supergiants, due to the intense radiation field, such that the solution of the statistical rate equations (non-LTE) is necessary. Secondly, the simplifying assumption of plane-parallel geometry is no longer valid for blue and red supergiants, so the scale heights of their atmospheres are no longer negligible with respect to their stellar radii. It is the combination of requiring non-LTE plus spherical geometry that has prevented the routine study of OB star atmospheres until recently.

3.1   LTE atmospheres

Effective temperatures of early-type stars, essential for subsequent determinations of radii and luminosities, are derived from a comparison between observed photometry and/or spectroscopy and models. Surface gravities also require comparison between observed line profiles and models.

LTE model atmospheres developed by Robert Kurucz during the 1970s and 1980s account very thoroughly for metal line blanketing and are widely employed for both early- and late-type stars. Line blanketing is the influence of thousands to millions of bound–bound spectral lines on the atmospheric structure. LTE means that the ionization state of the gas and the populations of the atomic levels can be obtained from the local T_e and n_e via the Saha–Boltzmann distribution, under the assumption that collisional processes dominate over radiative processes. Consequently, Kurucz models do not adequately apply to early-type supergiants, due to their deviations from LTE conditions.

The dominant source of continuum opacity in normal, solar-type stars is bound–free opacity from the H^- ion. Higher effective temperature A stars are dictated by bound–free opacity from atomic hydrogen, producing the well known Lyman (UV), Balmer (visual) and Paschen (near-IR) continuum jumps in such stars. In sufficiently hot stars, hydrogen and helium are completely ionized such that bound–free processes play a lesser role. Indeed, the continuum opacity of O stars is dictated largely by electron (Thompson) scattering (see below), although bound–free and free–free processes contribute to the continuum forming region of early-type stars. If the force from electron scattering were to exceed gravity above the sonic point, the surface of the star could not remain bound.

Amongst early-type stars – beyond the direct approach of Code *et al.* (1976) and Underhill *et al.* (1979) discussed in Chapter 2 – the simplest case to consider is that of mid- to late-B dwarfs which can be analyzed using continuum energy distributions, via low dispersion spectroscopy or photometry. Strömgren $uvby\beta$ photometry, coupled with LTE line blanketed model atmospheres, provides a powerful technique for the determination of T_{eff} in mid- to late-B stars via the Balmer discontinuity. As indicated in Fig. 3.1, the Strömgren photometric system provides an excellent indicator of the Balmer jump in non-supergiant B stars. The intrinsic UV–optical–IR spectral energy distributions of early B to early O stars vary only very subtly and are masked by uncertainties in interstellar reddening, since electron scattering dominates their continuous opacity, such that they do not display a Balmer jump.

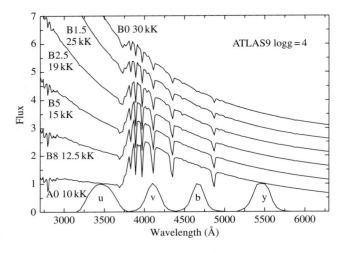

Fig. 3.1 Comparison between ATLAS9 (Kurucz 1991) line blanketed LTE models for B dwarfs with Strömgren photometric systems. From Crowther (1998).

Consequently, the effective temperatures of such stars are derived via analysis of their line spectra, generally optical photospheric lines of helium (O-type), silicon (B supergiants) or magnesium (A supergiants) For example, Venn (1995a) derived physical parameters for a sample of Galactic A supergiants using the Mg I-II and Hγ line spectra together with Kurucz model atmospheres. In A-type stars, Balmer line wings depend upon both effective temperature and surface gravity.

3.2   Non-LTE atmospheres

Generally, radiative processes dominate over collisional processes in the outer atmospheres of hot stars, so it is necessary to solve the equations of statistical equilibrium everywhere, i.e. non-LTE. The problem with this is that a determination of populations uses rates which are functions of the radiation field, itself a function of the populations. Consequently, numerically it is necessary to solve for the radiation field and populations simultaneously, which is computationally demanding, and requires an iterative scheme to obtain consistency. Unfortunately, the problem is too complex for analytical solutions.

Considerable effort has gone into developing realistic non-LTE model atmospheres for early-type stars in recent years by a number of independent groups. Consistently treating metal "line blanketing" in extended non-LTE atmospheres is computationally demanding. Consequently, the earliest attempts by Dimitri Mihalas in the late 1960s suffered from simplifications that only became surmountable two decades later, at which time large amounts of atomic data became available from the Opacity Project, which was led by Mike Seaton.

The solution of the equations of statistical equilibrium for complex model atoms is, by necessity, an iterative process, in which one may either start from LTE or a previous non-LTE model. For conventional "lambda iteration" techniques a very large number of iterations are required to arrive at a converged solution when scattering is important. Further, lambda iteration tends to *stabilize*, so it is difficult to judge when the process can be considered to be suitably converged.

Consequently, the mathematical concept of approximate lambda iteration (ALI) was developed, initially by Scharmer (1981). Nevertheless, the problem of "line blanketing", i.e. accounting for the effect of millions of spectral lines upon the emergent atmospheric structure and emergent spectra in non-LTE model atmospheres, remains challenging to the present date. Approximations involving either the calculation of mean opacities over a predefined spectral range in the ultraviolet or a restriction upon the atomic data considered represent the most common approaches, at least for cases in which spherical geometry is considered.

From such techniques, effective temperatures of hot stars are obtained from suitable diagnostic lines of adjacent ionization stages. Naturally, determinations of T_{eff} from model atmospheres are critically dependent on the assumptions used in such models, plus the accuracy with which surface gravities and elemental abundance ratios are determined.

Optical He I–II lines generally provide effective temperature diagnostics for O stars (e.g. Herrero *et al.* 1992) while Si II–IV lines are used for B stars (e.g. Becker & Butler 1990). The typical range of temperatures spanned is $T_{\text{eff}} \simeq 50\,000$ K at the earliest O2 V subtypes to $T_{\text{eff}} \simeq 30\,000$ K at B0 V, and $T_{\text{eff}} \simeq 10\,000$ K at A0 V.

Up until the mid-1990s, detailed analyses of OB stars necessitated the assumption of plane-parallel geometry, for which neither metal line blanketing nor spherical extension were considered. The most widely used non-LTE models were DETAIL/SURFACE written by Keith Butler and Jack Giddings, and TLUSTY by Ivan Hubeny and Thierry Lanz. Metal line blanketing has only recently been included in non-LTE analyses, for which the present version of TLUSTY (Lanz & Hubeny 2003) provides the most sophisticated approach for normal hot stars in which stellar winds are neglected.

To illustrate the effect of line blanketing, we consider the case of 10 Lac (O9 V). This star possesses a very weak wind, so we may isolate the role of line blanketing alone. It is common to construct a so-called "fit-diagram" in which for O stars the principal HeI–II and HI line strengths are compared to model predictions, enabling the determination of effective temperature and surface gravity. Herrero *et al.* (1992) determined the parameters of 10 Lac using an unblanketed non-LTE model, revealing a temperature of 37 500 K and surface gravity of $\log g = 4.00$ (cgs). An equivalent study was carried out by Herrero, Puls, & Najarro (2002) in which metal line blanketing *was* now included. The corresponding fit-diagram is illustrated in Fig. 3.2, from which a lower temperature of 35 500 K and surface

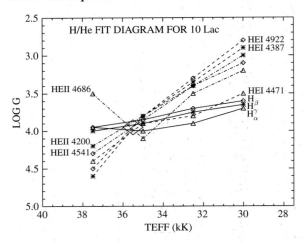

Fig. 3.2 Fit diagram for 10 Lac (O9 V) using line blanketed non-LTE model analyses, implying $T_{eff} = 35.5\,$kK, $\log g = 3.95$ and N(He)/[N(H)+N(He)]=0.09 by number. From Herrero *et al.* (2002).

gravity of $\log g = 3.95$ is determined. Accounting for line blanketing reduces the derived stellar temperature by 2000 K, due to back-warming, with the spectroscopic surface gravity barely affected in this instance.

Line blanketing has the effect of blocking UV radiation, which then emerges at optical and longer wavelengths. Some of the blocked photons are back-scattered, causing a back-warming effect, which leads to an enhanced ionization in the inner atmosphere. An excellent description is presented by Repolust, Puls, & Herrero (2004). In 10 Lac, blanketing affects the ionization structure of He, causing the change in the relative strengths of He I to He II lines, although this is not always the case.

In reality, OB stars possess stellar winds, which may contaminate ("fill-in") photospheric absorption lines, causing effective temperatures to be overestimated still further in the standard plane-parallel geometry assumption (Schaerer & Schmutz 1994). It has now been known for several decades that strong *wind blanketing* reduces temperature determinations by more than 10 %. The combination of stellar winds and line blanketing together can cause downward revisions in T_{eff} by up to 20 % in extreme OB supergiants relative to the standard approach. A plane-parallel, unblanketed model fit to optical photospheric lines of the extreme HDE 269698 (O4 If$^+$) is presented in Fig. 3.3, for which an effective temperature of 46 500 K is deduced. This star clearly possesses a strong wind, from the emission signatures at HeII 4686 and Hα.

Fortunately, a number of stellar atmosphere models have been developed for OB stars in which both spherical geometry and metal line blanketing have been incorporated, including codes developed by Adi Pauldrach (WM-Basic), Wolf–Rainer Hamann and Götz Gräfener (Po-WR), John Hillier (CMFGEN), and Joachim Puls (FASTWIND). Of these Po-WR and CMFGEN were originally developed for Wolf–Rayet stars, and have the advantage that they allow for the interaction between overlapping metal lines explicitly, albeit at the expense of being computationally time consuming. In contrast, WM-Basic and FASTWIND were specifically developed for O stars, and permit the rapid calculation of stellar atmosphere

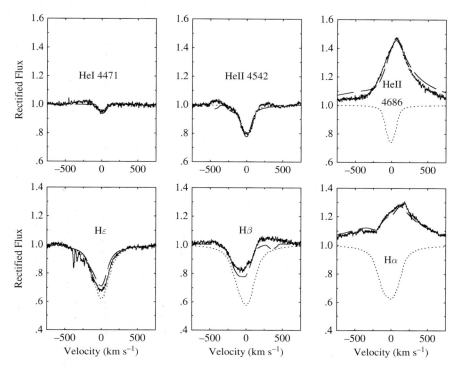

Fig. 3.3 Comparison between optical line profiles of the extreme O4 If$^+$ supergiant HDE 269698 in the LMC indicated as solid lines, and non-LTE, unblanketed plane-parallel model fits shown as dotted lines (spherical, line blanketed model fit shown as dashed lines) revealing $T_{\rm eff} = 46.5$ kK (40 kK), log $L_*/L_\odot = 6.25$ (6.0) and log $g = 3.7 (3.6)$, plus $\dot{M} = 0 (8.5 \times 10^{-6}) M_\odot$ yr^{-1}. Reproduced from Crowther *et al.* (2002b) by permission of the AAS.

models, now at the expense of an approximate, wavelength-averaged line opacity. Allowing for blanketing and spherical geometry produces an excellent match to both photospheric and wind lines for HDE 269698 (O4 If$^+$) for a reduced effective temperature of $T_{\rm eff} = 40\,000$ K (Fig. 3.3).

The currently adopted spectral type–effective temperature scale of galactic OB dwarfs and supergiants is presented in Table 3.1, updated from Martins, Schaerer, & Hillier (2005) and Crowther, Lennon, & Walborn (2006a). Individual temperature calibrations should be reliable to ± 2000 K for early O stars and ± 1000 K for early B stars, with the exception of the earliest subtypes which may approach $50\,000$ K at O2. Stellar luminosities follow from the calibration of absolute magnitude versus spectral type presented in Table 2.1. If the stellar temperature and luminosity are known, one may obtain the number of hydrogen ionizing photons $Q_0 = (N(\mathrm{LyC}),$ photon s$^{-1})$. This property is critical when considering the properties of associated HII regions in Chapter 9. Line blanketed CMFGEN models are used to estimate ionizing fluxes of OB stars, drawn in part from the calibration of $q_0 = Q_0/(4\pi R_*^2)$ from Martins *et al.* (2005).

Table 3.1 *Temperature scale, stellar masses, luminosities and ionizing fluxes* (N(LyC)) *of OB stars from Martins, Schaerer, & Hillier (2005), Crowther (2005), and Crowther, Lennon, & Walborn (2006a).*

Sp type	Dwarf				Supergiant			
	T_{eff} K	M_* $M_\odot$	$\log L_*$ $L_\odot$	$\log N(\text{LyC})$ ph s^{-1}	T_{eff} K	M_* $M_\odot$	$\log L_*$ $L_\odot$	$\log N(\text{LyC})$ ph s^{-1}
				O-type				
2	48 000	80	5.96	49.75	45 000	75	6.12	49.95
3	45 000	74	5.88	49.65	42 000	70	6.04	49.8
4	43 000	64	5.77	49.5	40 500	60	6.00	49.7
5	41 000	51	5.57	49.25	38 500	56	5.98	49.7
6	39 000	41	5.39	49.0	37 000	52	5.93	49.65
7	37 000	36	5.25	48.8	35 000	45	5.86	49.55
8	35 000	31	5.10	48.55	33 000	30	5.63	49.2
9	33 000	25	4.91	48.2	31 500	24	5.58	49.05
9.5	31 500	23	4.78	48.0	29 000	20	5.58	48.8
				B-type				
0	29 500	19	4.61	47.4	27 500	20	5.59	48.6
0.5	28 000	18	4.48	46.8	26 000	20	5.58	48.2
0.7					22 500	18	5.49	47.65
1	26 000	14	4.24	46.3	21 500	18	5.49	47.5
1.5	24 000	12	3.96	46.0	20 500	18	5.48	47.45
2	21 000	9	3.59	45.2	18 500	17	5.43	47.3
2.5	19 000	7.5	3.34	44.6	16 500	15	5.33	47.05
3	17 500	6	3.06	44.0	15 500	14	5.27	46.55
5	15 400	5	2.79		13 500	12	5.12	
8	12 300	4	2.35		11 400	9	4.84	

Table 3.1 shows that N(LyC) varies by five orders of magnitude between the hottest main sequence O2–3 V stars and B2.5 V stars. Historically, a "typical" O7 V star has been considered to possess an ionizing flux of 10^{49} photon s^{-1}, and for convenience we shall adopt this value here.[1] However, the recent downward T_{eff} calibration for galactic O stars suggests a somewhat lower ionizing flux of N(LyC)(O7V) $= 10^{48.86}$ photon s^{-1}. Predicted UV spectral energy distributions from line blanketed TLUSTY models for O stars are presented in Fig. 3.4. Approximately 3/5 of the bolometric luminosity from an O2 star originates in the Lyman continuum, compared to just 1/4 of the bolometric luminosity from an O7 star, and 1/10 for a O9.5 dwarf.

The role of metal line blanketing in O stars implies that the effective temperature at a given spectral type is dependent upon the metallicity of the star. For example, metal line blanketing for an O7 star within a low metallicity environment such as the Small Magellanic Cloud would play a lesser role than for an O7 star within the Milky Way. Indeed, recent spectroscopic results

---

[1] Panagia (1973) provides a primary reference for ionizing fluxes from O stars from which N(LyC)(O7V) $= 10^{48.86}$ photon s^{-1}. A substantially higher output of $10^{49.12}$ photon s^{-1} was reported by Vacca *et al.* (1996) on the basis of their high temperature calibration.

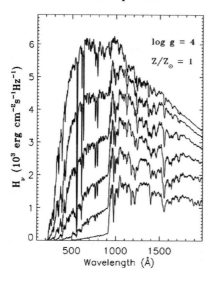

Fig. 3.4 Predicted emergent flux from solar metallicity O dwarf models (55, 50, 45, 40, 35, 30 kK from top to bottom) using the non-LTE, line blanketed, plane-parallel model atmosphere code TLUSTY. Reproduced from Lanz & Hubeny (2003) by permission of the AAS.

for O stars within the SMC (and LMC) suggest a higher temperature scale than their galactic counterparts, closer to previous calibrations in which line blanketing was not considered (e.g. Mokiem *et al.* 2006). Consequently, typical O7V stars in the low metallicity SMC likely possess ionizing fluxes somewhat higher than their galactic counterparts. Indeed, perhaps the historically adopted 10^{49} photon s^{-1} for an O7 star is reasonable for moderately metal-poor environments.

Stellar temperatures for W-R stars are more difficult to characterize, since the geometric extension in these stars is comparable to the stellar radius. Atmospheric models for W-R stars are typically parameterized by the radius of the inner boundary, R_*, at large optical depth, typically $\tau_{\mathrm{Ross}} = 10$–100. Of course, only the optically thin part of the atmosphere is seen by the observer, so R_* relies upon an adopted velocity structure for the invisible, optically thick, supersonic part of the wind. The optical continuum radiation originates from an "effective" photosphere, within the wind, where the Rosseland optical depth reaches 2/3. Typical W-R winds have reached a significant fraction of their terminal wind velocity, v_∞, before they become optically thin in the continuum. In any case, with the radius at an optical depth of 2/3, $R_{2/3}$ lies at supersonic velocities, well beyond the hydrostatic domain. For typical W-R stars possessing relatively dense winds, $R_{2/3}$ may exceed R_* by a factor of $\sim$3. A comparison between the stellar radii and line formation regions of O and W-R stars is shown in Fig. 3.5.

In contrast, stellar radii R_{evol}, deduced from evolutionary models, relate to the hydrostatic core. Theoretical corrections are based upon fairly arbitrary assumptions, which especially relate to the velocity structure. Consequently, a direct comparison between temperatures for W-R stars from evolutionary calculations and empirical atmospheric models is not useful, except that one requires $R_{2/3} > R_{\mathrm{evol}}$, with the difference attributed to the extension of the supersonic region.

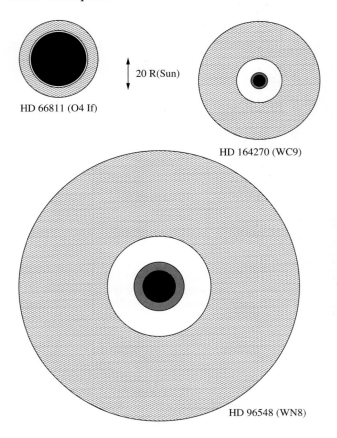

HD 66811 (O4 If)

20 R(Sun)

HD 164270 (WC9)

HD 96548 (WN8)

Fig. 3.5 Comparisons between stellar radii at Rosseland optical depths of 20 (= R_*, black) and 2/3 (= $R_{2/3}$, grey) for ζ Pup (HD 66811, O4 If), HD 96548 (WN8) and HD 164270 (WC9), shown to scale, together with the region corresponding to the primary optical wind line forming region, $10^{11} \leq n_e \leq 10^{12}$ cm^{-3} (hatched), plus $n_e \geq 10^{12}$ cm^{-3} (white) in each case, illustrating the highly extended winds of W-R stars. From Crowther (2007).

Analogous to O stars, stellar temperatures of W-R stars are determined using emission lines from adjacent ions of nitrogen or helium (WN stars) or carbon (WC stars). Stellar temperatures of W-R stars, relating to R_*, range from 30 000 K amongst late-type WN stars to well in excess of 100 000 K for WO stars, although differences in wind density complicate accurate determinations. The highly extended, hot photospheres but cool winds were first recognised as such by Cherepashchuk, Eaton, & Khaliullin (1984) and Hillier (1989). This is illustrated in Fig. 3.6 for a WC9 atmospheric model, demonstrating the variation of temperature, density, velocity, and ionization balance with Rosseland optical depth. In some respects the structure of W-R atmospheres is more reminiscent of ionized nebulae than normal stellar atmospheres.

The Lyman continuum ionizing flux distributions of WR stars in general extend those for O stars to higher temperature, and so significant He I continua are generally obtained, with strong He II continua in a few high temperature, low density cases, such as WN3 stars. Typical properties of Galactic W-R stars are listed in Table 3.2, where we have separated strong and

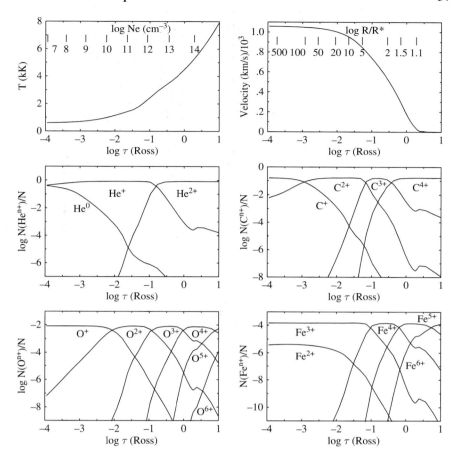

Fig. 3.6 Ionization structure of an atmospheric model for HD 164270 (WC9). Reproduced from Crowther, Morris, & Smith (2006b) by permission of the AAS.

weak-lined WN3–6 stars, which have very different wind properties, and included parameters for LMC WC4 and WO stars (following Crowther 2007).

Luminosities of He-burning W-R stars in the Milky Way range from $\sim 100\,000\ L_\odot$ for WC stars to $250\,000\ L_\odot$ for early WN stars, and are a factor of 2–3 times higher for late WN stars. Intrinsically luminous late-type WNha stars, which are thought to be H-burning stars with strong winds, have luminosities of order $10^6 L_\odot$. Note that LMC WC stars are a factor of three times more luminous than their Milky Way counterparts, as expected since they are thought to be descended from higher initial mass stars.

W-R wind properties are discussed in Chapter 4. In general, Lyman continuum ionizing fluxes are typical of mid-O stars. Due to the low ratio of N(W-R)/N(O) in star forming regions, W-R stars are expected to play only a minor role in the ionizing budget. Luminous H-rich WN stars, observed in young massive clusters, do significantly contribute to the ionising budget, since their output is comparable to the O3 supergiants from Table 3.1.

Turning to physical properties of LBVs, their excursions back and forth in the H-R diagram are generally considered to occur at fairly constant bolometric luminosity. This is as a result of

Table 3.2 *Physical and wind properties of Milky Way (and LMC) W-R stars, adapted from Crowther (2007). Mass-loss rates are shown assuming a volume filling factor of $f = 0.1$.*

Sp type	T_* K	M_v mag	$\log L_*$ $K_\odot$	$\log \dot{M}$ $M_\odot \mathrm{yr}^{-1}$	v_∞ $\mathrm{km\,s}^{-1}$	$\log N(\mathrm{LyC})$ $\mathrm{ph\,s}^{-1}$
			WN subtypes			
3w	85	−3.1	5.3	−5.3	2200	49.2
4s	85	−4.0	5.3	−4.9	1800	49.2
5w	60	−4.0	5.2	−5.2	1500	49.0
6s	70	−4.1	5.2	−4.8	1800	49.1
7	50	−5.4	5.5	−4.8	1300	49.4
8	45	−5.5	5.4	−4.7	1000	49.1
9	32	−6.7	5.7	−4.8	700	48.9
6ha	45	−6.8	6.2	−5.0	2500	49.9
9ha	35	−7.1	5.9	−4.8	1300	49.4
			WC and WO subtypes			
(3)	(150)	(−2.8)	(5.2)	(−5.0)	(4100)	(49.0)
(4)	(90)	(−4.5)	(5.5)	(−4.6)	(2750)	(49.4)
5	85	−3.6	5.1	−4.9	2200	48.9
6	80	−4.0	5.1	−4.9	2200	48.9
7	75	−4.5	5.3	−4.7	2200	49.1
8	65	−4.0	5.1	−5.0	1700	49.0
9	50	−4.6	4.9	−5.0	1200	48.6

the sensitive dependence of BC upon T_{eff}. Stahl *et al.* (2001) present a spectroscopic analysis of AG Carinae between 1988 and 2000 (recall Fig. 2.20), revealing a change in effective temperature between 26 000 K at visual minimum (corresponding to a WN11 star) and 8 000 K at visual maximum (A supergiant). The physical radius varies between 50 and 500 $R_\odot$ over this S Dor-type cycle, as illustrated in Fig. 3.7. Typical quiescent stellar luminosities are $10^{5.8\pm0.4} L_\odot$ although distance uncertainties to galactic LBVs are considerable. Only η Car and perhaps the Pistol star in the Quintuplet cluster at the galactic center are believed to possess higher quiescent luminosities. In external galaxies, the SMC W-R binary system HD 5980 underwent a brief LBV giant eruption during the early 1990s, and NGC 2363–V1 in the low metallicity galaxy NGC 2366 is also undergoing a giant eruption. Stellar luminosities during giant eruptions exceed $10^{6.5} L_\odot$.

Complications occur for cases such as η Car in which the star is veiled by the Homunculus bipolar reflection nebula produced in the nineteenth century "Giant Eruption". η Car is currently losing mass at such an exceptional rate of $10^{-3} M_\odot \mathrm{yr}^{-1}$ that the effective temperature of the underlying star cannot be well determined (Hillier *et al.* 2001). The current bolometric luminosity of η Car is believed to be $10^{6.7} L_\odot$, as derived from its re-radiated IR luminosity. For an assumed surface composition, the minimum mass inferred for the system is approximately $120 M_\odot$ from the Eddington limit. If a putative secondary component of η Car has a mass of $\leq 30 M_\odot$, the primary must be at least $90 M_\odot$.

Finally, turning to red supergiants, their high luminosities and low temperatures conspire to produce enormous physical radii with highly extended atmospheres. For example, Betelgeuse

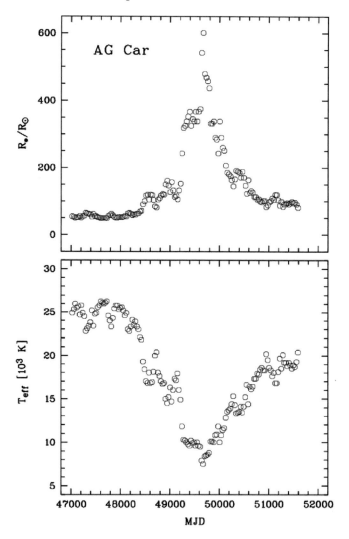

Fig. 3.7 Variation in effective temperature and radius for the LBV AG Carinae between 1988 and 2000. From Stahl *et al.* (2001).

has a radius of $645 \pm 129 R_\odot$ (3 AU!) according to interferometric data from the IOTA (Infrared-Optical Telescope Array) and its Hipparcos parallax, from which an effective temperature of 3640 K and luminosity of $\log L_*/L_\odot = 4.8$ may be derived (Perrin *et al.* 2004). In the optical, M supergiants are classified by their strong TiO (and VO) molecular bands, as illustrated in Fig. 3.8, the opacity of which has only recently been incorporated into atmospheric models for late-type stars, leading to an increased temperature–spectral type calibration for mid-M subtypes. Consequently, accurate temperatures and luminosities have eluded quantitative studies of RSGs until recently (e.g. Levesque *et al.* 2005, 2006).

Progenitor masses of $M_{\rm init} \leq 20\text{–}25\, M_\odot$ are inferred for normal RSG from comparison with the latest evolutionary models allowing for rotational mixing. These rely upon assumed

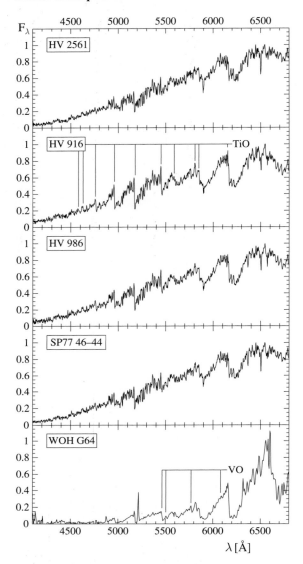

Fig. 3.8 Optical spectra of M supergiants showing the strong TiO, with spectral types in the range M1.5–2.5 except for WOH G64 (M 7.5) which has strong VO bands. From van Loon *et al.* (2005).

initial rotational velocities for which initial rates of $v_{init} = 300\,\mathrm{km\,s^{-1}}$ are commonly adopted, yet average *empirical* rotational velocities are substantially lower, e.g. $175\,\mathrm{km\,s^{-1}}$ for O stars in NGC 346 (SMC) according to Mokiem *et al.* (2006). As a consequence, the actual upper mass limit for stars evolving through an RSG phase may be $M_{init} \sim 30$–$35 M_\odot$.

The temperature scale of lower mass red giants is well constrained from interferometric datasets; there are insufficient nearby RSGs for this approach, for which Betelgeuse represents

Table 3.3 *Temperature scale and bolometric correction of Milky Way and Magellanic Cloud RSG from Levesque et al. (2005, 2006).*

Sp type	Milky Way		LMC		SMC	
	T_{eff} K	BC mag	T_{eff} K	BC mag	T_{eff} K	BC mag
K1–K1.5	4100	−0.79	4300	−0.70	4196	−0.73
K2–K3	4015	−0.90	4050	−0.80	4020	−0.92
K5–M0	3840	−1.16	3850	−1.09	3788	−1.27
M0	3790	−1.25	3738	−1.31	3625	−1.62
M1	3745	−1.35	3695	−1.45	3625	−1.61
M1.5	3710	−1.43	3654	−1.59	..	..
M2	3660	−1.57	3625	−1.69	3475	−2.07
M2.5	3615	−1.70	3545	−1.99	..	..
M3	3605	−1.74	3542	−2.01	..	..
M3.5	3550	−1.96	..	..	..	..
M4–M4.5	3535	2.03	3450	−2.18	..	..
M5	3450	−2.49	..	..	..	..

the best calibrator. Instead, historically, temperature scales have typically employed broadband colors, from a calibration of those with measured diameters, plus a blackbody continuum distribution. This approach suffers from the role of gravity on the intrinsic B–V color and the lack of continuum windows, such that a model atmosphere approach would be preferred.

A minimum requirement for late-type supergiants is spherically symmetric geometry in LTE, allowing for all relevant opacities. However, molecular opacities, from e.g. TiO, need to be incorporated into model atmosphere codes, which have been generally neglected. Laboratory molecular datasets are typically obtained at a much lower temperature than those of M stars, such that theoretical line lists have been necessary. These have only recently been calculated (e.g. Harris, Polyansky, & Tennyson 2002). The improved treatment for molecular opacities within suitable model atmosphere codes – such as MARCS (Gustafsson *et al.* 1975) – has permitted a revised temperature scale and bolometric calibration for Milky Way K and M supergiants. This is presented in Table 3.3, and agrees well with the interferometrically derived temperature for Betelgeuse (M2 Iab), although the validity of LTE within the low densities of red supergiants may be called into question. Similar calibrations have been presented for RSG in the Magellanic Clouds.

3.3 Surface gravities and masses

The most widely used method of spectroscopically deriving a hot star's mass is via a determination of the surface gravity from fitting line profile wings of Balmer hydrogen absorption lines, incorporating Stark broadening theory (e.g. Vidal, Cooper, & Smith 1973). In early-type stars, line broadening is due to (microscopic) thermal broadening, (macroscopic) rotational broadening, both with characteristic Gaussian line profiles, plus linear Stark pressure broadening with a Lorenzian profile. Typical hydrogenic (and He I) lines possess Gaussian line cores and Lorenzian wings.

The spectroscopic mass is determined from this gravity (pressure) determination, using

$$g = \frac{GM_*}{R_*^2} \tag{3.1}$$

and is expressed as $\log g$ in cgs units. In reality, the gravity determined for hot stars is affected by uncertainties in the radiation pressure in the line-wing forming regions (see next chapter). Typical values for OB dwarfs are $\log g = 4$, with lower gravities implying more extended atmospheres, for supergiants, e.g. $\log g \sim 3.5$ in O supergiants, $\log g \sim 3$ for early B supergiants, or $\log g \sim 2$ for early A supergiants. High dispersion optical spectroscopy is needed in order to model the Balmer line wings accurately enough.

Herrero *et al.* (1992) first established a "mass discrepancy" between the spectroscopic mass determinations of Galactic OB stars and those resulting from comparison with evolutionary tracks. Evolutionary masses from the location in the H-R diagram were up to a factor of two higher than spectroscopically derived masses. Potential problems lay with both the stellar luminosities obtained from model analyses, due to uncertain distances for galactic objects, and in the evolutionary models. This has not been fully resolved, although improvements in both stellar atmosphere and interior models have helped to reduce the discrepancy. Similar results have been identified for O stars in the Magellanic Clouds. Interior models will be discussed in Chapter 5.

Kudritzki, Bresolin, & Przybilla (2003) utilized the near constant $g/T_{\rm eff}^4$ (flux weighted gravity) among A and B supergiants to derive a relationship between absolute bolometric magnitude and flux weighted gravity, i.e.

$$-M_{\rm Bol} = a \log(g/T_{\rm eff}^4) + b \tag{3.2}$$

where $a = -3.71$ and $b = 13.49$ for Local Group AB supergiants plus counterparts in NGC 300 and NGC 3621. Since AB supergiants are visually the brightest stars in external galaxies, this has the potential for providing independent distances to galaxies at up to 10 Mpc.

Strong stellar winds mask the stellar photospheres of W-R stars so alternative techniques to spectroscopic gravities need to be applied. If bolometric luminosities of H-free W-R stars can reliably be inferred, one may apply the theoretical mass–luminosity relation (see Eq. 5.1). More directly, surface gravities may be inferred for W-R stars whose masses are known from binary orbits and radii are derived from spectroscopic analysis. For example, the WC component of γ Velorum has been studied from which a stellar mass of $7M_\odot$ is obtained, somewhat lower than $9.5\,M_\odot$ resulting from an orbital solution (De Marco *et al.* 2000).

3.4   Surface composition

The French philosopher Auguste Comte famously, and inaccurately, predicted of stars in 1835 that "... *we would never know how to study by any means their chemical composition* ..." Fortunately, theoretical knowledge of stellar atmospheres and atomic data, together with high-resolution spectrographs, does now permit abundance determinations for most normal stars, including hot, luminous OB stars.

This can only readily be carried out using optical photospheric lines of H, He and metals, based on the same non-LTE techniques described above that are used to derive $T_{\rm eff}$ and $\log g$. Consequently, solely hydrogen–helium abundances have been widely determined for a relatively large number of O stars. Most main sequence stars show normal He abundances, He/H ≈ 0.1 by number, as expected. Rapidly rotating OB stars appear to display enriched He at their surfaces, suggesting that rotational mixing has brought products of core nucleosynthesis to the surface while the star is still on the Main Sequence.

CNO elemental abundances in blue supergiants have generally been lacking until recently, with the notable exception of Venn (1995b, 1999) who studied Milky Way and SMC A supergiants. Great care should be taken when analysing OB stars for abundance determinations since the neglect of "micro-turbulent" line broadening may imply erroneous He-enrichments. There is both qualitative and quantitative evidence for enhanced nitrogen, accompanied with decreased carbon from studies of optical photospheric lines, from Venn for A supergiants and Trundle & Lennon (2005) and Crowther, Lennon, & Walborn (2006a) for B supergiants, as shown in Table 3.4 (Milky Way) and Table 3.5 (SMC). Typical nitrogen enrichments are of a factor of 10 with respect to carbon i.e. $\log(N/C) - \log(N/C)_{HII} = [N/C] \sim 1$, and a factor of 5 with respect to oxygen i.e. $[N/O] \sim 0.7$. Similar conclusions are reached from analysis of UV spectral lines, as carried out by Pauldrach, Hoffmann, & Lennon (2001) for the O supergiants ζ Pup and α Cam.

Table 3.4 *CNO elemental abundances* ($\log X/H + 12$) *in Milky Way blue supergiants (Venn 1995b; Crowther, Lennon, & Walborn 2006a; Pauldrach, Hoffmann, & Lennon 2001) versus the solar photosphere (Asplund et al. 2005) and the Orion HII region (Esteban et al. 1998).*

	Sun	HII	A Iab	B Iab	O Iab[a]
C	8.4	8.4	7.8 ± 0.1	7.9 ± 0.3	8.2/7.4
N	7.8	7.8	7.9 ± 0.2	8.4 ± 0.3	8.9/8.0
O	8.7	8.6	..	8.5 ± 0.3	8.8/8.4
log (N/C)	−0.6	−0.6	+0.1	+0.5	+0.7/ + 0.6
log (N/O)	−0.9	−0.8	..	−0.1	+0.1/ − 0.4

(a) ζ Pup and α Cam, respectively.

Table 3.5 *CNO elemental abundances* ($\log X/H + 12$) *in SMC blue supergiants (Venn 1999; Trundle & Lennon 2005; Evans et al. 2004a) versus SMC HII regions (Kurt et al. 1999).*

	HII	A Iab	B Iab	O Iab
C	7.5	..	7.3 ± 0.1	7.4 ± 0.2
N	6.6	7.5 ± 0.1	7.7 ± 0.3	8.1 ± 0.3
O	8.1	8.1 ± 0.1	8.1 ± 0.1	8.0 ± 0.2
log (N/C)	−0.9	..	+0.4	+0.7
log (N/O)	−1.5	−0.6	−0.4	+0.1

The current consensus amongst stellar atmosphere model studies indicates that pristine CNO abundances are observed in OC and BC stars with weak nitrogen lines. In "normal" OB stars, moderate enrichment of nitrogen is found at the expense of carbon. Large nitrogen enrichments are observed in ON stars and rare BN stars with strong nitrogen lines. The high degree of enrichment at relatively early evolutionary phases – much greater than those predicted by standard interior models – led to the incorporation of rotational mixing in stellar evolutionary models, as we shall see in Chapter 5. Indeed, significant CNO processing *is* now predicted at the surfaces of blue supergiants evolving towards the red part of the H-R diagram, with more extreme abundances expected for those returning from a red supergiant phase, owing to convective dredge-up.

Relatively few luminous blue variables have been subject to detailed abundance analysis, due to the need for complex, non-LTE, spherical model atmospheric tools. Those that have been studied indicate rather uniform surface helium abundances of He/H $\sim$ 0.4 by number, intermediate between values observed in OB stars and late-type WN stars, adding to the evidence for an evolutionary link between the two groups. Most recently, high quality UV and optical spectra of η Car have been obtained with HST and analyzed to reveal enriched nitrogen by at least a factor of ten, whilst carbon and oxygen are depleted. ζ^1 Sco (HD 152236), one of two early B hypergiants studied by Crowther, Lennon, & Walborn (2006a), possesses a nitrogen enhancement of [N/O] = +1.3 which is more typical of nebulae associated with luminous blue variables, e.g. [N/O] = +1.3 to +1.9 dex for AG Carinae and R127 (Lamers *et al.* 2001). Such ejecta nebulae will be discussed further in Chapter 8.

For W-R stars, it has long been suggested that abundances represented the products of core nucleosynthesis (Gamov 1943; Paczynski 1967). However, it was only from detailed analysis of recombination lines that WN stars are indeed found to be depleted in H, C, and O, with He and N enriched, and conversely WC stars are enhanced in He, C, and O, with H and N absent (Smith 1973; Smith & Hummer 1988). These are presented in Table 3.6. WN abundance patterns are consistent with material processed by the CNO nuclear burning cycle, whilst WC stars reveal products of He burning as we shall see in Chapter 5. Elemental abundances in rare intermediate WN/C stars are in good agreement with these stars occupying a brief transition stage between the WN and WC stages.

Table 3.6 *Elemental abundances in Galactic W-R stars, by number, as derived from non-LTE model atmosphere models (Massey 2003).*

	H/He	N/He	C/N	(C+O)/He
WNL	$\leq$4	0.005	0.04	..
WNE	$\leq$0.5	0.005	0.04	0.0004
WN/C	0.0	0.005	2–3	0.025
WC	0.0	0.0	∞	0.1$-$3
Cosmic	10	0.001	5	0.015

Recombination line studies, using theoretical coefficients for different transitions, are most readily applicable to WC stars, since they show a large number of lines in their optical spectra. Atomic data are most reliable for hydrogenic ions, such as high lying transitions of C IV and O VI, so early-type WC and WO stars can be most readily studied (e.g. Kingsburgh, Barlow, & Storey 1995). In reality, optical depth effects come into play, so detailed abundance determinations for all subtypes can most reliably be carried out using non-LTE model atmospheres.

One successful method of determining elemental abundances in WC stars introduced by Hillier (1989) is shown in Fig. 3.9 for an LMC WC4 star, in which non-LTE models of varying carbon abundance, 15 %, 25 %, and 35 % by mass, are compared with the optical HeII 5412 and CIV 5471 recombination lines, from which an abundance of C/He ~ 0.13 (~ 25 % by mass) is determined. Such optical and near-IR recombination lines are formed at high densities of 10^{11} cm^{-3} at radii of 3–30 R_* (recall Fig. 3.6) within their inner dense stellar winds.

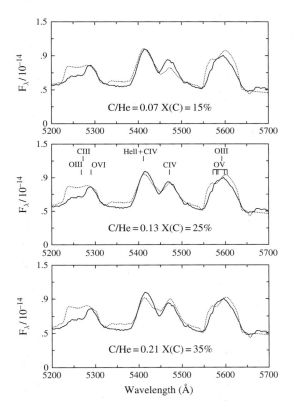

Fig. 3.9 Illustration of the determination of C/He abundances in WC stars, using model fits (dotted) to the de-reddened optical spectrum of HD 32125 (WC4, LMC, solid), from which C/He ~ 0.13 by number, or a carbon content of 25 % by mass is derived. Adapted from Crowther *et al.* (2002a).

In addition, Wolf–Rayet stars also show fine-structure mid-IR lines of neon and sulphur, formed at critical densities of 10^5 cm^{-3}, corresponding to hundreds of stellar radii. The [Ne II] 12.8 μm and [Ne III] 15.5 μm stellar lines were first investigated by Barlow, Roche, & Aitken (1988) for γ Vel (WC8+O) using an analytical approach. These fine-structure lines are also commonly seen in HII regions, in which narrow *nebular* lines are observed, if the ultraviolet flux distributions of stars ionizing the nebula are sufficiently "hard". In contrast, broad *stellar* lines originate due to the high density winds from Wolf–Rayet stars. Barlow *et al.* (1988) came to the conclusion that neon was not greatly enhanced in this WC star. This was unexpected, since interior evolutionary models predicted substantial enrichment of neon (~ 2 % by mass), as we shall see in Chapter 5. This question was re-visited when ISO was launched, but the line fluxes measured by Barlow *et al.* were supported. It was only when the clumped nature of W-R winds was taken into consideration, as we shall see in Chapter 4, that neon was found to be enhanced by a factor of ∼ 10, with ∼1 % by mass. Similar enrichments were measured for other WC stars from recent mid-IR Spitzer spectroscopic observations, and are shown for two WC stars in Fig. 3.10, illustrating strong, broad [Ne II] 12.8 μm and/or [Ne III] 15.5 μm emission lines. In the case of rapidly rotating low metallicity stars, the abundance of ^{22}Ne is predicted to be much higher than the original CNO abundance, via primary nitrogen production (see Chapter 5). Such predictions remain to be observationally tested.

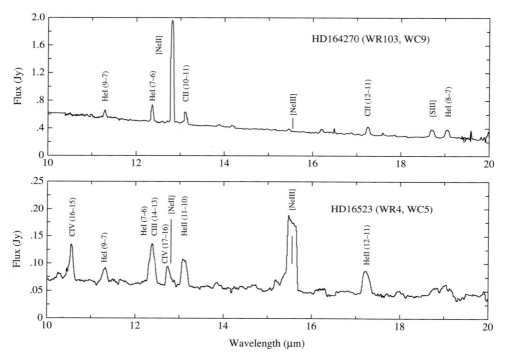

Fig. 3.10 Mid-IR Spitzer IRS spectroscopy of two Milky Way WC stars, revealing strong [Ne II] 12.8 μm or [Ne III] 15.5 μm emission lines formed in the low density outer wind, in addition to He and C recombination lines formed in the high density inner wind (see Crowther *et al.* 2006b).

For a subset of hot, luminous stars an alternative approach to the determination of elemental abundances is possible via study of their circumstellar environment, as will be discussed in Chapter 8. In this way one determines the surface abundances during the epoch at which the material was ejected through its stellar wind, typically 10^4 years previously. An added complication is that abundances obtained in this way may be mixed with the local ISM. However, since this technique is most readily applied to stars with dense HII regions, generally luminous blue variables and a subset of W-R stars, their local environment will generally have been evacuated by the stellar wind. Chapter 8 will discuss the immediate environment of massive stars in greater detail.

4

Stellar winds

Stellar winds are ubiquitous amongst massive stars, although the physical processes involved depend upon the location of the star within the H-R diagram. Mass-loss crucially affects the evolution and fate of a massive star (Chapter 5), while the momentum and energy expelled contribute to the dynamics and energetics of the ISM (Chapter 8). The interested reader is referred to the monograph by Lamers & Cassinelli (1999) on the topic of stellar winds, or Kudritzki & Puls (2000) for a more detailed discussion of mass-loss from OB and related stars.

The existence of winds in massive stars was first proposed by Beals (1929) to explain the emission line spectra of Wolf–Rayet stars. This gained observational support in the 1960s when the first rocket UV missions revealed the characteristic P Cygni signatures of mass-loss from CIV $\lambda 1550$, SiIV $\lambda 1400$, and NV $\lambda 1240$ in O stars (Morton 1967). A theoretical framework for mass-loss in hot stars was initially developed by Lucy & Solomon (1970) involving radiation pressure from lines, and refined by Castor, Abbott, & Klein (1975), thereafter known as CAK theory. The observational characteristics of stellar winds are velocity and density. The former can be directly observed, whilst the latter relies on a varying complexity of theoretical interpretation.

4.1   Radiation pressure

When a photon is absorbed or scattered by matter, it imparts its energy, $h\nu$, and momentum, $h\nu/c$, where h is Planck's constant and c is the velocity of light. Consequently, radiation is a very inefficient carrier of momentum. Nevertheless, we will show that continuum and line radiation pressure is capable of driving an outflow from early-type stars. Radiation pressure may be neglected for low temperature stars such as the Sun, but becomes increasingly important for luminous high temperature stars (e.g. Underhill 1949). Stellar winds are primarily characterized by mass-loss rate $\dot{M} = dM/dt$ and terminal wind velocity v_∞. For a spherical, stationary wind at a distance r, the mass-loss rate relates to the velocity field, $v(r)$, and density, $\rho(r)$, via the equation of continuity

$$\dot{M} = 4\pi r^2 \rho(r) v(r) \tag{4.1}$$

in which $v_\infty = \lim_{r\to\infty} v(r)$.

Electron (Thompson) scattering dominates the visible continuum opacity in hot stars. This is independent of wavelength, so we may consider a so-called "gray" atmosphere in which the total radiation pressure, P_R, is

$$P_{\mathrm{R}} = \frac{4\sigma}{3c}\,T_{\mathrm{eff}}^4, \tag{4.2}$$

where σ is the Stefan–Boltzmann constant and T_{eff} is the effective temperature. We may estimate the radiative acceleration, g_{e}, i.e.

$$g_{\mathrm{e}} = \frac{4\pi}{c}\,\frac{q\sigma_{\mathrm{e}}}{m_{\mathrm{H}}}\,\sigma\,T_{\mathrm{eff}}^4, \tag{4.3}$$

where the number of free electrons per atomic mass unit is q, σ_{e} is the Thompson cross-section and m_{H} is the mass of a hydrogen atom. We can then write the ratio of radiative acceleration due to electron scattering to the surface gravity as Γ_{e}, the so-called Eddington parameter,

$$\Gamma_{\mathrm{e}} = \frac{4\pi}{c}\,\frac{q\sigma_{\mathrm{e}}}{m_{\mathrm{H}}}\,\sigma\,T_{\mathrm{eff}}^4\,\frac{R_*^2}{GM_*} = \frac{q\sigma_{\mathrm{e}}}{cm_{\mathrm{H}}G}\,\frac{L_*}{M_*} = 10^{-4.5}\,q\,\frac{L_*/L_\odot}{M_*/M_\odot}. \tag{4.4}$$

For the Sun, $\Gamma_{\mathrm{e}} \ll 1$, but massive stars approach the limit $\Gamma_{\mathrm{e}} = 1$, since $L \propto M_*^{2.5}$ for hot, luminous stars. For a given stellar mass, there exists a maximum (Eddington) luminosity, L_{Edd}, above which no stable star may exist, i.e.

$$\frac{L_{\mathrm{Edd}}}{L_\odot} = 10^{4.5}\,q^{-1}\,\frac{M_*}{M_\odot}. \tag{4.5}$$

We may write the Eddington parameter, Γ_{e}, in terms of effective temperature and gravity, so the Eddington limit corresponding to $\Gamma_{\mathrm{e}}^{\mathrm{Max}} = 1$ becomes a straight line in the $\log T_{\mathrm{eff}}$, $\log g$ diagram,

$$\log \Gamma_{\mathrm{e}}^{\mathrm{Max}} = -15.12 + \log q + 4\log T_{\mathrm{eff}} - \log g = 0, \tag{4.6}$$

as shown in Fig. 4.1 for a pure H atmosphere.

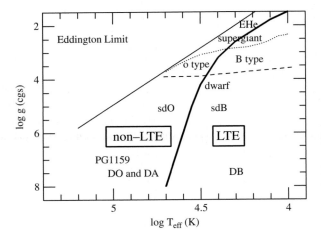

Fig. 4.1 Regimes in effective temperature and gravity in which non-LTE versus LTE models operate. The Eddington limit for a pure hydrogen star is indicated (thin line) as are the OB dwarf (dashed) and supergiant (dotted) scales. From Crowther (1998).

Although electron scattering dominates the *continuum* opacity in hot stars, the basic mechanism by which hot star winds are driven is the transfer of photospheric photon momentum to the outer stellar atmosphere through absorption by spectral *lines*. Bound electrons provide much more opacity than free electrons. It is the combination of a plethora of line opacity within the same spectral region (extreme UV) as the maximum flux of photospheric radiation which allows for efficient driving of winds by radiation pressure (Milne 1926).

In a static atmosphere, the photospheric radiation will only be efficiently absorbed or scattered in the lower layers of the atmosphere, weakening any radiative acceleration from the outer layers. However, if the outer layers are expanding, the velocity gradient allows the atoms to see the photosphere unimpeded as red-shifted radiation. The Doppler shift allows atoms in the outer atmosphere to absorb undiminished continuum photons in their line transitions, as shown in Fig. 4.2. Sobolov (1960) first realized that radiation emitted from the star at a wavelength blueward of the line moves freely until it is redshifted by the accelerating outflow into a local line resonance.

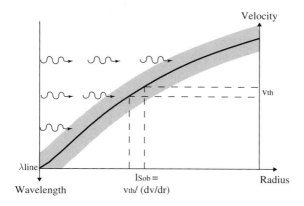

Fig. 4.2 Doppler-shifted resonance line absorption in an accelerating wind. Photons with a wavelength just shortward of a line propagate freely until – in the reference frame of the outward flowing medium – they are redshifted into a narrow line resonance (whose width is represented by the shaded region due to thermal line broadening). Reprinted from Owocki (2004) with permission. Copyright 2004 by EDP Sciences.

Using the notation of Owocki (2001), the geometric width of this resonance is $v_{\text{th}}/(\mathrm{d}v/\mathrm{d}r)$, a Sobolev length, where the ion thermal speed is v_{th}. In a supersonic outflow, the Sobolev length is of order $v_{\text{th}}/v \ll 1$ smaller than the scale of a variation in the typical flow. Consequently, line scattering can be described by local conditions at any radius. For example, the total optical depth of radiation passing through the line resonance may be approximated in terms of the *local* density and velocity gradient:

$$\tau = \frac{\kappa \rho v_{\text{th}}}{\mathrm{d}v/\mathrm{d}r} \tag{4.7}$$

where ρ is the density and κ the line opacity. This leads to a straightforward expression for the line acceleration,

$$g_{\text{line}} = g_{\text{thin}} \frac{(1 - e^{-\tau})}{\tau}. \tag{4.8}$$

In the optically thin limit, $\tau \ll 1$, the line acceleration resembles the electron scattering case,

$$g_{\text{thin}} = \frac{\kappa v_{\text{th}} v_0 L_\nu}{4\pi r^2 c^2} \approx \frac{\kappa}{\kappa_e} \frac{v_{\text{th}}}{c} g_e, \tag{4.9}$$

where the line frequency v_0 is assumed to lie near the peak of the stellar luminosity L_ν, i.e. $v_0 L_\nu \approx L_*$. Conversely, in the limit of an optically thick line with $\tau \gg 1$,

$$g_{\text{thick}} \approx \frac{g_{\text{thin}}}{\tau} = \frac{L_*}{4\pi r^2 \rho c^2} \frac{dv}{dr} = \frac{L_*}{\dot{M} c^2} v \frac{dv}{dr}, \tag{4.10}$$

where we use the definition of the wind mass-loss rate from Eq. 4.1. The force from optically thick lines is independent of the opacity, and scales with the velocity gradient. Neglecting gravity for the moment, the steady-state acceleration from a single optically thick line near the peak of the continuous energy distribution can be approximated as

$$v \frac{dv}{dr} = g_{\text{line}} \approx \frac{L_*}{\dot{M} c^2} v \frac{dv}{dr}. \tag{4.11}$$

As the acceleration cancels out, $\dot{M} \approx L_*/c^2$, i.e. the mass-loss scales with the stellar luminosity. In reality, hot-star winds are driven by many optically thick lines. If these are independent, the total mass-loss scales with the number of optically thick lines, N_{thick}, i.e.

$$\dot{M} \approx N_{\text{thick}} \frac{L_*}{c^2}. \tag{4.12}$$

If the wind has a terminal velocity v_∞ there can be at most $N_{\text{thick}} \approx c/v_\infty$ non-overlapping thick lines spread through the spectrum. This implies the so-called "single-scattering limit" for mass-loss,

$$\dot{M} v_\infty \leq \frac{L_*}{c}. \tag{4.13}$$

Winds of hot luminous stars do not generally exceed this limit. Extreme O supergiants and Wolf–Rayet stars do exceed the single-scattering limit, which requires a more elaborate multiple scattering treatment that we shall discuss below.

Of course, the properties of hot star winds depend on both the number of lines available to absorb photon momentum, and on their ability to absorb, i.e. their optical thickness and their frequential position. Ions which provide the radiative acceleration in hot stars constitute a very small fraction of the wind, perhaps 10^{-5}. Most of the wind is composed of H and He which barely contribute to the acceleration, as they are almost fully ionized. Ions which absorb strong metal line photons will typically increase their velocity by ~ 20 cm s^{-1}. Consequently, the effective increase in momentum per absorption is 2×10^{-3} cm s^{-1}, since the momentum gained by the ion is shared with all the constituents of the wind, via collisions with surrounding protons, ions, and electrons due to the charge of the ion. The velocity reached far from the stellar surface, say at $\sim 50 R_*$, tends towards a maximum (so-called "terminal") value, v_∞. Typically $v_\infty \sim 2000$ km s$^{-1} = 2 \times 10^8$ cm s^{-1}, therefore of order 10^{11} absorptions are required. The time available to accelerate the gas is $50 R_*/v_\infty \sim 1.5 \times 10^5$ s for a typical O7 dwarf with radius $R_* = 10 R_\odot$. Ions which provide the acceleration have to absorb typically $10^{11}/1.5 \times 10^5 = 7 \times 10^5$ photon s^{-1}. Indeed, this is typical of strong UV metal lines, which are plentiful due to their complex electronic structure.

Castor, Abbott, & Klein (1975) (CAK) developed a self-consistent solution of the wind properties, in which the flux-weighted number distribution of lines is approximated by a power law in the line opacity, i.e.

$$\frac{dN}{d\kappa} = \frac{1}{\kappa_0}\left(\frac{\kappa}{\kappa_0}\right)^{\alpha-2}, \tag{4.14}$$

where $0 < \alpha < 1$ and κ_0 is defined such that $\kappa_0 \, dN/d\kappa = 1$. The cumulative force can be obtained by integrating the single-line result over this opacity distribution. The CAK line force lies between the optically thin ($\alpha = 0$) and optically thick ($\alpha = 1$) extremes.

Consequently, a hot star wind overcomes gravity ($g = GM_*/r^2$) through a combination of continuum opacity ($g_e = g\Gamma_e$) and line opacity (g_{line}) as follows:

$$v\frac{dv}{dr} = -g + g_e + g_{\text{line}} = g_{\text{line}} - \frac{GM_*}{r^2}(1 - \Gamma_e). \tag{4.15}$$

From CAK theory we obtain a velocity law of the following form:

$$v(r) = v_\infty \left(1 - \frac{R_*}{r}\right)^\beta, \tag{4.16}$$

where $\beta = 0.5$. The terminal velocity is given by

$$v_\infty = v_{\text{esc}}\left[\frac{\alpha}{(1-\alpha)}\right]^{1/2} \tag{4.17}$$

in which v_{esc} is the (effective) surface escape velocity

$$v_{\text{esc}} = (2g_{\text{eff}}R_*)^{0.5}. \tag{4.18}$$

Inserting a typical value of $\alpha = 0.6$ for O stars implies $v_\infty/v_{\text{esc}} \sim 1.2$ which is far too low in comparison with observations. Pauldrach, Puls, & Kudritzki (1986), hereafter PPK, revised the CAK approach, revealing a slightly non-linear dependence of v_∞ on v_{esc}. In the above equation the (effective) gravity

$$g_{\text{eff}} = \frac{GM_*}{R_*^2}(1 - \Gamma_e), \tag{4.19}$$

i.e. the stellar gravity corrected for the reducing effect of radiation pressure (via Thompson scattering) on the gravitational potential via the Eddington parameter Γ_e. Owocki (2003) expresses the solution to the CAK mass-loss rate in terms of the Eddington parameter,

$$\dot{M} \propto \frac{L_*}{c^2}\left[\frac{\Gamma_e}{1 - \Gamma_e}\right]^{-1+\frac{1}{\alpha}}. \tag{4.20}$$

Smith & Conti (2008) have applied this formulation to explain the luminous H-rich WN stars at a relatively early post-main sequence phase. On the main sequence, the stellar luminosity increases, while the mass decreases due to the stellar wind, causing an increase in Γ_e and consequently in the mass-loss rate.

From the above, there is expected to be a close relationship between the terminal velocity and the escape velocity, and it is clear that the determination of stellar escape velocities depends on knowledge of stellar masses and radii. These results apply under the idealized assumption that the stellar radiation is radially streaming from a point source. If the finite angular extent of the disk is taken into account (Friend & Abbott 1986; PPK) the velocity law in Eq. (4.16) is flatter, with $\beta \sim 0.8$ for hot O stars with $40\,000\,\text{K} \leq T_{\text{eff}} \leq 50\,000\,\text{K}$.

Abbott (1982) originally developed the concept of force multiplier parameters. α controls the fraction of optically thick versus optically thin lines (recall Eq. 4.14), k controls the number of lines with strengths larger than a critical value, and δ describes the change in ionization. The radiative acceleration, g_{line}, may be written in terms of the electron scattering acceleration g_e via $\Upsilon(\alpha, k, \delta)$ involving these force multipliers as follows:

$$g_{\text{line}} = g_e \Upsilon(\alpha, k, \delta) = k \left(\frac{\sigma_e^{\text{ref}} v_{\text{th}} \rho}{dv/dr} \right)^{-\alpha} \left(\frac{n_e}{W(r)} \right)^{\delta}, \tag{4.21}$$

where σ_e^{ref} is a reference value of the electron scattering opacity, v_{th} is the mean thermal velocity of protons in a wind whose temperature is equal to the effective temperature of the star, n_e is the electron density, and $W(r)$ is the radiative "dilution factor",

$$W(r) = \frac{1}{2} \left[1 - \left(1 - \frac{R_*}{r} \right)^{1/2} \right]. \tag{4.22}$$

Finally, the equation of motion, Eq. 4.15, may be written in the form

$$v \frac{dv}{dr} = \frac{GM}{r^2} \left[\Gamma_e (1 + \Upsilon(\alpha, k, \delta)) - 1 \right]. \tag{4.23}$$

It is clear that both Γ_e and $\Upsilon(\alpha, k, \delta)$ need to be large for an early-type star to possess a wind, i.e. $v\, dv/dr \geq 0$. We will discuss recent developments involving a Monte Carlo approach later in this chapter.

Since metal lines drive the winds of OB stars, their properties are predicted to depend on the metal content, Z (e.g. Kudritzki, Pauldrach, & Puls 1987). Modern revisions are discussed by Kudritzki & Puls (2000) such that

$$\dot{M} \propto Z^{(1-\alpha)/(\alpha-\delta)}. \tag{4.24}$$

Inserting typical O star values of α (= 0.6 to 0.7) and δ (= 0.0 to 0.1) suggests an exponent of ~0.6 for the metallicity dependence of mass-loss rates for O stars. Specifically, Vink *et al.* (2001) predicted an exponent of 0.69 ± 0.10. An empirical comparison between winds of O stars in the Milky Way, LMC, and SMC presented in Fig. 4.3 suggests an exponent of 0.83 ± 0.16 (Mokiem *et al.* 2007), i.e. mass-loss rates of O stars in the solar neighborhood are on average a factor of four times stronger than those of equivalent SMC O-type stars. Wind properties of OB stars in the Magellanic Clouds will be compared to Milky Way counterparts later in this chapter.

Finally, a widely used method of comparing wind properties of stars in different metallicity environments is through the so-called wind "momentum–luminosity relation". Kudritzki *et al.* (1999) introduced a theoretical relationship between the bolometric luminosity of an early-type star and the wind momentum, $\dot{M} v_\infty$, modified by the square-root of the stellar radius as follows:

$$\dot{M} v_\infty (R_*/R_\odot)^{0.5} \propto L_*^{1/\alpha_{\text{eff}}}, \tag{4.25}$$

where $\alpha_{\text{eff}} = \alpha - \delta \sim 0.6$ for typical force multiplier values appropriate for O stars. The wind momentum–luminosity relationship provides a potentially independent method of determining distances to galaxies beyond the Local Group, although it is primarily of use for lower

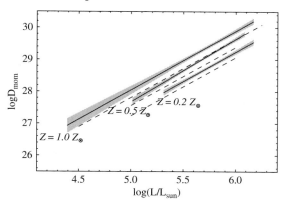

Fig. 4.3 Empirical wind momenta ($D_{mom} = \dot{M}v_\infty R_*^{0.5}$ in cgs units) of Galactic, LMC and SMC O stars (solid lines), together with theoretical predictions (dotted lines) of Vink *et al.* (2001). Mass-loss rates are not corrected for possible wind clumping. From Mokiem *et al.* (2007).

luminosity A supergiants which are visually the brightest hot stars in galaxies, as shown in Fig. 4.4. Kudritzki *et al.* (1999) obtained $\alpha_{eff} \sim 0.4$–0.5 for A and mid-B supergiants, lower than for O stars, as theoretically expected for driving by lower ionization iron peak lines. The flux-weighted gravity–luminosity relation (Eq. 3.2) provides an alternative method from which distances may be derived using AB supergiants without recourse to mass-loss rates.

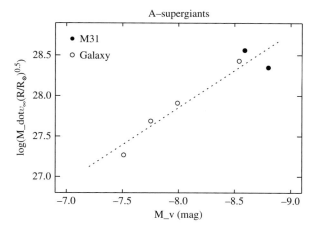

Fig. 4.4 Wind momenta of galactic and M31 A supergiants as a function of absolute visual magnitude. The dashed line is a linear regression to the data, i.e. $\log(\dot{M}v_\infty R_*^{0.5}) = 20.52 - 0.916M_V$. From Kudritzki *et al.* (1999).

Calculations by Puls, Springmann, & Lennon (2000) and Vink, de Koter, & Lamers (2000, 2001) reveal that light, relatively simple elements (e.g. CNO) are the principal line drivers for the outer, supersonic part of the wind, whilst heavy, complex elements (e.g. Fe) are responsible for the inner, subsonic part. CNO elements determine the wind velocity, and the Fe-peak elements dictate the mass-loss rate, at least for stars with compositions close to solar metallicity.

4.2   Wind velocities

Ultraviolet P Cygni profiles, ubiquitous in O-type stars, provide a direct indication of stellar winds (Morton 1967; Walborn, Nichols-Bohlin, & Panek 1985). Such profiles involve the absorption of photons by an atom and rapid re-emission by spontaneous de-excitation, and are generally associated with resonance transitions from abundant ions, such as C^{3+}, N^{4+}, Si^{3+}. Figure 4.5 illustrates the formation of a P Cygni line profile. Wind material approaching the observer within a column in front of the star is blueshifted by the Doppler effect. Scattering of radiation out of this direction causes a reduction in the observed flux on the blue side of the profile. From the regions on either side of this column wind material may scatter radiation towards the observer. This may occur from either the approaching (blue-shifted) or receding (red-shifted) hemisphere, the extra flux seen by the observer, originating from this scattered radiation, is a symmetric emission component on both sides of the profile. The overall effect is asymmetric with blue absorption and red emission. If the column density of the absorbing ion is small, solely a blue-shifted absorption line will be observed.

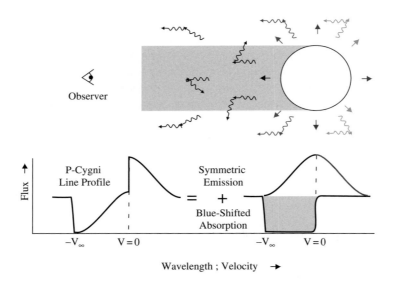

Fig. 4.5 Schematic showing the formation of a P Cygni line profile in a stellar wind. Reprinted from Dessart (2004) with permission. Copyright 2004 by EDP Sciences.

OB stars

The wavelength of the blue edge of the "black" absorption provides a measure of the asymptotic wind velocity. Accurate wind velocities of OB and W-R stars can be readily obtained in this way from HST or IUE observations of saturated CIV λ1550, or SiIV λ1400 P Cygni profiles, as illustrated in Fig. 4.6. Differences between the maximum (v_{edge}) and asymptotic velocities (v_∞), for which $v_\infty \sim 0.85 v_{edge}$, are thought to be related to the shocks (e.g. Prinja *et al.* 1990). Hot star winds are intrinsically unstable, as we will discuss shortly. Terminal wind velocities may either be directly measured, or inferred from UV line profile modelling using the Sobolev with exact integration (SEI) technique (Lamers,

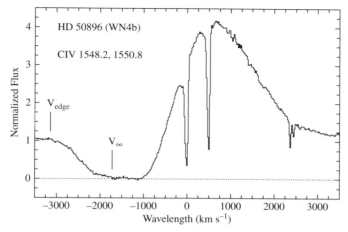

Fig. 4.6 C IV 1550 P Cygni profile of the WN4b Wolf–Rayet star HD 50896, illustrating the asymptotic wind velocity, v_∞, and maximum extent, v_{edge} (see Prinja *et al.* 1990). Narrow absorption features are interstellar features.

Cerruti-Sola, & Perinotto 1987). Recall, the radial dependence of velocities from O stars are expected to follow a β-law with $\beta = 0.8$ (Eq. 4.16).

Table 4.1 presents wind velocities for galactic O stars and B supergiants. Wind velocities from the earliest O stars approach $v_\infty \sim 3000$ km s^{-1}, i.e. one percent of the speed of light. Wind velocities of 1000 km s^{-1} are typical of early B supergiants, decreasing to 100 km s^{-1} for A supergiants.

Alas, no reliable diagnostics of O star wind velocities exist at optical wavelengths. In contrast, optical Hα line profiles from late B and A supergiants are P Cygni in nature, allowing wind velocities to be directly measured without resort to UV spectroscopy (Kudritzki *et al.* 1999). The switch from pure emission Hα in O supergiants to P Cygni Hα in A supergiants is as a result of hydrogen remaining ionized in O stars but recombining to a neutral form in the wind of A supergiants. In this case, the Lyman continuum and the Lyman lines become extremely optically thick, and the $n = 2$ level (lower state of Hα) becomes the effective ground-state, causing Hα to behave more like a resonance line than as a recombination line.

Radiatively driven wind theory predicts terminal velocities to exceed the escape velocity by a uniform factor for O stars (recall Eq. 4.17). Prinja *et al.* (1990) compare escape velocities, v_{esc}, to terminal wind velocities, v_∞, for a large sample of O stars and B supergiants. These are shown in Fig. 4.7 and reveal a mean of $v_\infty/v_{esc} = 2.36 \pm 0.51$ from 151 normal O stars. Recall from PPK that terminal velocities are predicted to exceed the escape velocity by a factor of $\sim$3. Lower α force multipliers are predicted for AB supergiants, suggesting a smaller ratio of $v_\infty/v_{esc} \sim 1$.

The number of OB stars observed in our Galaxy, principally with IUE, far exceeds that from external galaxies due to their relative UV brightness. Nevertheless, sufficient extragalactic hot stars have now been observed, principally with HST, for us to be able to make useful comparisons. Wind velocities of LMC O stars differ little from galactic counterparts. A more

Table 4.1 *Terminal wind velocities of Milky Way OB stars from the empirical measurements of Prinja, Barlow, & Howarth (1990), updated to account for the early O star classification scheme of Walborn et al. (2002).*

Sp type	Supergiant		Giant		Dwarf	
	v_∞ km s^{-1}	N	v_∞ km s^{-1}	N	v_∞ km s^{-1}	N
O2	3150	(1)	–			
O3	–		–			
O3.5	–		–		3265	(2)
O4	2325	(3)	–		2965	(6)
O5	1885	(1)	2810	(4)	2875	(4)
O5.5					1960	(1)
O6	2300	(1)	2560	(2)	2570	(12)
O6.5	2180	(3)	2545	(4)	2455	(10)
O7	2055	(4)	2600	(3)	2295	(10)
O7.5	1980	(1)	2175	(7)	1975	(3)
O8	1530	(3)	2125	(1)	1755	(7)
O8.5	1955	(4)	2255	(1)	1970	(2)
O9	1990	(5)	1875	(9)	1500	(7)
O9.5	1765	(7)	1505	(4)		
O9.7	1735	(7)				
B0	1535	(11)				
B0.5	1405	(14)				
B0.7	1155	(1)				
B1	1065	(20)				
B1.5	750	(4)				
B2	790	(7)				
B2.5	490	(2)				
B3	590	(5)				

prominent effect *is* observed in the SMC; Milky Way early O dwarf stars possess much stronger P Cygni absorption and higher wind than comparable SMC stars (see Kudritzki & Puls 2000). Indeed, α is predicted to be a subtle function of metallicity, $v_\infty \propto Z^{0.1}$ for OB stars.

Bistability jump

The term "bistability jump" was first discussed with respect to hot massive stars by Pauldrach & Puls (1990) for the case of the LBV P Cygni (B1.5 Ia$^+$). P Cygni has a very slow (185 km s^{-1}), dense, stellar wind. Subtle changes in predicted ionization lead to one of two quite different regimes, with hydrogen either ionized in its outer stellar wind, producing pure emission Balmer lines, or neutral causing strong P Cygni-type Balmer lines (see also Langer *et al.* 1994). Indeed, a closely related extreme B supergiant He 3–519 has been observed to switch between these states in the past decade, as shown in Fig. 4.8.

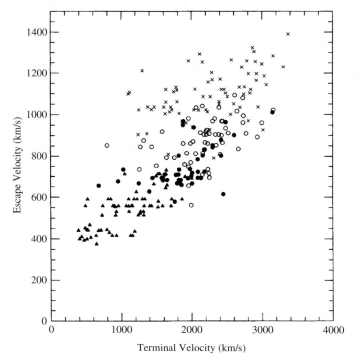

Fig. 4.7 Comparison between the escape and terminal velocities for a large sample of Milky Way O stars and B supergiants. B supergiants are shown as filled triangles, O supergiants are filled circles, O giants are open circles and O dwarfs are crosses. Reprinted from Prinja *et al.* (1990) by permission of the AAS.

Lamers, Snow, & Lindholm (1995) highlighted an apparent bistability jump in the wind properties of normal supergiants around $T_{\text{eff}} \sim 21\,000$ K, corresponding to B1.5 subtype, with $v_{\infty}/v_{\text{esc}} \simeq 2.65$ at higher temperatures, and $v_{\infty}/v_{\text{esc}} \simeq 1.4$ at lower temperatures. This sudden decrease is predicted to result from the change in ionization of elements contributing to the line force, according to theoretical wind models. Recent SMC and Milky Way studies of B supergiants confirm a trend to lower $v_{\infty}/v_{\text{esc}}$ for later B subtypes, as illustrated in Fig. 4.9, although there is no evidence in favor of a sharp discontinuity around $\sim$22 000 K once uncertainties in stellar masses are taken into account. Morphologically there is a distinction between B0.5 and B0.7 supergiants at UV wavelengths. The primary cause of this change is the substantially lower temperatures of B0.7 supergiants with respect to B0.5 supergiants (recall Table 3.1), causing the appearance of AlIII and CII lines at lower ionization.

W-R stars

In contrast with O stars, for which solely UV P Cygni resonance metal lines are available, the greatly increased wind densities of W-R stars permits alternative diagnostics. Optical and near-IR P Cygni lines of He I may be used. Alternatively, mid-IR fine-structure lines (recall Fig. 3.10) may be employed since they are formed far out in the stellar wind at the asymptotic flow velocity.

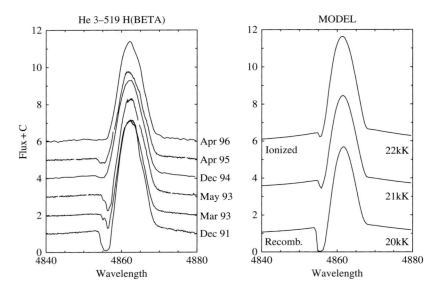

Fig. 4.8 Comparison between observed Hβ line profiles of the extreme B supergiant He3–519 (left) and theoretical models (right) in which a strong P Cygni absorption is predicted if H is recombined in the outer wind, and a pure emission profile is predicted if H remains ionized. From Crowther (1997).

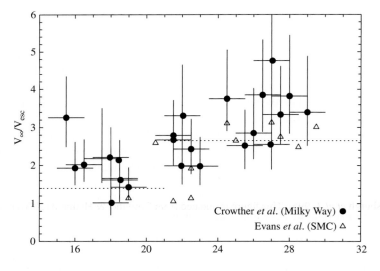

Fig. 4.9 Comparison between the ratio v_∞/v_{esc} versus effective temperature (in kK) for B supergiants in the SMC and Milky Way, together with the Kudritzki & Puls (2000) scaling formula (dotted line) based upon earlier results. From Crowther *et al.* (2006a).

Of course, space-based UV and IR observations are not available for the great majority of W-R stars. In addition, many W-R winds are highly ionized, causing He I to be weak or absent. More generally, it would be preferable to measure wind velocities from strong optical recombination lines. Such lines are typically formed interior to the asymptotic flow velocity, especially for W-R stars exhibiting relatively weak winds, so reliable wind velocity measurements from such diagnostics should ideally be based on stellar atmosphere models. Regardless, lower wind velocities are observed for late-type WN and WC stars with respect to early-type W-R stars (recall Table 3.2). The current record holders amongst non-degenerate stars for the fastest stellar wind are WO stars, for which wind velocities of $\sim$5500 km s^{-1} have been obtained from optical recombination lines (Kingsburgh *et al.* 1995). WO stars have been identified in a number of external galaxies, in which reduced line widths (wind velocities) are observed at reduced metallicities. This is illustrated in Fig. 4.10 and is thought to occur for other W-R spectral types, although numbers are small.

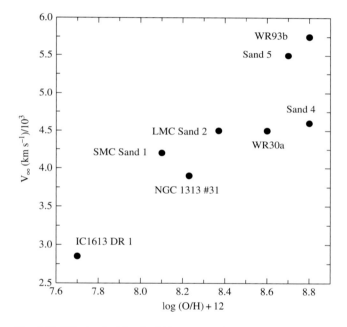

Fig. 4.10 Wind velocities for WO stars as a function of metallicity, following the approach of Kingsburgh *et al.* (1995). This figure is updated from Crowther & Hadfield (2006) to include NGC1313 #31 from Hadfield & Crowther (2007).

4.3   Mass-loss rates

Observationally, estimates of mass-loss rates for early-type stars can be obtained from radio continuum fluxes, optical/IR line profiles, or UV wind lines. In extreme OB supergiants, LBVs, and W-R stars, $\dot{M}$ may exceed 10^{-5} $M_{\odot}$ yr^{-1} but values 10–100 times lower are more typical of normal O stars. Complications arise due to the radiatively unstable winds producing compressions and rarefactions in their outflows, such that winds are clumped in nature.

IR–radio continua

Winds in hot stars can be readily observed at IR–mm–radio wavelengths via the free–free (Bremsstrahlung) "thermal" excess caused by the stellar wind, under the assumption of homogeneity and spherical symmetry. Mass-loss rates can be determined via application of relatively simple analytical relations which reveal that the continuum flux, S_ν in a spherically symmetric, isothermal envelope which is expanding at constant velocity scales with ν^α, with $\alpha \simeq 0.6$. The emergent flux is set by the distance to the star d, mass-loss rate $\dot{M}$ and terminal velocity v_∞ as follows:

$$S_\nu \propto \left(\frac{\dot{M}}{v_\infty}\right)^{4/3} \frac{\nu^\alpha}{d^2}, \tag{4.26}$$

following Wright & Barlow (1975) and Panagia & Felli (1975). Other factors, including differences in composition, ionization balance, and electron temperature, play a minor role, except in the case of W-R stars. This approach, together with IR (free–free) excesses, provided the first robust mass-loss rates for a large sample of galactic OBA supergiants (Barlow & Cohen 1977). Complications arise if recombination takes place in the outer wind, which would lead to a change in the spectral index from 0.6 to $\sim$1.3 if the temperature distribution is insensitive to radius in the outer wind. Indeed, the observed spectral energy distribution for P Cygni is indicated in Fig. 4.11. The spectral index increases from the mm to the radio. Recombination of H does not occur for most early-type supergiants.

Radio continuum measurements of hot stars have been carried out by David Abbott and collaborators with the Very Large Array (VLA), and by Claus Leitherer and collaborators with the Australia Telescope Compact Array (ATCA). OB stars with relatively weak winds do not show a strong free–free IR excess or radio flux, so mass-loss rates from this technique have solely been for nearby hot stars with relatively, dense winds (Bieging, Abbott, & Churchwell 1989; Leitherer, Chapman, & Koribalski 1995).

The typical region sampled in the radio corresponds to $\sim$100R_* for OB stars, such that the mass-loss rate obtained from the radio continuum samples a quite different physical region from UV/optical lines. Early radio results for OB stars formed the basis for the empirical calibration of mass-loss rates by de Jager, Nieuwenhuijzen, & van der Hucht (1988) which have been widely applied in evolutionary calculations until recently.

Nugis, Crowther, & Willis (1998) estimated mass-loss rates for W-R stars from radio observations, allowing for the clumped nature of their winds (see Section 4.4), which were coupled to parameters derived from spectroscopic analysis to provide empirical mass-loss scaling relations (Nugis & Lamers 2000). For their combined WN and WC sample,

$$\log \dot{M}/(M_\odot \text{yr}^{-1}) = -11.00 + 1.29 \log L_*/L_\odot$$
$$+ 1.74 \log Y + 0.47 \log Z, \tag{4.27}$$

where Y and Z are the mass fractions of helium and metals, respectively.

As we will show in Chapter 6, collisions between stellar winds from stars in a binary system will cause non-thermal (synchrotron) radio emission, so care needs to be taken against

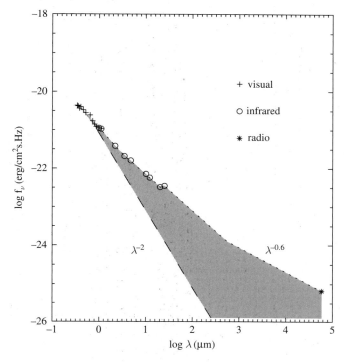

Fig. 4.11 A comparison between the de-reddened UV to radio energy distribution for P Cygni (dotted line) with that expected for a hydrostatic photosphere without a wind (dashed line). The gray region is the excess due to free–free emission. Reproduced from Lamers & Cassinelli (1999) with permission of Cambridge University Press.

overestimating mass-loss rates from (apparently) single stars in this way. Generally this is accomplished by carrying out observations at multiple radio frequencies, to ensure the radio exponent is consistent with free–free emission.

Optical and IR line profiles

If an ion in a stellar wind collides with a free electron, it can recombine. In principle, the most likely recombination is directly to the ground state. However, it may recombine to an excited state, and cascade downwards by photo de-excitation. Each de-excitation results in the emission of a line photon. In hot stars, optical lines appear in emission if mass-loss rates are very high, for which Hα has long been recognized as the prime source of mass-loss information in early-type stars. Since recombination involves the combination of ion and electron density, the strength of recombination wind lines such as Hα scale with the *square* of the density. Consequently, Hα is formed at the highest electron densities immediately above the photosphere in O stars, at $\sim 1.1 R_*$, in stark contrast to $\sim 100 R_*$ from the radio continuum.

Initial attempts at exploiting the wind-sensitivity of Hα were restricted to attempts to measure its line luminosity, after corrections for the underlying Hα photospheric line and blend with He II $\lambda 6560$ were taken into account. This was carried out by Lamers & Leitherer (1993) for a sample of O stars for which radio continuum derived mass-loss rates were known. A comparison between the two approaches revealed good consistency, as shown in Fig. 4.12.

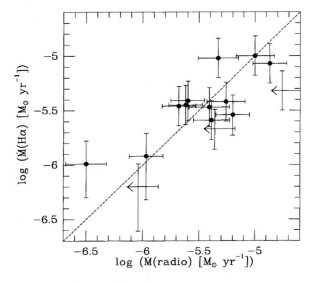

Fig. 4.12 Comparison between radio and Hα derived mass-loss rates of OB stars. Reproduced from Lamers & Leitherer (1993) by permission of the AAS.

Their sample was dominated by nearby O supergiants, but demonstrates the typical mass-loss rates of such stars as $10^{-6} \leq \dot{M}/(M_\odot \ \mathrm{yr}^{-1}) \leq 10^{-5}$. A comparison of empirical Hα and radio O star mass-loss rates with theoretical predictions from Pauldrach *et al.* (1986), as shown in Fig. 4.13, revealed agreement at low wind densities. For O supergiants and late-type

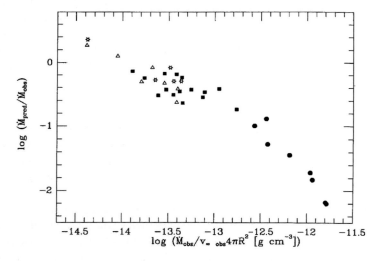

Fig. 4.13 Comparison between the ratio of predicted to observed mass-loss rates for O stars (squares: supergiants; asterisks: giants; triangles: dwarfs) and late-type WN stars (solid circles) versus mean wind density. This demonstrates that the standard theory of radiatively driven winds (which neglects the effects of line-overlap, e.g. Pauldrach, Puls, & Kudritzki 1986) does not reproduce the strong winds of O supergiants and WN stars. Reproduced from Lamers & Leitherer (1993) by permission of the AAS.

WN stars, possessing higher wind densities, predicted mass-loss rates were too small by up to 1–2 orders of magnitude for W-R stars.

Subsequently, Puls *et al.* (1996) studied a large sample of O stars in the Milky Way and Magellanic Clouds, now including profile fits to Hα lines for more reliable mass-loss rates. Nevertheless, the results of Lamers & Leitherer were supported, in the sense that theory was unable to reproduce the observed high mass-loss rates. In the standard CAK approach, optically thick lines are assumed not to overlap within the wind. In fact, this is rarely true in the extreme UV where spectral lines are very tightly packed and the bulk of the line driving originates. Consequently, allowance for multiple scattering needs to be considered.

Multiple scattering

A Monte Carlo technique which treats multi-line transfer in O and W-R atmospheres has been developed by Abbott & Lucy (1985) using photon "packets". Figure 4.14 illustrates the scattering history for a typical extreme UV photon in regions of closely spaced spectral lines in their model for the O4 supergiant ζ Pup. The photon undergoes line scatterings with many distinct transitions, such that its path may be described as a random-walk. For this typical photon, Fig. 4.15 shows the cumulative change of its rest frequency (in units of $v_0 v_\infty/c$ where v_0 is its initial rest frequency), plus the cumulative momentum deposition in units of $h v_0/c$. In this case, the net effect of these scatterings is the imparting of energy $\sim 4 h v_0/c$ to the wind, i.e. multiple scattering.

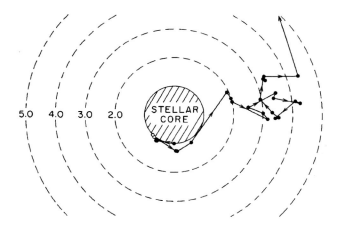

Fig. 4.14 Path of an extreme UV photon experiencing multiple scattering within the wind of an early O supergiant. Filled circles indicate interactions with lines while open circles indicate electron scatterings. The transfer of momentum is purely radial and the photon's path is shown projected onto a plane. Reproduced from Abbott & Lucy (1985) by permission of the AAS.

For solar metallicity, the Monte Carlo simulations of Vink, de Koter, & Lamers (2000) provide a theoretical mass-loss rate recipe of

$$\log \dot{M} = -6.70 + 2.19 \log(L_*/10^5) - 1.31 \log(M_*/30)$$
$$-1.23 \log\left[(v_\infty/v_{\rm esc})/2\right]$$
$$+0.93 \log(T_{\rm eff}/40000) - 10.92(\log T_{\rm eff}/40000)^2$$

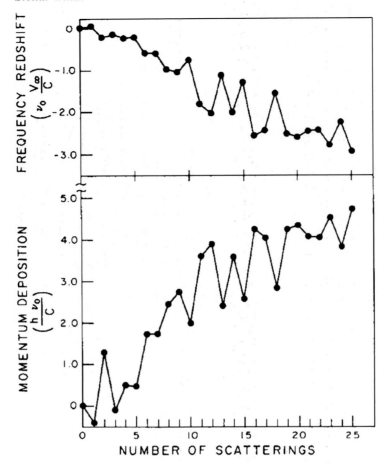

Fig. 4.15 Cumulative frequency redshift and momentum transfer at each scattering of the photon from the previous figure. Reproduced from Abbott & Lucy (1985) by permission of the AAS.

for O stars ($27\,500\,\mathrm{K} < T_{\mathrm{eff}} \leq 50\,000\,\mathrm{K}$), where $\dot{M}$ is in $M_\odot\,\mathrm{yr}^{-1}$, L and M are in solar units, and T_{eff} is in K, and

$$\log \dot{M} = -6.69 + 2.21 \log(L_*/10^5) - 1.34 \log(M_*/30)$$
$$-1.60 \log\left[(v_\infty/v_{\mathrm{esc}})/2\right]$$
$$+1.07 \log(T_{\mathrm{eff}}/20000)$$

for mid-type B supergiants ($15\,000\,\mathrm{K} < T_{\mathrm{eff}} \leq 22\,500\,\mathrm{K}$). There is no single prediction at $T_{\mathrm{eff}} = 25\,000 \pm 2500\,\mathrm{K}$ due to varying wind ionization, attributed to the bistability jump. Representative wind properties of OB dwarfs and supergiants predicted from these relations are presented in Table 4.2.

Table 4.2 *Representative wind properties of galactic OB dwarfs and supergiants based upon the mass-loss calibration of Vink et al. (2000).*

Sp type	Dwarf				Supergiant			
	T_{eff} K	v_∞ km s^{-1}	log $\dot{M}$ $M_\odot$ yr^{-1}	$\dot{M}v_\infty$ (L_*/c)	T_{eff} K	v_∞ km s^{-1}	log $\dot{M}$ $M_\odot$ yr^{-1}	$\dot{M}v_\infty$ (L_*/c)
O3	45 000	3200	−5.4	0.8	42 000	3200	−5.2	0.8
O5	41 000	2900	−5.9	0.5	38 500	2200	−5.2	0.7
O7	37 000	2300	−6.3	0.3	35 000	2100	−5.4	0.5
O9	33 000	1500	−6.8	0.15	31 500	2000	−5.9	0.3
B0	29 500	1200	−7.4	0.06	27 500	1500	−6.0	0.2

Figure 4.16 compares the O star results of Puls *et al.* (1996) and previous Hα and radio mass-loss rates with the Vink *et al.* predictions, which now show good agreement. Within the past decade, significant progress has also been achieved regarding the development of atmospheric models in which non-LTE effects, line blanketing, and spherical geometry are now possible (recall Chapter 3). Such techniques have already been touched upon with regard to effective temperatures of hot stars (recall Fig. 3.3). For O stars, stellar temperatures are reduced but derived mass-loss rates are not greatly affected since the ionization structure of hydrogen is largely unaffected. Good agreement is achieved for early to mid-type dwarfs and giants. However, empirical mass-loss rates of O supergiants with strong winds need to be reduced to match theory, which might be attributed to clumping in the inner wind, where Hα forms. One added complication is that late-type, low luminosity O dwarfs possess much weaker winds than is predicted by radiatively driven wind theory.

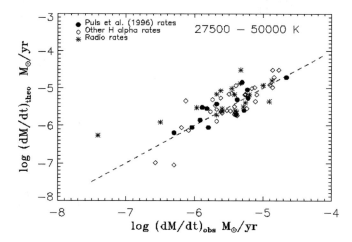

Fig. 4.16 Comparison between observed mass-loss rates of O stars from Hα and radio determinations and predictions from wind models. From Vink *et al.* (2000).

For Milky Way O stars uncertainties in distance severely hinder empirical calibrations of mass-loss versus physical parameters. Magellanic Cloud O stars of course do not suffer from such issues and provide us with the means of comparing empirical mass-loss properties with predictions at lower Z than the Milky Way. Although the number of stars analyzed in detail for the Magellanic Clouds remains somewhat limited, the mass-loss rates of luminous LMC and SMC O stars are indeed lower than galactic counterparts (recall Fig. 4.3) for which $\dot{M} \propto Z^{0.83}$, in reasonable accord with theoretical predictions (Mokiem *et al.* 2007).

Turning to B supergiants, the inclusion of line blanketing and winds has only a subtle effect upon effective temperatures but causes significant revisions to Hα derived mass-loss rates for mid-B spectral types. A comparison between mid-B supergiants in the SMC and Milky Way again reveals weaker winds for the former, but observed mass-loss rates for stars in both galaxies are up to a factor of ten times lower than predicted by theory. If we account for the possible clumped nature of their inner winds, this would exacerbate the comparison further. Detailed comparisons for A supergiants have yet to be published.

Finally, on the topic of recombination line diagnostics of OB stars, moderate to high spectral resolution near-IR observations to date are rare. However, analogous wind diagnostics to optical lines are available (recall Fig. 2.15). This provides the prospect of the determination of mass-loss properties for OB stars that are obscured at optical wavelengths using infrared diagnostics. Brα at 4 μm is accessible from the ground while Pα 1.87 μm lies between the H- and K-band ground-based windows, and so generally requires space-based observations. This will be discussed in detail in Chapter 7.

Turning to the still higher wind density of W-R stars, spectroscopic analysis requires the simultaneous mass-loss rate and stellar temperature determinations from non-LTE model atmospheres, since their atmospheres are highly stratified (recall Fig. 3.5). This adds extra parameters regarding spectral synthesis of W-R stars with respect to O stars. Consequently, Schmutz, Hamann, & Wessolowski (1989) introduced the so-called transformed radius, R_t:

$$R_t = R_* \left[\frac{v_\infty}{2500 \text{ km s}^{-1}} \right]^{2/3} \left[\frac{\dot{M}}{10^{-4} \ M_\odot \text{yr}^{-1}} \right]^{-2/3}, \tag{4.28}$$

for which similar line equivalent widths are predicted for a particular wind density R_t, permitting the parameters of W-R stars spanning a range of wind velocities and mass-loss rates to be efficiently determined. Metals such as C, N, and O provide efficient coolants, as in HII regions, so that the outer wind electron temperature is typically 8000 K to 12 000 K in W-R atmospheres, in dramatic contrast to very high stellar temperatures.

Lucy & Abbott (1993) and Springmann (1994) produced Monte Carlo wind models for W-R stars in which multiple-scattering was achieved by the presence of multiple ionization stages in the wind. However, a velocity structure was adopted a priori, with the inner wind acceleration unexplained. Schmutz (1997) first tackled the problem of driving W-R winds from radiatively driven winds consistently, but succeeded only in maintaining a strong outflow in the outer (clumpy) wind, thus failing to initiate a sufficiently powerful acceleration in the deep atmospheric layers.

Historically, the properties of W-R winds were considered to be independent of metallicity. However, Crowther *et al.* (2002a) noted that single LMC WC stars possessed wind strengths $\sim$0.2 dex lower than Milky Way counterparts, as shown in Fig. 4.17. Indeed, their result also provided a basis for the different WC subtype distributions for the LMC and Milky Way, in the sense that all known LMC WC stars are WC4 stars (negligible CIII λ5696 emission) whilst Milky Way WC stars span WC5–9 (weak to strong CIII λ5696 emission). This result followed from the fact that CIII emission is extremely sensitive to mass-loss rate.

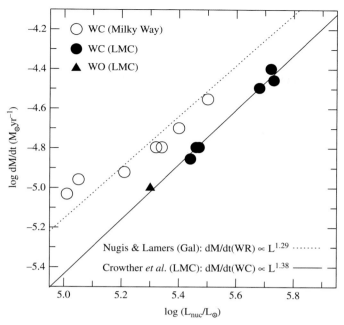

Fig. 4.17 Comparison between the mass-loss rates and luminosities of LMC WC4 (filled squares), WO stars (filled triangle), and galactic WC5–9 stars (open squares) at known distances. The lines are from the calibration of Nugis & Lamers (2000, dotted) assuming $C/He = 0.2$ and $C/O = 4$ by number, plus a linear fit (solid) to the LMC stars. From Crowther *et al.* (2002a).

A significant theoretical advance was achieved by Gräfener & Hamann (2005) who incorporated opacities from highly ionized FeIX–XVII lines, which provided the necessary driving deep in the WR atmosphere. A self-consistent wind solution was achieved for the WC5 star HD 165763 (WR111), in which the wind acceleration due to radiation and gas pressure matches the mechanical and gravitational acceleration. The velocity structure from their hydrodynamical model closely matches a typical $\beta = 1$ velocity law in the inner supersonic wind (recall Eq. 4.16), plus a slower law in the extreme outer wind. The critical parameter involving the development of strong W-R winds may be their proximity to the Eddington limit.

Theoretically, both the hydrodynamical models of Gräfener & Hamann (2005) and recent Monte Carlo wind models for W-R stars by Vink & de Koter (2005) argue in favor of radiation pressure through metal lines as being responsible for the multiple scattering observed in W-R stars. The latter study – carried out for W-R stars with low stellar temperatures for an assumed Eddington parameter – suggests a metallicity scaling that is equivalent to O stars for WN subtypes, with an exponent of 0.86 for $10^{-3} \leq Z/Z_\odot \leq 1$. The higher metal content of WC atmospheres suggests a weaker dependence with exponent 0.66 for $10^{-1} \leq Z/Z_\odot \leq 1$ and 0.35 for $10^{-3} \leq Z/Z_\odot \leq 10^{-1}$.

The strength of W-R winds also impacts upon their ionizing radiation. Atmospheric models for W-R stars in which mass-loss rates are high leads to a relatively soft ionizing flux distribution in which extreme UV photons are redistributed to longer wavelength by the opaque stellar wind (Schmutz, Leitherer, & Gruenwald 1992). In contrast, relatively weak winds predict a hard ionizing flux distribution, in which extreme UV photons pass through the relatively transparent wind unimpeded. Consequently, the shape of the ionizing flux distribution of WR stars is as dependent on the wind density as the stellar temperature. This is illustrated in Fig. 4.18, where we compare the predicted Lyman continuum ionizing flux distribution from four 100 000 K WN models, in which only the low metallicity, low mass-loss rate models predicts prodigious photons below the He$^+$ edge at 228Å. One expects evidence of hard ionizing radiation from W-R stars (e.g. nebular HeII 4686) solely at low metallicies, which is generally borne out by observations of HII regions.

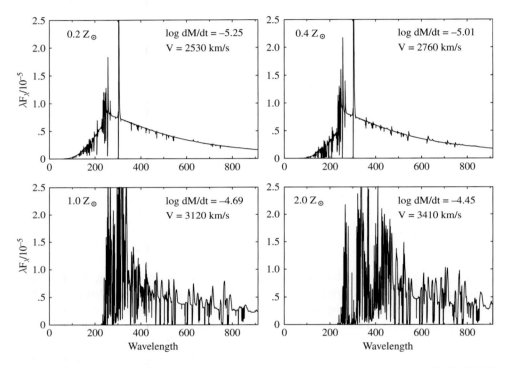

Fig. 4.18 Comparison between the Lyman continuum ionizing fluxes of early WN CMFGEN models with fixed parameters (100 kK, $\log L/L_\odot = 5.48$), except that the mass-loss rates and wind velocities depend upon metallicities according to Smith, Norris, & Crowther (2002), such that solely the low wind density models predict a significant flux below the He$^+$ edge at 228Å. From Crowther (2007).

UV P Cygni profiles

In addition to radio continuum and optical/near-IR recombination line techniques, ultraviolet P Cygni profiles from metal resonance lines also permit determinations of mass-loss rates in hot stars. Unfortunately, mass-loss determinations require knowledge of elemental abundances, the degree of ionization of the ion producing the line, and knowledge about the form of the velocity law. Additionally, observation of UV lines require low dust extinctions (problematic for most Milky Way sight-lines) although they represent the most sensitive probe of mass-loss of all the available techniques.

Analyses of hot star winds via UV spectroscopy are generally carried out using the Sobolev with exact integration (SEI) method (Lamers, Cerruti-Sola, & Perinotto 1987), named after the pioneer of stellar wind studies, V. Sobolev, in the late 1950s. Hamann (1981) has used an identical method to study mass-loss from UV spectral lines. In the SEI approach, UV P Cygni line profiles are fitted to reveal $\dot{M}q_i$, where q_i is the, a priori unknown, fractional population of a particular ionization stage within a particular element.

From IUE or HST, direct ionization information is limited to observed ions: generally N^{4+}, C^{3+}, and S^{3+}. Therefore total mass-loss rates may only be estimated from an assumed ionization balance, generally from non-LTE atmospheric models. Moreover, N V 1240 and C IV 1550 are often saturated in galactic OB stars, so that only lower limits for mass-loss rates can generally be derived. In view of these difficulties, up until recently it was necessary to combine column density information from unsaturated UV P Cygni profiles with independently determined mass-loss rates (from radio or Hα) to derive empirical ionization fractions.

Fortunately, a more direct approach is available for early-type stars by incorporating far-UV spectroscopy with one of the Copernicus, HUT, or FUSE satellites, which affords many more resonance lines from many more ionization stages, e.g. CIII, NIII, OVI, SIV, SVI, PIV–V. SEI modeling of the unsaturated PV doublet in O stars, for which one anticipates $q_i \sim 1$ for mid-O subtypes, argue in favor of significant clumping in the winds of LMC O-type stars, from comparison with theoretical mass-loss rates. SEI UV modeling of late-type O stars in the Milky Way and Magellanic Clouds also reveals substantially lower mass-loss rates than predicted theoretically.

Red supergiants

Finally in this section, mass-loss determinations may be derived for dust forming cool supergiants in the Milky Way and Magellanic Clouds via their mid-IR dust continuum emission. However, this approach necessitates an assumed dust-to-gas conversion factor to calculate gas mass-loss rates (Bowers, Johnston, & Spencer 1983; van Loon *et al.* 2005). Dust is formed well beyond the stellar photosphere of M supergiants, so radial pulsations are generally assumed to provide the mechanism of transporting material to sufficient distance that dust absorption of continuous radiation expels the wind. Outflow velocities and total (gas) mass-loss rates of nearby Milky Way O-rich cool supergiants may be derived from CO emission lines, in which CO serves as a proxy for the total gas. Outflow velocities of RSG are typically 10–40 km s^{-1} as measured from CO emission lines (Jura & Kleinmann 1990).

From an analysis of LMC dust enshrouded RSG and O-rich AGBs, van Loon *et al.* (2005) obtained

$$\log \dot{M} = -5.65 + 1.05 \log(L_*/10\,000 L_\odot) - 6.3 \log(T_{\rm eff}/3500K), \tag{4.29}$$

where $\dot{M}$ has units $M_\odot$ yr^{-1}, based on the assumption of a gas-to-dust ratio of 500 (versus 200 for typical solar metallicity giants). This compares fairly well with galactic AGBs and dust enshrouded RSGs. Optically bright RSGs possess much lower mass-loss rates, most likely due to their higher temperatures and consequently an inability to form dust. Overall, the level of consistency between different IR/sub-mm techniques for RSG mass-loss rates remains poor. As such calibrations of mass-loss rates for cool luminous supergiants is problematic when applied to stellar evolutionary models.

4.4   Structure and clumping

There is extensive observational evidence that hot star winds are not smooth, steady outflows. Intensive monitoring of unsaturated UV P Cygni lines with IUE led to the detection of extensive variability within the absorption components. To produce such variability, the wind structure must be on a large scale, covering a substantial fraction of the disk. In contrast, emission components of P Cygni profiles were found to be rather less variable, since they form from a more global average of the wind. Optical spectroscopic studies have also led to absorption and emission line variability.

Structured winds

Observational and theoretical evidence suggests that OB winds are inhomogeneous, exhibiting both large- and small-scale structures. In particular, several early-type stars were continuously monitored for 16 days in the ultraviolet as part of the IUE MEGA campaign (Massa *et al.* 1995). Deep-seated regular structures originating from so-called co-rotating interaction regions (CIRs) were observed in the early B supergiant HD 64760 (see Fig. 4.19) related to its probable 4.8 day rotational period, and characterized by a velocity plateau plus

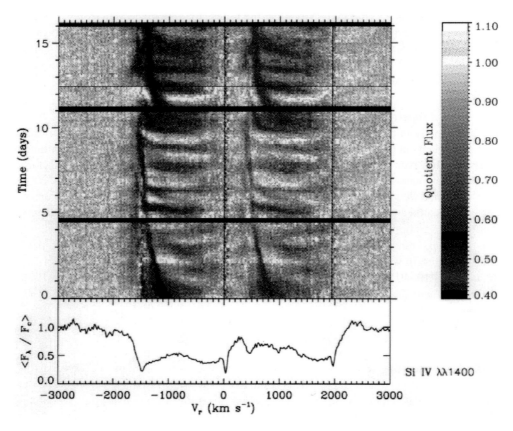

Fig. 4.19 IUE MEGA continuous 16-day time-series observations of the effect of CIRs on the SiIV 1393–1402 doublet in the fast rotating (4.8 day period) B0.5 Ib star HD 64760. Reproduced from Massa *et al.* (1995) by permission of the AAS.

density compression. These are believed to be formed at the interface between fast and slow streams in the wind. Either stellar pulsation or magnetic fields are thought to be responsible for such photospheric modulation in early-type stars, although star spots may also induce CIRs. The schematic equatorial wind density structure for a CIR is illustrated in Fig. 4.20.

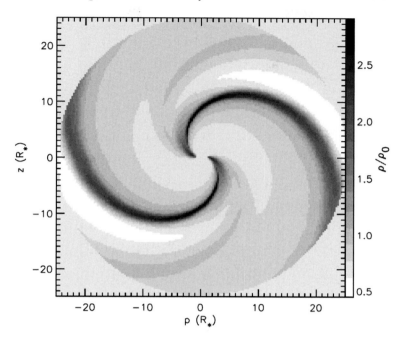

Fig. 4.20 Equatorial wind density structure for a CIR model, normalized to the unperturbed case. From Dessart & Chesneau (2002).

There is also indirect evidence that OB winds have a turbulent structure. From the IUE MEGA programme, the well known O4 I(n)f supergiant ζ Pup displays continuously varying discrete absorption components (DACs), formed at $0.3-0.5\,v_\infty$ (see Fig. 4.21). Saturated P Cygni profiles have extended black troughs, which are thought to be a signature of a highly non-monotonic wind. Soft X-ray emission is also observed, which is thought to originate from embedded wind shocks, for which $L_X/L_{Bol} \sim 10^{-7}$ (Chlebowski, Harnden, & Sciortino 1989).

Shocks are also thought to be the means by which high ionization stages are observed, most notably NV and OVI, in stars for which the T_{eff} is too low from normal means. This is via a process known as super-ionization. Sensitive X-ray satellites such as Chandra now permit the observation of spectrally resolved line profiles from OB stars. For example, X-ray spectroscopy of ζ Pup is presented in Fig. 4.22. The forbidden, intercombination and resonance lines of He-like ions (e.g. OVII, NeIX, MgXI, SiXIII, SXV) permit the location of X-rays to be addressed, using the ratio of forbidden to intercombination lines. A very small fraction of the wind is in the form of such very hot gas.

The most likely explanation for this turbulent structure is theoretical evidence for a strong instability of line driving to small-scale velocity perturbations (Lucy & Solomon 1970; Owocki, Castor, & Rybicki 1988). There is a strong potential in line scattering to drive wind

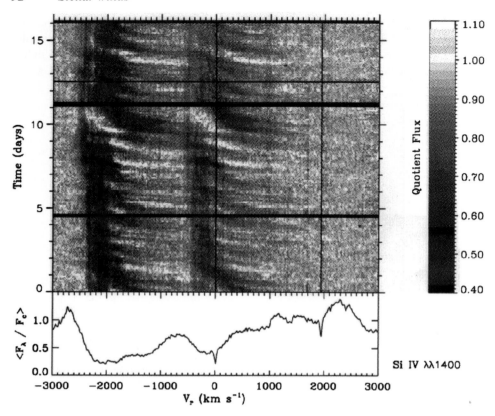

Fig. 4.21 IUE MEGA continuous 16-day time-series observations of DACs in the Si IV 1393–1402 doublet in the fast rotating (~5 day period) O4 I(n)f star ζ Pup. Reproduced from Massa *et al.* (1995) by permission of the AAS.

material with accelerations that greatly exceed the mean outward acceleration. Simulations demonstrate that this instability may lead naturally to a highly structured flow dominated by multiple shock compressions.

Mass-loss rates – corrected for clumping?

If hot star winds are structured, or clumped, it remains to be seen whether derived mass-loss rates of OB stars, as derived from optical or radio measurements, represent over-estimates of true values. Overall consistency between, e.g., Hα and radio determinations is rather good (recall Fig. 4.12), suggesting that if winds are highly clumped, they are similar on scales of $\leq 1.5 R_*$ (Hα) and ~1000 R_* (radio). Time-averaged line profiles of OB stars that have been intensively monitored do remain remarkably stable on timescales of years.

Recent spectroscopic evidence from unsaturated UV P Cygni lines observed with FUSE argue strongly in favor of clumping. First, non-LTE model atmosphere line profiles in which mass-loss rates have been obtained from Hα strongly overestimate the UV absorption strength of unsaturated P Cygni profiles such as PV in O stars and SIV in B supergiants. The only means of resolving these discrepancies is either by assuming artificially low elemental abundances

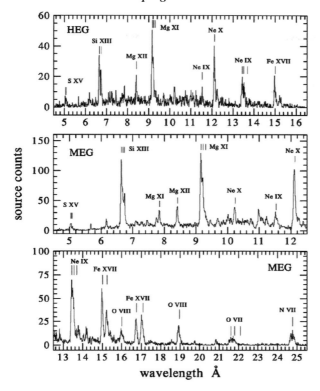

Fig. 4.22 Chandra X-ray spectrum of ζ Pup (O4I(n)f). X-rays appear to form in hot material embedded in much cooler gas within the inner wind of OB stars, from just above the photosphere to $\sim 10 R_*$. Ions responsible for the strongest line emissions are identified. Reproduced from Cassinelli *et al.* (2001) by permission of the AAS.

for P or S, or by accounting for clumping (Crowther *et al.* 2002b; Evans *et al.* 2004b). The second, related technique involves comparing SEI line profile fits to the PV unsaturated lines revealing $\dot{M}q_i$ and comparing these with predictions from radiatively driven wind theory across all O subtypes (Fullerton, Massa, & Prinja 2006). One would expect that P^{4+} is the dominant ion for some O subtypes, i.e. $q_i = 1$, yet the *maximum* value of q_i inferred is 0.1, based upon current radiatively driven wind theory. Either this is seriously in error, or O stars winds are strongly clumped. Similar conclusions were reached by multi-wavelength continuum studies of OB stars (Puls *et al.* 2006).

In contrast with O stars, evidence for highly clumped winds for W-R stars appears overwhelming. Line profiles show small-scale structures or "blobs" (Moffat *et al.* 1988). Clumping has also been long suspected in W-R stars since electron scattering wings, which scale linearly with density, are strongly overestimated in homogeneous models that match the (density-squared) recombination lines (Hillier 1991). In this manner typical volume filling factors of $f \sim 0.1$ have been obtained, which are alternatively characterized as clumping factors, i.e. $f_{\rm cl} = 1/f \sim 10$.

In addition, one may directly determine W-R mass-loss rates for close binary systems using the variation of linear polarization – described by the two Stokes parameters – as a function of orbital phase. The phase-dependent modulation of linear polarization due to the relative motion of the companion originates via Thomson scattering of photons from the companion by the free electrons in the W-R wind. This technique has been applied to several W-R binaries including V444 Cyg (HD 193576, WN5+O) by St-Louis *et al.* (1993).

The effect of a reduced wind strength during the OB phase is illustrated in Fig. 4.23(a). Relative to a uniform high mass-loss rate (homogeneous case) for an initial $120M_\odot$ star, a clumped OB wind has the effect of a more modest reduction in stellar mass with age. It is likely that the most massive stars undergo an extended H-rich WN phase (recall Chapter 2). Nevertheless, the greater role of giant LBV eruptions is illustrated, in order that the star becomes a classical W-R star with a mass in the range of that observed (see Chapter 6). An alternative case is shown in Fig. 4.23(b) which allows for the possibility that the most massive stars explode during the LBV phase as Type IIn SN before reaching the W-R phase.

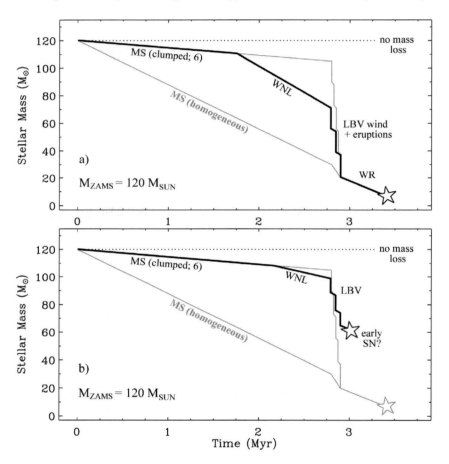

Fig. 4.23 Illustration of reduced mass-loss during OB phase due to wind clumping (MS clumped) relative to the standard case (MS homogeneous) for an initial $120M_\odot$ case. In (a) an extended H-rich WN phase is taken into account which equates to that lost during LBV eruptions, prior to the classical W-R phase, whilst in (b) a shorter H-rich WN phase occurs, plus the possibility that a supernova explosion occurs during the LBV phase. Reproduced from Smith (2008) with permission of Cambridge University Press.

Owocki, Gayley, & Shaviv (2004) have proposed that the "porosity" resulting from radiative instabilities in hot, luminous stars close to, or in breach of, the Eddington limit may reduce the coupling between matter and radiation. Potentially, this permits mass-loss rates to be driven far higher than those achieved using conventional line-driven radiation pressure. Specifically, this could have potential application in giant LBV outbursts, such as that experienced by η Car during the nineteenth century (Davidson & Humphreys 1997). The time-averaged mass-loss rate during the giant eruption of η Car was $\sim 0.2\, M_\odot \, \mathrm{yr}^{-1}$ and the radiative luminosity was of order $10^{7.3} L_\odot$, substantially above its Eddington limit. Radiative transport within a spatially inhomogeneous atmosphere may favor low density channels between the high density regions. This has the effect of lowering the effective driving opacity in the deep, denser layers, and is in effect a super-Eddington, continuum-driven wind.

4.5 Influence of stellar rotation

Results presented so far assume that winds are spherically symmetric, with rotation neglected. For a rotating star, the total gravity (g_{tot}) is the sum of the gravitational (g_{grav}), centrifugal (g_{rot}) and radiative (g_{rad}) accelerations, i.e.

$$g_{tot} = g_{eff} + g_{rad} = g_{grav} + g_{rot} + g_{rad}. \tag{4.30}$$

Conventionally, rotation is neglected when considering the Eddington limit, with $\Gamma_e^{Max} = 1$ (Eq. 4.1), i.e. $g_{rad} + g_{grav} = 0$. Alternatively, we may consider the break-up velocity as an $\Omega\, (= v_e/v_{crit})$ limit, for equatorial rotation velocity v_e and critical velocity v_{crit} defined in Eq. 2.17. If radiation pressure effects are neglected in Eq. 4.31, then $g_{eff} = g_{grav} + g_{rot} = 0$ at the break-up or Ω limit. Together, the so-called $\Omega\Gamma_e$ limit, introduced by Maeder & Meynet (2000), is reached when the total gravity $g_{tot} = 0$, due to rotation and radiation.

Langer (1998) introduced the concept of the Ω limit, in which *critical* rotational velocities are achieved before reaching the Eddington limit, since

$$v_{crit}^2 = \frac{GM_*}{R_e} (1 - \Gamma_e) \tag{4.31}$$

for a rotating star with equatorial radius R_e. This is true so long as the brightness of the rotating star is uniform over its surface, which contradicts von Zeipel's theorem. Accounting for this leads to $g_{tot} = g_{eff} (1 - \Gamma_e^\Omega)$ where Γ_e^Ω is a modified Eddington parameter including a term involving angular velocity, Ω. In this case, two solutions to $g_{tot} = 0$ exist – the standard case (Eq. 2.17) for which radiation is neglected, plus a second solution if the Eddington parameter $\Gamma_e > 0.64$, due to the effects of rotation and radiation. A critical velocity is reached if Γ_e approaches unity, in agreement with Langer (1998).

Two-dimensional radiatively driven wind models of Friend & Abbott (1986) predict a modest increase in mass-loss rate of only 30% for an equatorial rotational velocity $v_e = 350\, \mathrm{km\, s}^{-1}$, which may be parameterized as follows:

$$\dot{M}(v) = \dot{M}(v = 0) \left(\frac{1}{1 - \dfrac{v}{v_{crit}}} \right)^\zeta \tag{4.32}$$

where $\zeta = 0.43$. This approach does not account for the von Zeipel theorem and diverges at break-up velocity. Bjorkman & Cassinelli (1993) first considered deviations from spherical symmetry. Adopting azimuthal symmetry, they considered a purely radial CAK line force in the equations of motion, which predicted a polar deflection of wind towards the equatorial plane, owing to the combination of centrifugal, gravitational, and radiative acceleration. For extreme rotation rates of $\Omega = v_e/v_{crit} \geq 0.5$ an oblate "wind-compressed disk" is predicted, produced by the collision of wind material from each hemisphere, with a high density contrast, ρ_e/ρ_p, of up to 20–30.

Initially, this was considered as an explanation for the origin of circumstellar disks in B[e] supergiants (recall Section 2.3). Subsequently, Owocki, Cranmer, & Gayley (1996) additionally considered the effects on the wind dynamics from the *non-radial* components of the CAK line force, plus von Zeipel gravity darkening. They found that wind material is redistributed towards the poles from consideration of non-radial components of the line force, plus an enhanced photon flux over the poles due to gravity darkening. In contradiction to the wind-compressed disk model, a prolate density structure is obtained, with a dense, fast outflow at the poles and a slow, thin wind at the equator.

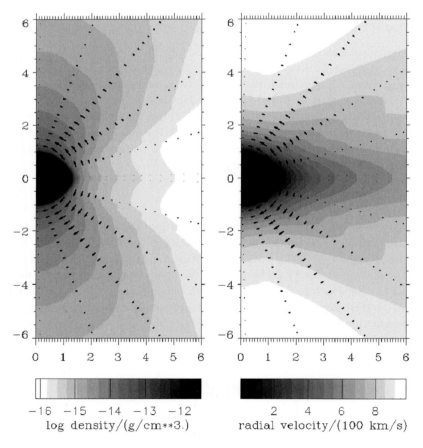

Fig. 4.24 Density (left) and radial velocity (right) predicted for the 2D wind model of a mid-B supergiant, calculated with consistent radial and non-radial CAK force multiplier parameters and gravity darkening. From Petrenz & Puls (2000).

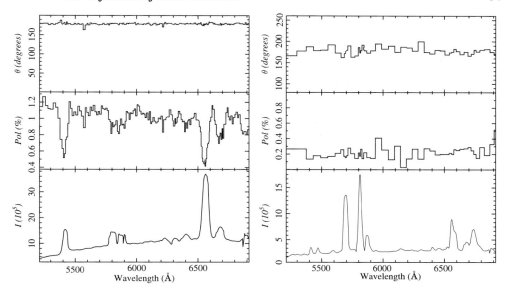

Fig. 4.25 Representative polarization data for WR134 (WN6, left) and WR135 (WC8, right) showing the position angle of polarization (top), the percentage polarization (middle) showing a strong line effect for WR134 but not WR135, and intensity spectrum (bottom). From Harries, Hillier, & Howarth (1998).

For rotating O stars, gravity darkening favors polar ejection, since the polar caps of rotating stars are hotter than the equatorial region, although spectropolarimetric observations favor globally distorted winds in only a few cases. Petrenz & Puls (2000) relaxed the previously assumed *uniform* CAK force multiplier parameters – since the density and ionizing radiation field are functions of latitude and distance from the stellar surface – with results in agreement with Owocki *et al*. By way of example, a mid-B dwarf rotating at 85% of its critical velocity was found by Petrenz & Puls to have a strongly prolate structure, with $\rho_p/\rho_e \leq 15$, and a polar (equatorial) terminal velocity of 1030 (730) km s^{-1}, as illustrated in Fig. 4.24. The influence of rotation on the global wind properties is that the total-mass loss rate deviates from the 1D value by at most 10–20%, even for the highest rotation rates of up to 0.85 v_{crit}.

Classical Be stars, for which wind-compressed disk models were originally considered appropriate, are now thought to be rotating sufficiently close to break-up that their circumstellar disks may be produced via pulsation or gas pressure (Townsend, Owocki, & Howarth 2004). A subset of non-supergiant early-type B stars are known to pulsate, on timescales of hours, known as β Cephei stars, with masses in the range 10–15$M_\odot$. The origin of this variability has been attributed to the classical κ-mechanism caused by the hot iron opacity peak at 2×10^5 K. Lower mass late-type B stars pulsate on timescales of days, known as slowly pulsating B stars (SPBs), attributed to the same mechanism.

The situation for B[e] supergiants, LBVs and W-R stars is less straightforward. High rotational velocities of 200–500 km s^{-1} have been reported in two cases, although the association of the measured features with the W-R core is unclear since one is a binary, and the absorption lines of the other are formed within the stellar wind. Recent spectropolarimetric evidence suggests that the winds of LBVs are rather more asymmetric than W-R and O supergiants, despite a statistically small sample to date. Harries, Hillier, & Howarth (1998) studied spectropolarimetric datasets for 29 W-R stars, from which just four single WN stars and one WC binary show a strong line effect, indicative of significant departures from spherical symmetry, as indicated in Fig. 4.25.

Radiative transfer calculations suggest that the observed continuum polarizations can be matched by models with equator to pole density ratios of 2–3. W-R stars with non-spherical winds have amongst the highest mass-loss rates of all Milky Way W-R stars. Based upon line profile variability studies, Harries, Hillier, & Howarth estimated equatorial rotation velocities for the stars with non-spherical winds and concluded that the ratio of rotational to critical velocity for such systems did not exceed $v_e/v_{crit} \sim 0.25$.

5

Evolution of single stars

In this chapter we discuss the evolution of massive stars ($\geq 8M_\odot$) from their Main Sequence through their chemically evolved supergiant phases, to their ultimate demise as supernovae. The birth of massive stars is deferred until Chapter 7, and gamma ray bursts are discussed in Chapter 11.

The physical scale of stellar masses results from gravity and stellar structure considerations, whilst the ability for stars to maintain their hydrostatic equilibrium over millions or billions of years is as a consequence of the properties of their atomic nuclei. For non-degenerate stars the energy radiated is extracted from either their gravitational or nuclear reservoir. Gravitational contraction maintains mechanical equilibrium over a Kelvin–Helmholtz timescale, t_{KH}, whilst the nuclear energy source can sustain the stellar luminosity during much longer times. Schematically, there exist four ranges of stellar mass, those that undergo (i) no nuclear burning (brown dwarfs, $\leq 0.08\ M_\odot$); (ii) hydrogen burning (0.08–0.5 $M_\odot$); (iii) hydrogen and helium burning (0.5–8 $M_\odot$); (iv) beyond helium burning ($\geq 8\ M_\odot$). Of these, solely the final mass range is discussed here, since low and intermediate mass stars are widely discussed in the literature (e.g. Iben 1967). A modern review of massive stellar evolution is provided by Maeder & Meynet (2000).

5.1   Nucleosynthesis

Arnould & Takahashi (1999) provide a modern review of nuclear astrophysics. During the Main Sequence, massive stars generate energy in their high temperature cores through the conversion of hydrogen to helium via the CN cycle, which produces sufficient pressure to perfectly balance the gravitational attraction of the outer layers. In hydrogen fusion CNO elements are used as catalysts, i.e.

$$\begin{aligned}
{}^{12}\text{C} + \text{p} &\rightarrow {}^{13}\text{N} + \gamma \\
{}^{13}\text{N} &\rightarrow {}^{13}\text{C} + \text{e}^+ + \nu_\text{e} \\
{}^{13}\text{C} + \text{p} &\rightarrow {}^{14}\text{N} + \gamma \\
{}^{14}\text{N} + \text{p} &\rightarrow {}^{15}\text{O} + \gamma \\
{}^{15}\text{O} &\rightarrow {}^{15}\text{N} + \text{e}^+ + \nu_\text{e} \\
{}^{15}\text{N} + \text{p} &\rightarrow {}^{12}\text{C} + {}^4\text{He}.
\end{aligned}$$

The reaction begins and ends with a carbon nucleus, in which the addition of four protons produces a helium nucleus. The slowest reaction in the CN cycle is the capture of a proton by

^{14}N, which acts as a bottleneck, The ON cycle is an offshoot from the main cycle, producing ^{14}N as follows:

$$
\begin{aligned}
{}^{15}\text{N} + \text{p} &\rightarrow {}^{16}\text{O} + \gamma \\
{}^{16}\text{O} + \text{p} &\rightarrow {}^{17}\text{F} + \gamma \\
{}^{17}\text{F} &\rightarrow {}^{17}\text{O} + e^+ + \nu_e \\
{}^{17}\text{O} + \text{p} &\rightarrow {}^{14}\text{N} + {}^{4}\text{He}.
\end{aligned}
$$

When the core reaches a temperature of 10^8K, the thermal energy is sufficient to allow helium nuclei to tunnel through their Coulomb barrier. The binding energy of two He nuclei – known as alpha particles – is negative, so the resulting ^{8}Be nucleus is very unstable to fission and would decay back to two He nuclei very rapidly (10^{-16} s). However, this timescale is much greater than the transit time of two α particles (10^{-19}s) within the core, such that α capture by ^{8}Be permits the formation of stable ^{12}C as follows:

$$
\begin{aligned}
{}^{4}\text{He} + {}^{4}\text{He} &\leftrightarrow {}^{8}\text{Be} \\
{}^{8}\text{Be} + {}^{4}\text{He} &\leftrightarrow {}^{12}\text{C}^* \rightarrow {}^{12}\text{C} + \gamma
\end{aligned}
$$

which leads to the production of ^{16}O via α-capture:

$$
{}^{12}\text{C} + {}^{4}\text{He} \rightarrow {}^{16}\text{O} + \gamma.
$$

Significant uncertainties remain in this nuclear reaction rate, for which different authors disagree by a factor of $\sim$3 (e.g. Kunz *et al.* 2002). This reaction competes with the triple-alpha reaction during helium burning to determine the ratio of carbon to oxygen at the onset of carbon burning, which explains why oxygen and carbon are the third and fourth most abundant elements in the Universe after hydrogen and helium.

Core He burning in massive stars also has the effect of transforming nitrogen to neon and magnesium:

$$
\begin{aligned}
{}^{14}\text{N} + {}^{4}\text{He} &\rightarrow {}^{18}\text{F} + \gamma \\
{}^{18}\text{F} &\rightarrow {}^{18}\text{O} + e^+ + \nu_e \\
{}^{18}\text{O} + {}^{4}\text{He} &\rightarrow {}^{22}\text{Ne} + \gamma \\
{}^{22}\text{Ne} + {}^{4}\text{He} &\rightarrow {}^{25}\text{Mg} + n
\end{aligned}
$$

and serves as the main neutron source for the s-process in massive stars. Generally the production of s-process elements is dominated by the AGB phase in lower mass stars.

Since the cross section of ^{16}O$(\alpha, \gamma)^{20}$Ne is tiny because of missing resonances close to the relevant He-burning energies, further nucleosynthesis reactions in stars more massive than $\sim 8 M_\odot$ produce ^{20}Ne, ^{24}Mg via carbon-burning at extremely high temperatures of 10^9K, as follows:

$$
\begin{aligned}
{}^{12}\text{C} + {}^{12}\text{C} &\rightarrow {}^{20}\text{Ne} + {}^{4}\text{He} \\
&\rightarrow {}^{23}\text{Na} + \text{p} \rightarrow {}^{24}\text{Mg} + \gamma,
\end{aligned}
$$

producing a core which is now dominated by ^{16}O and ^{20}Ne. The dominant loss of energy now shifts from electromagnetic radiation to neutrinos, *dramatically* shortening the evolutionary timescales. Oxygen is not the next fuel to burn since it is extremely stable. Instead, neon burning occurs at yet higher temperatures of 1.5×10^9 K.

$$^{20}\text{Ne} + {}^4\text{He} \quad \rightarrow \quad {}^{24}\text{Mg} + \gamma$$
$$^{24}\text{Mg} + {}^4\text{He} \quad \rightarrow \quad {}^{28}\text{Si} + \gamma.$$

O-burning then occurs when the temperature reaches 2×10^9 K:

$$^{16}\text{O} + {}^{16}\text{O} \quad \rightarrow \quad {}^{28}\text{Si} + {}^4\text{He}$$
$$\rightarrow \quad {}^{32}\text{S} + \gamma$$
$$\rightarrow \quad {}^{31}\text{P} + \text{p}$$
$$\rightarrow \quad {}^{31}\text{S} + \text{n}.$$

After O-exhaustion, the stellar core is composed of ^{28}Si and either ^{32}S or ^{38}Ar for sufficiently massive stars. Ultimately, there is a competition between photodisintegration processes and α-captures which progressively shifts the equilibrium towards the Fe-peak elements, producing an iron–nickel core, surrounded by α-element shell burning. The evolution of the temperature and density for the center of two massive stars of initial mass 15 and $25 M_\odot$ are shown in Fig. 5.1, illustrating the continued contraction to higher density and temperature until the core undergoes final collapse. During most of the evolution, the density is proportional to the cube of the temperature, as expected for the non-adiabatic contraction of a star obeying the ideal gas law through successive stages of hydrostatic equilibrium. An alternative schematic view of the interior evolution of a $25 M_\odot$ star is illustrated in Fig. 5.2. Shell burning takes place at the boundary of the former convective core, where that fuel has been exhausted. An "onion-skin" structure is developed, with heavier nuclei dominating the

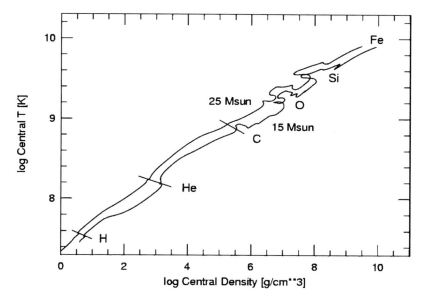

Fig. 5.1 The evolution of the temperature and density for the center of 15 and $25 M_\odot$ stars, with the location of successive nuclear reactions indicated. From Woosley & Janka (2005).

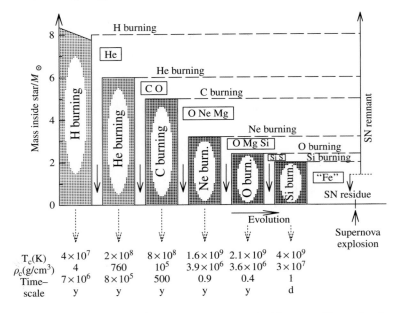

Fig. 5.2 Schematic view of the interior evolution of a $25M_\odot$ star illustrating the corresponding central temperatures, densities and timescales. From Arnould & Goriely (2003) with permission from Elsevier Limited.

composition as one moves from the surface to the center. At this stage the star has arrived at the end of its "quiescent" life.

One final aspect relating to nucleosynthesis is the issue of primary nitrogen production at low metallicity. ^{14}N is mostly synthesized within the CNO cycle from carbon/oxygen initially present in the star, i.e. the increase in nitrogen scales with the initial carbon and oxygen content, known as "secondary" nitrogen. If, instead, nitrogen is built directly from hydrogen and helium, it scales with that of other primary heavy elements and is known as "primary" nitrogen. Observational evidence for the creation of primary nitrogen first came from the study of old, metal-poor stars, in which the ratio of N/O declines until it reaches a plateau for decreasing metallicity (see e.g. Edmunds & Pagel 1978).

Theoretically, in a star which has both a He burning core and a H burning shell, some newly synthesized carbon from the core is transported into the H burning shell, where the CNO cycle converts it into primary ^{14}N. Mixing induced by rotation may provide the necessary driving mechanism in intermediate mass stars at low metallicity. The CNO burning occurs at much higher temperatures at low metallicities, causing the H-burning shell to lie closer to the edge of the He-burning core, permitting easier transport of ^{12}C from the core to the shell.

5.2   Evolution to a red supergiant

Stellar lifetimes are very steeply dependent upon mass. For most stars, around 10% of the initial mass is involved in nuclear reactions, so the main-sequence lifetime will be proportional to the mass and inversely proportional to the luminosity, i.e. $\tau_{\rm MS} \propto M_*/L_*$. From Chapter 6, $\tau_{\rm MS} \propto M_*^{-3.7}$ for solar-type stars for which $L_* \propto M_*^{4.7}$, and $\tau_{\rm MS} \propto M_*^{0.75}$ for stars more massive than $30M_\odot$ for which $L_* \propto M_*^{1.75}$. The main-sequence lifetime for

Table 5.1 *Main-sequence lifetimes, τ_{MS}, of massive stars at solar metallicity according to Massey & Meyer (2001).*

Mass ($M_\odot$)	τ_{MS} (Gyr)	$T_{\text{eff}}(K)$	Sp. type	$\log L_*/L_\odot$
120	0.003	53 300	O2–3 V	6.2
60	0.003	48 200	O4 V	5.7
25	0.006	37 900	O6.5 V	5.3
12	0.016	28 000	B1 V	4.0
5	0.095	17 200	B5 V	2.7
2.5	0.585	10 700	B9 V	1.6
1.25	4.91	6 380	F5 V	0.3
1.0	9.84	5 640	G8 V	−0.2
0.8	25.0	4 860	K2 V	−0.6
0.5	100	3 890	M0 V	−1.4
0.2	4 000	3 300	M4 V	−2.2

stars of low to high mass is presented in Table 5.1 from Massey & Meyer (2001). Focusing upon stars initially above $8M_\odot$, the highest mass O dwarfs evolve from the main sequence after only a couple of Myr, while the main sequence lifetimes of early B dwarfs is counted in tens of Myr, an order of magnitude longer, but still incredibly rapid in comparison with low-mass stars. The H and He burning lifetimes for stars initially spanning 1–100 $M_\odot$ are presented in Fig. 5.3. Solar and metal-poor cases are shown, illustrating that lifetimes are relatively weak functions of metallicity.

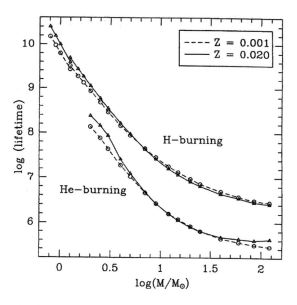

Fig. 5.3 The logarithm of the lifetime in years for the H and He burning phases versus initial mass, expressed in solar units for $Z = 0.02$ (solar) and $Z = 0.001$ (metal-poor). From Schaller *et al.* (1992).

The various inputs to stellar interior evolutionary models come from either laboratory experiments (e.g. opacities, nuclear reaction rates) or astronomical observations (e.g. mass-loss properties, rotation rates). Extensive evolutionary models for a range of metallicities using radiative opacities from OPAL (Iglesias & Rogers 1991) – later supported by independent Opacity Project results (Seaton *et al.* 1994) – are available from the Geneva (e.g. Schaller *et al.* 1992) and Padova (e.g. Bressan *et al.* 1993) groups.

Both sets of grids, covering initial masses up to 120 $M_\odot$, employ the following empirical inputs for high-mass stars. Mass-loss rates originate from de Jager, Nieuwenhuijzen, & van der Hucht (1988) for most evolutionary phases, with an adopted metallicity scaling of $\dot{M} \propto Z^{0.5}$ for OB stars. A mass-dependent mass-loss rate from Langer (1989) was adopted for all W-R stars in the Padova group, while a uniform mass-loss rate is adopted for W-R stars containing surface hydrogen in the Geneva group calculations. Geneva tracks have been most widely applied massive stars (e.g. isochrones from Lejeune & Schaerer 2001), and Padova grids (e.g. isochrones from Bertelli *et al.* 1994) used to study low and intermediate mass stars.

Until recently, the solar heavy-metal content was thought to be close to 2% by mass. However, more sophisticated solar model atmosphere studies have indicated that the oxygen abundance for the Sun has been substantially overestimated. Since oxygen is astrophysically the most abundant element, the downward reduction in metallicity is significant. The inferred current solar metallicity is 1.2% (Asplund *et al.* 2005), or a solar neighborhood metallicity of 1.4% after correction for depletion of heavy elements in the 4.5 Gyr old Sun (Meynet, 2008). This revision reconciles previous (low) oxygen abundances for the Orion nebula with the previously (high) oxygen abundance for the solar photosphere. Consequently, the widely used solar metallicity models with $Z = 0.02$ in fact refer to a somewhat metal-rich environment. Figure 5.4 presents a theoretical H-R diagram for $Z = 0.02$ from Schaller *et al.* (1992), illustrating the core H and He burning phases as hatched regions.

Mixing and convection

Numerous aspects of stellar models depend on the way chemical elements are mixed in their interior. Mixing may occur within the core and in the radiative envelopes, driven by various instabilities. For example, shear mixing is due to the turbulence induced by the friction of stellar layers rotating at different velocities.

On the Main Sequence, the composition of the core is slowly changing from hydrogen to helium. Evolution proceeds slowly, with the luminosity increasing and surface temperature decreasing. The central temperature of high-mass stars is a factor of two higher than solar-mass stars, with densities and pressures an order of magnitude lower. Nevertheless, the very strong temperature dependence of the CNO cycle means that the energy generation of high-mass stars is highly concentrated in the center. The resulting temperature gradient is unstable to convection, so that massive stars possess *convective* cores. In contrast, energy transport is radiative for the solar core. One may define a boundary of the convective core, known as the Schwarzschild radius, R_s, when the radiative gradient equals the adiabatic gradient. In reality, convective motions may occur beyond this point and "overshoot". Such core overshooting may be parameterized by an overshooting parameter, α, in which the overshooting distance is αR_s. As α increases, the core H-burning timescale and stellar luminosity also increase.

Convection has the effect of efficiently *mixing* the material in the core, bringing a fresh supply of hydrogen into the center, and distributing the newly formed helium throughout the core. This has the effect of keeping the core composition uniform, such that once the hydrogen

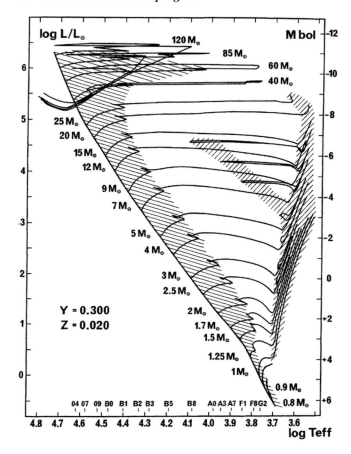

Fig. 5.4 Theoretical H-R diagram for $Z = 0.02$ (solar) metallicity models, including the regions of core H and He burning. Spectral types are approximate, and originate primarily from Böhm-Vitense (1981). From Schaller *et al.* (1992).

is exhausted energy production ceases throughout the core. The envelopes of hot massive stars are radiative, such that the internal structure of massive stars is the reverse of cool stars, including red supergiants, for which cores are radiative and envelopes are convective. As the mass of the convective core decreases, there will be a varying molecular weight gradient within the stellar interior. Within massive stars this zone may be dynamically stable but vibrationally unstable (Kato 1966) and initiate "slow" mixing. Convection may either be treated as an efficient, fast process (the Schwarzschild criterion) or an inefficient, slow process (the Ledoux criterion). Semi-convection refers to the zone between these extremes, subject to mild mixing, and is usually treated by a diffusion process. Observational evidence may be used to distinguish between the two extreme cases (Langer & Maeder 1995).

There are various predictions that very massive stars are pulsationally unstable on the core hydrogen burning and core helium main sequence (e.g. Kiriakidis, Fricke, & Glatzel 1993), although these have not been observed. Langer *et al.* (1994) present $60 M_\odot$ evolutionary models which somewhat speculatively assume that pulsations drive enhanced mass-loss during the Main Sequence and post-Main Sequence phases. Stothers & Chin (1993) predict a dynamical instability for luminous post-main sequence stars, which they attributed to the LBV phase.

From H to Si burning

The abrupt end of core hydrogen burning in massive stars causes the star to undergo a slow contraction. This has the effect of heating up the core in which the central density and temperature increase. Eventually, the hydrogen-rich shell surrounding the core is heated up, leading to a thin shell of hydrogen burning, causing the star to expand and cool. Overall contraction ceases, and the energy from the core contraction is fed into an expansion of the envelope, causing the star to rapidly move redward in the H-R diagram. One expects, and observes, very few massive stars in the yellow part of the H-R diagram. As the star expands and cools towards a red supergiant, convection becomes more efficient than radiation transport in the outer layers. Due to conservation of angular momentum, red supergiants rotate very slowly ($\sim$10 km s^{-1}). The evolution of the internal structure of a $15M_\odot$ star is shown in Fig. 5.5. Cloudy regions represent convective zones, heavy diagonals indicate regions of nucleosynthesis, with evolution proceeding from a main sequence O phase to a blue supergiant and ultimately a red supergiant.

Core helium burning, during the red supergiant phase for moderately high mass stars, operates for $\sim$10% of the main-sequence lifetime (recall Fig. 5.3), with shell H-burning occuring surrounding the core. If the star has retained its massive envelope, the outer layers of the red supergiant have no time to respond to core exhaustion and ignition during the late evolutionary phases since the stellar Kelvin–Helmholtz timescale is larger than the evolutionary timescale of the core. Consequently, although the core is increasing in temperature, observationally the star remains in the red part of the H-R diagram.

After He-exhaustion, the exceptionally high temperatures ($> 8 \times 10^8$ K) and densities provides a means of producing copious neutrinos. Neutrino–anti-neutrino pairs are produced

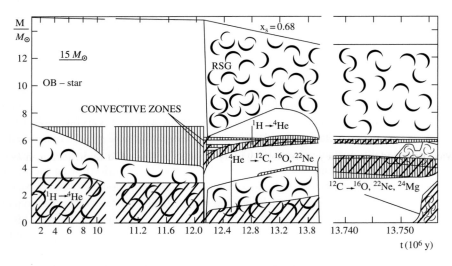

Fig. 5.5 Evolution of the interior structure of an initial $15M_\odot$ star up to central C-exhaustion. Cloudy regions represent convective zones, whilst heavy diagonals indicate regions of high nuclear energy generation. From Maeder & Meynet (1987).

by electron and positron annihilation. Since neutrinos interact only weakly with matter, they escape from the stellar core, taking with them the majority of the available thermal energy produced by nuclear reactions or gravitational contraction. To compensate for the neutrino losses, the core has to increase its nuclear energy production by further contraction. As a result, the late evolutionary phases are greatly accelerated. The time between C-ignition and the final supernova is less than 0.1% of the corresponding main-sequence lifetime, due to neutrino losses and the decrease in yield from exothermic reactions of increasingly heavy nuclei. Once iron – the most stable element with the highest binding energy per nucleon – is produced in the core, it would be necessary to add energy to produce heavier elements (endothermic). Therefore, no further energy may be extracted from nucleosynthesis, leading to a catastrophic core-collapse, as we shall discuss later in this chapter. The uncertain $^{12}C(\alpha, \gamma)^{16}O$ reaction rate (e.g. Kunz *et al.* 2002) affects the final core mass, since a smaller rate gives a larger carbon abundance after helium burning, and so a smaller iron core mass.

High mass stars typically undergo classical Fe core-collapse during either the RSG or W-R phase. In contrast, low and intermediate mass stars possess degenerate CO cores at the end of their life, entering the white dwarf cooling sequence after the AGB stage. Indeed, the 40 Myr old massive star cluster NGC 1818 in the LMC hosts a luminous white dwarf, setting a lower initial mass limit of $M_{init} \sim 7.6\ M_\odot$ for stars undergoing core-collapse and producing neutron stars instead of white dwarfs (Elson *et al.* 1998).

Between these two extremes, stars with initial masses of 8–9 $M_\odot$ produce massive Super-Asymptotic Giant Branch (S-AGB) stars, which may explode as electron capture supernovae. Such stars undergo carbon burning and a thermal pulse, causing the core to grow until either the entire envelope is ejected (producing a massive white dwarf) or it encounters electron captures on ^{20}Ne and ^{24}Mg (triggering a supernova). Indeed, perhaps up to 10% of core-collapse SNe may be due to S-AGB stars at solar metallicity, with a yet higher proportion in metal-poor environments. The Crab nebula has been identified as a possible remnant from an exploding 8–10 $M_\odot$ electron capture supernova.

5.3 Evolution to the Wolf–Rayet stage

The effect of mass-loss has a dominant effect upon stellar models for the most massive stars, albeit hindered by uncertainties both in blue and (especially) red parts of the H-R diagram (Woosley, Langer, & Weaver 1993). Up until the last decade, rotational mixing was neglected. Indeed, revised Geneva group tracks were produced in which mass-loss rates were artificially increased relative to empirical calibrations of de Jager, Nieuwenhuijzen, & van der Hucht (1988) in order to predict W-R populations that were in better agreement with observations (Maeder & Meynet 1994). No equivalent set of Padova evolutionary tracks was produced, since they were most widely applied to low- and intermediate-mass stellar populations.

The predicted evolutionary path of very massive stars at solar metallicity is indicated in Fig. 5.6, according to Schaller *et al.* (1992). The evolution of such stars is dictated by mass-loss rather than convection (see also Woosley, Langer, & Weaver 1993). Standard mass-loss rates following de Jager, Nieuwenhuijzen, & van der Hucht (1988) for the pre-W-R evolution are adopted in the upper panel, whilst main-sequence mass-loss rates which are enhanced by a factor of two are presented in the lower panel. Additional calculations were performed for which W-R mass-loss rates were also artificially enhanced by a factor of two (Meynet *et al.* 1994). The observed upper H-R diagram is reasonably well matched by such models,

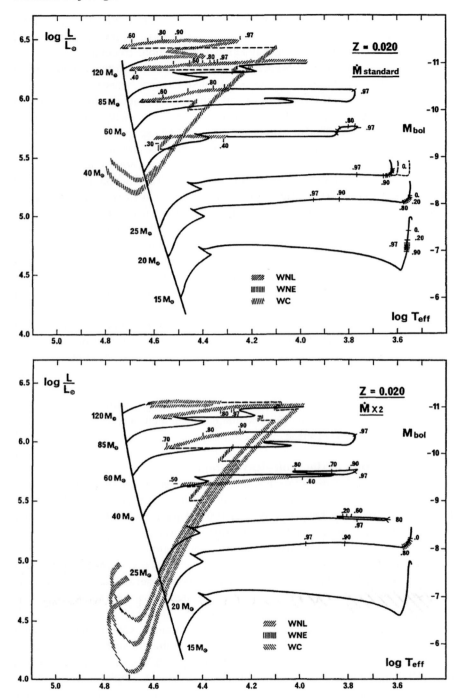

Fig. 5.6 Theoretical H-R diagram for $Z = 0.02$ (solar) metallicity models, with standard (upper plot) versus a factor of two enhanced de Jager *et al.* (1988) mass-loss rates (lower plot). From Schaller *et al.* (1992).

in which moderately massive stars spend the majority of their post-main sequence life as either blue or red supergiants. The intermediate yellow supergiant phase is rapidly passed through by the star evolving from blue to red. For stars of higher initial mass, the RSG phase is circumvented due to extreme mass-loss during the luminous blue variable stage, in agreement with the observed absence of luminous red supergiants above the Humphreys & Davidson (1979) limit (recall Fig. 2.2).

If mass-loss during the blue and/or red supergiant phase is sufficient to remove much of the H-rich envelope whilst H-shell burning, the stellar core will evolve blueward in the H-R diagram to ignite He-burning and become a Wolf–Rayet star, remaining in the blue part of the H-R diagram until final core-collapse. The evolution of the interior structure of a star of initial mass $60M_\odot$ is shown in Fig. 5.7, for which the evolution of the surface of the star is indicated, relating to O, WN, and WC phases. According to such non-rotating models the lower mass limits for the formation of W-R stars ranges from $M_{\rm init} \sim 60-80M_\odot$ at $\leq 0.1\,Z_\odot$ to $M_{\rm init} \sim 21-25M_\odot$ at $3\,Z_\odot$.

Schaerer & Maeder (1992) provided a mass–luminosity relation for H-free W-R stars:

$$\log \frac{L_*}{L_\odot} = 3.032 + 2.695 \log \frac{M_*}{M_\odot} - 0.461 \left(\log \frac{M_*}{M_\odot} \right)^2 . \tag{5.1}$$

This expression is effectively independent of the chemical composition since the opacity is purely electron scattering.

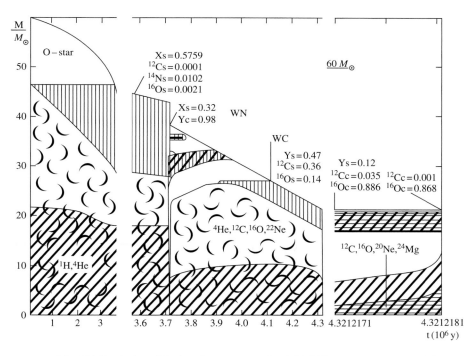

Fig. 5.7 Evolution of the interior structure of an initial $60M_\odot$ star up to central C-exhaustion. Cloudy regions represent convective zones, whilst heavy diagonals indicate regions of high nuclear energy generation. From Maeder & Meynet (1987).

In addition to the extensive Geneva and Padova predictions, Langer *et al.* (1994) also calculated evolutionary models for very massive $60M_\odot$ stars at solar metallicity, including the theoretical pulsational instability driven mass-loss of Kiriakidis, Fricke, & Glatzel (1993). Depending upon the efficiency, ξ, of such instabilities, which has a maximum strength of $3\times10^{-5}M_\odot$ yr^{-1} for $\xi=1$ at $T_{\rm eff} \sim 34\,000$ K, the star remains on the blue side of the instability region in the H-R diagram during its main-sequence phase, which Langer *et al.* identify as H-rich WN stars. Such high mass-loss produces a hotter, less luminous star with a lower mass convective core, ~ 0.3 dex less luminous than Geneva group predictions of Schaller *et al.* (1992). Evolution then proceeds as normal, in which overall contraction is terminated when the H-burning shell ignition causes the stellar radius to increase, causing the star to cross the pulsational instability regime. For surface temperatures below 20 000 K, stronger instabilities are predicted, causing a very high mass-loss rate, which Langer *et al.* associate with the LBV phase, and adopt $5 \times 10^{-3} M_\odot$ yr^{-1} leading to the removal of $6M_\odot$ over a short timeframe, comparable to the giant eruption experienced by η Carinae in the nineteenth century.

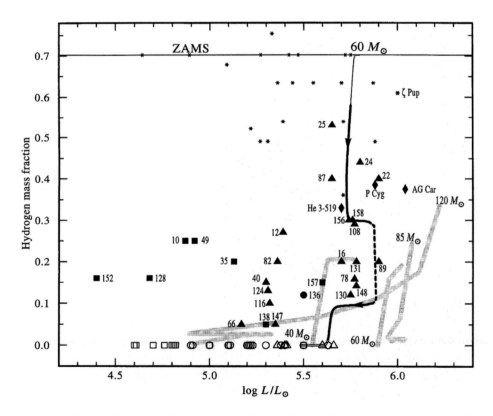

Fig. 5.8 Variation of hydrogen mass-fraction for the $60M_\odot$ evolutionary model of Langer *et al.* (1994). The location of the LBV P Cygni, WNL (triangles), strong-lined WNE (circles), weak-lined WNE (squares) and WC (diamonds) stars from unblanketed analyses by Hamann *et al.* (1993) and Koesterke & Hamann (1995) are presented. Filled-in symbols relate to cases where hydrogen as been detected. The late WN phase for Meynet *et al.* (1994) models are also indicated as shaded lines. From Langer *et al.* (1994).

Once the star returns from its brief excursion to lower temperatures central He burning has just started. A hydrogen-rich envelope of 2.5 $M_\odot$ remains on the star, which is quickly removed due to high Wolf–Rayet mass-loss rates, for which the mass-dependent rates of Langer (1989) have been adopted by Langer *et al.* They emphasized that two late WN phases are anticipated – one during core H-burning, and another following the LBV excursion. Crowther *et al.* (1995) came to a similar conclusion for the Carina H-rich WN stars, confirming the suspicion of Walborn (1973) from spectral morphology (recall Fig. 2.17). For very massive stars, the sequence of

$$O \rightarrow WN(H - rich) \rightarrow LBV \rightarrow WN(H - poor) \rightarrow WC \rightarrow SN\ Ic$$

from Chapter 2 is reinforced in Fig. 5.8 where the surface H-mass fractions for Of supergiants, LBVs, and W-R stars are compared with the evolutionary track of Langer *et al.* plus the late WN phase of the Meynet *et al.* (1994) models.

Metallicity impacts upon interior models through a combination of lower mass-loss rates and a smaller radius (higher temperature) on the ZAMS due to reduced opacity. The effect is a shift to the blue by 0.04–0.1 dex in log T_{eff} for a factor of 10 reduction in metallicity. From the enhanced mass-loss rate models of Meynet *et al.* (1994) good agreement between the observed W-R populations for metallicities ranging from the SMC to solar metallicity were obtained. W-R lifetimes and N(W-R)/N(O) number ratios increase with increasing metallicity. However, as we have already discussed, both O star and W-R mass-loss rates have been *reduced* due to clumping in recent years, such that some other process – which is now believed to be rotational mixing – must operate.

5.4   Rotation and mass-loss

Mass-loss and rotation are intimately linked for the evolution of massive stars. Rapid rotation may produce enhanced mass-loss, as is observed in η Car (Smith *et al.* 2003), while stellar winds will cause a loss in angular momentum, leading to spin down. Since metallicity influences the strength of stellar winds, the spin down of massive stars is anticipated to be rapid at solar metallicity, erasing initial rotational velocity properties within a few million years. In metal-poor environments, initial conditions remain preserved during the main sequence lifetime of O-type stars.

Various deficiencies in Geneva group evolutionary models for which rotational mixing was neglected led to its inclusion. These included: (1) poor agreement between interior models and observations of OB stars exhibiting significant surface enrichment of nitrogen relative to carbon (and oxygen); (2) the observed blue to red supergiant ratio; (3) the presence of intermediate WN/C stars; (4) the observed W-R population at low metallicity.

Centrifugal acceleration causes the surface temperature of a rotating star to depend upon latitude, as a consequence of the von Zeipel (1924) theorem. In addition, stellar winds are no longer isotropic, such that one expects higher mass-loss from the poles than the equator due to its higher effective gravity (recall Chapter 3).

Rotational mixing has been incorporated into evolutionary calculations, following the theoretical treatment of Zahn (1992) who derived the equations describing the transport of chemical species and angular momentum in the shellular rotation law. This approach seems reasonable, and indeed applies within the solar interior. Two variations have been developed by Meynet & Maeder (1997) and by Heger, Langer, & Woosley (2000) to introduce these equations, although the same physical framework is applied.

Meynet & Maeder (1997) describe the transport of angular momentum in the stellar interior through the shear and meridional instabilities. Models involve a parameterization for the timescale and efficiency of the shear instability, with the meridional currents transporting angular momentum to the surface from the core, or vice versa. On the main sequence, there are two cells of meridional currents, one close to the surface, transporting angular momentum outwards, the other just above the convective core, transporting angular momentum inwards. Alternatively, Heger, Langer, & Woosley (2000) include magnetic fields – which may be considered as a consequence of rotation, in which differential rotation in the radiative zone acts as a dynamo amplifying the initial magnetic field – such that the redistribution of angular momentum in the stellar interior during core H-burning is close to rigid rotation, such that momentum is transported radially from the core to the surface.

Rotation introduces three primary effects.

1. Chemically enhanced layers are transported outwards from the convective core, providing early enrichment of products from the CNO cycle.
2. Rotating cores are larger and more massive than non-rotating cores, such that the effect of rotation is similar to that of convective core overshooting during core H-burning. Models which rotate faster imply a longer core H-burning timescale and higher luminosities.
3. Stars which rotate above a critical rate may follow tracks which are essentially those of homogeneous evolution (Maeder 1987b), i.e. evolution proceeds directly to the Wolf–Rayet phase, circumventing the red part of the H-R diagram.

Maeder & Meynet (2000) find that the most massive stars at solar metallicity slow down much more rapidly than lower-mass stars, due to mass-loss. Observationally, Main Sequence O stars are indeed observed to possess lower rotational velocities than B stars (recall Fig. 2.21). At solar metallicity, the initial rotation velocity v_{init} may produce quite different final supernova progenitor stars, either a red supergiant ($v_{\mathrm{init}} = 0$ km s^{-1}), blue supergiant ($v_{\mathrm{init}} = 200$ km s^{-1}) or a W-R star ($v_{\mathit{init}} = 300$ km s^{-1}) for a star of initial mass 20 $M_{\odot}$.

Rotating models evolve more rapidly in the red supergiant phase, during which they lose mass at a higher rate, favoring a subsequent bluewards evolution. In addition, more helium is present in the H-burning shell, leading to the H-shell becoming less active and leading to a less extended convective zone. This has the effect of increasing the fraction of the time spent in the red relative to the blue, in better agreement with the observed blue-to-red supergiant ratio, B/R, that is observed in young star clusters such as NGC 330 in the SMC (Maeder & Meynet 2005). The evolution of equatorial velocities and the ratio of the angular velocity to critical angular velocity at the surface of stellar models for a variety of metallicities are presented in Fig. 5.9.

Rotation also influences the early surface abundances of massive stars. For a representative 9 $M_{\odot}$ case, surface enrichment of N/C is only predicted to occur during the *red supergiant* phase for a non-rotating model, whilst enrichment in N/C is predicted during the *main sequence* phase for a solar metallicity model with $v_{\mathrm{init}} = 300$ km s^{-1}. Indeed, blue supergiants possess substantial N/C enhancement (recall Section 3.4).

Metallicity impacts upon interior models through lower mass-loss rates and a smaller radius. The former effect together with internal transport processes produce an increase in the rotational velocity for main-sequence stars according to Maeder & Meynet (2000). Even

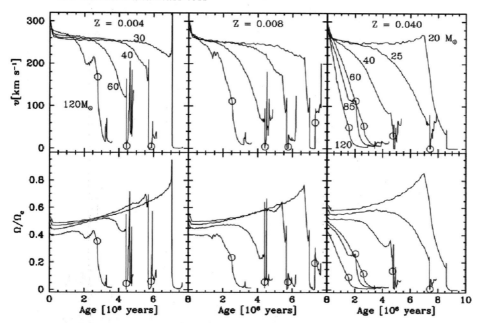

Fig. 5.9 Evolution of the equatorial velocities (upper panels) and the ratio of the angular velocity to the critical angular velocity (lower panels) at the surface of stellar models for various initial masses and metallicities for initial rotational velocity $v_{\text{init}} = 300 \text{ km s}^{-1}$. Empty circles are places where the star enters the W-R phase. From Meynet & Maeder (2005).

if the initial distribution of velocities on the ZAMS does not depend upon metallicity, one would expect the relative number of stars close to break-up velocity will increase at lower metallicity. Smaller radii for low metallicity stars also favours a larger early enrichment of N/C during the main sequence evolution of massive stars.

In the extreme cases of I Zw 18 and SBS0335-052E, nearby HII galaxies with a metallicity of only 1/30 of the Solar Neighborhood (see Chapter 9), Wolf–Rayet stars are predicted for only the most massive single stars ($\geq 90 M_{\odot}$) according to non-rotating models of de Mello *et al.* (1998), for which WN stars are expected to dominate the W-R population, with N(WC)/N(WN)$\leq$0.15. Indeed, W-R stars are observed in UV and optical spectroscopy of regions within I Zw 18, as illustrated in Fig. 5.10, and SBS 0335-052E. However only WC stars have been unambiguously identified. From Chapter 4 it appears that the strength of WN winds depends upon metallicity more sensitively than WC stars, the former will be extremely difficult to directly detect in remote metal-poor galaxies. More likely, since hot, weak-lined W-R stars are predicted to have hard ionizing flux distributions (recall Fig. 4.18) such stars may be indirectly indicated via the presence of strong nebular HeII 4686 emission, which is observed in I Zw 18 and other metal-poor star-forming galaxies. Consequently, large W-R populations are inferred in such galaxies, for which single star evolutionary models appear to have difficulty in reproducing. Close binary evolution may dominate the formation channel for such W-R stars.

The relationship between the final and initial mass for various stellar models at various metallicities is shown in Fig. 5.11. Rotation favors the evolution into the W-R phase at earlier

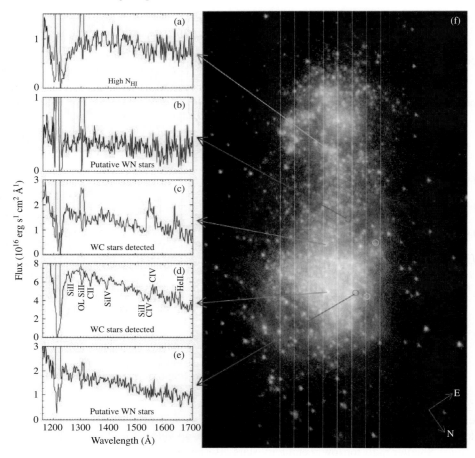

Fig. 5.10 Far-UV HST/STIS spectra of star clusters in the metal-deficient ($1/30Z_\odot$) blue compact dwarf galaxy I Zw 18 (shown as a composite far-UV/near-UV/V-band HST image in panel f), including two regions (c) and (d) observed to host WC stars. Reproduced from Brown *et al.* (2002) by permission of the AAS.

stages, adding to the W-R lifetime, plus allowing lower initial mass stars to enter the W-R phase (Meynet & Maeder 2005). For regions of constant star formation (as opposed to starbursts), representative of Local Group galaxies, one can estimate a theoretical number ratio of W-R to O stars from the average lifetime of W-R to O stars, weighted over the initial mass function (IMF, see Chapter 7).

For a Salpeter IMF slope for massive stars, the ratios obtained by rotating models are in much better agreement with $N(W\text{-}R)/N(O) \sim 0.01$ at SMC metallicity increasing to $N(W\text{-}R)/N(O) \sim 0.1$ at solar metallicity. Since the O star population is relatively imprecise, the observed WN to WC ratio for non-starburst regions are often used instead. Recalling Fig. 2.9, agreement between the observed distribution and evolutionary models is successful at low metallicity, with the exception of IC 10, whose number ratio remains controversial, since it lies in the galactic plane and so WN stars are harder to identify than WC stars (recall

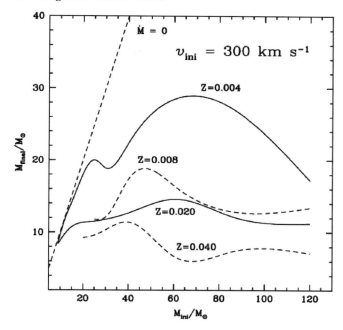

Fig. 5.11 Relation between the initial and final mass for rotating stellar models at a variety of metallicities. The slope labeled $\dot{M} = 0$ corresponds to the case without mass-loss. From Meynet & Maeder (2005).

Fig. 2.10). At high metallicity the empirical N(WC)/N(WN) ratio continues to increase while the rotating model predictions from Meynet & Maeder (2005) flatten out. Eldrige & Vink (2006) have calculated evolutionary models of massive stars in which W-R mass-loss rates are metallicity dependent, resulting in improved agreement for metal-rich W-R populations (Fig. 2.9).

5.5 Magnetic massive stars

Rapidly rotating massive stars will lose their angular momentum from magnetohydrodynamical processes, in which the magnetic field is generated in radiative layers by the dynamo due to differential rotation (Spruit 2002). This dynamo is based upon a predicted Tayler instability (Tayler 1973), although there is currently no proof of its existence. According to Spruit, even a weak horizontal magnetic field is subject to this instability, leading to a vertical field component which is wound up by differential rotation. Field lines become progressively closer, leading to a strong horizontal field. The magnetic field created by the dynamo does not reach the stellar surface. However, there are consequences which may be observed. The role of magnetic fields is to impose nearly solid body rotation. Consequently, magnetic fields favor higher rotational velocities during the main-sequence evolution than cases in which magnetic fields are not accounted for, as illustrated in Fig. 5.12. In addition, rotation with a magnetic field leads to a larger core at the end of the main-sequence phase, affecting the evolution in the H-R diagram. In addition, magnetic models (when the meridional circulation is included) lead to an enhanced N-enrichment, for which there is some observational support.

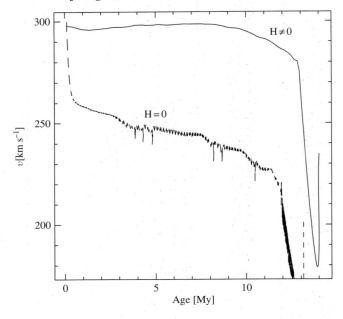

Fig. 5.12 Evolution of the surface rotational velocities for a $15M_\odot$ model initially rotating at $v_{\rm eq} = 300\ {\rm km\ s}^{-1}$ in which magnetic fields are excluded (H = 0) or included (H ≠ 0). From Maeder & Meynet (2005).

Table 5.2 *Periods of young pulsars from Heger, Woosley, & Spruit (2005) and references therein.*

Pulsar	Current period (ms)	Initial period (ms)
PSR J0537–6910 (N157B, LMC)	16	10
PSR B0531+21 (Crab)	33	21
PSR B0540–690 (LMC)	50	39
PSR B1509–58	150	20

Numerical studies of the late evolutionary phases of rotating massive stars have failed to reproduce the observed rotation rates of young neutron stars, predicting very rapidly rotating (∼1 ms) neutron stars. Observationally, initial rotation rates are in the 10–40 ms range, as shown in Table 5.2. The inclusion of magnetic torques is sufficient to reproduce the observed rotation rates of young neutron stars (Heger, Woosley, & Spruit 2005).

5.6   Core-collapse supernovae

D. Baade and F. Zwicky first suggested in the late 1930s that supernova explosions are related to the death of massive stars. However, the precise mechanism of core-collapse explosions remains uncertain to this day. Thielemann, Nomoto, & Hashimoto (1996) discuss core-collapse SNe and their ejecta, whilst Woosley & Janka (2005) consider more recent theoretical developments.

Observations of SNe

Historically, supernovae have been classified as Type I or II on the basis of the absence or presence of hydrogen in their optical spectra, respectively. Since the mid-1980s, Type I SNe have been subdivided into Ia, b, or c, according to whether silicon or helium lines are present, whilst Type II SNe are generally subdivided by the shape of their light curves, as summarized in Fig. 5.13. Representative spectra from these main types are presented in Fig. 5.14, at visual maximum, plus three weeks and one year later.

Type Ib/c and Type II SNe are, for the most part, associated with regions of recent star formation, whilst Type Ia SNe are discovered in all types of galaxies, suggesting a different type of progenitor. The current, widely held, view is that Type Ia SNe are associated with the thermonuclear explosion of a white dwarf due to accretion of material above the Chandrasekhar limit, whilst the other types involve the core-collapse of massive stars. Type Ia are fairly uniform in peak luminosity, with $M_B \sim -19$ mag, although there is an intrinsic dispersion in absolute magnitudes. This is well correlated to the rate of decline in B-band light curves (Phillips 1993), which has permitted their use as cosmological standard candles to probe the nature of the cosmological constant (Perlmutter *et al.* 1999).

A supernova may be bright either because it produces a lot of radioactive ^{56}Ni (e.g. Type Ia) or because it has a large radius and large envelope (e.g. Type II). A higher radioactivity produces more energy at late times, whilst a higher initial radius produces a higher luminosity at early times. Typical explosion energies of core-collapse SNe span an order of magnitude

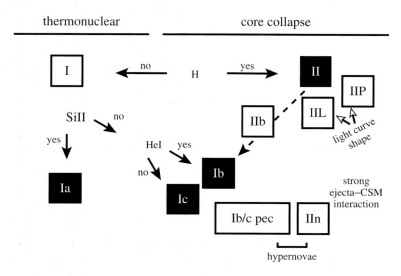

Fig. 5.13 Classification scheme for supernovae. Type Ia SNe are associated with the thermonuclear explosion of accreting white dwarfs, whilst other types are associated with the core-collapse of massive stars. Some bright, energetic Type Ib/c and IIn SNe are often called "hypernovae". From Turatto (2003), with kind permission of Springer Science and Business Media.

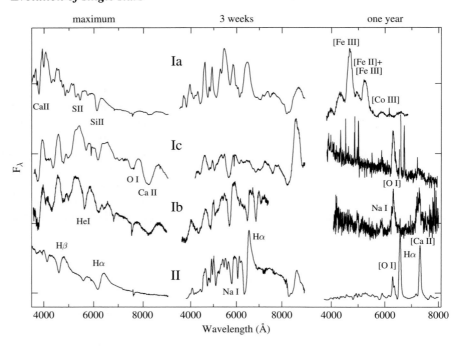

Fig. 5.14 Spectra of the main SN types at maximum light, plus three weeks and one year later. From Turatto (2003) with kind permission of Springer Science and Business Media.

from 10^{51} to $> 10^{52}$ erg, with peak absolute visual magnitudes of order $M_V \sim -17$ mag for local Type II SNe, and $M_V \sim -18$ mag for local Type Ib/c SNe.

Type II SNe are subdivided into several types, of which the bulk are either II-L (linear) which show an uninterrupted decline in brightness or II-P (plateau) which stops declining in brightness shortly after maximum light for 2–3 months. The difference between these subtypes is thought to be due to differences in envelope mass (lower for Type II-L SNe). SN 1987A, the brightest SNe since Kepler's supernova in 1604, will be discussed separately. Within the local Universe, within a volume limited sample, Type II-P SNe dominate the core-collapse statistics.

Type Ib (and Ic) SNe are conspicuous by their absence of H (and He) lines, suggesting that they are associated with very massive stars that have shed their outer H (He) envelopes, namely Wolf–Rayet stars. Alternatively, lower mass binaries may produce Type Ib (Type Ic) SNe in which H (both H and He) has been stripped away due to Roche lobe overflow and/or common envelope evolution. Indeed, a small fraction of Type Ib/c SNe does occur in elliptical/S0 host galaxies, favoring such lower mass progenitors, on the basis of their low star formation rates. Cappellaro, Evans, & Turatto (1999) present rates of 60%:30%:10% for Type II:Ia:Ibc SN type, although more recent results based upon increased statistics suggest a higher rate of Type Ibc SN among core-collapse SNe with Ibc $/(\text{Ibc} + \text{II}) \approx 29\%$ (J. Leaman, priv. comm.)

A subset of intrinsically luminous Type Ic SNe with extremely broad spectral lines have been termed "hypernovae" (HNe) in the literature (Nomoto *et al.* 2000), a number of which have been observationally associated with several nearby GRBs (e.g. SN1998bw = GRB980425, Galama *et al.* 1998; SN2003dh = GRB030329, Hjorth *et al.* 2003).

Theory of core-collapse SNe

Once the core of a massive star is dominated by Fe-peak elements, further core contraction and increase in temperature cannot release nuclear energy to sustain the internal pressure since reactions involving stable nuclei at the peak of the binding energy curve are endothermic. The iron core collapses on a very rapid timescale (milliseconds) precipitated by a loss of energy through neutrino production. Photodisintegration of Fe nuclei leads to α particles, i.e.

$$\gamma + {}^{56}\text{Fe} \rightarrow 13\,{}^4\text{He} + 4\text{n}$$

which subsequently decays to

$$\gamma + {}^4\text{He} \rightarrow 2\text{p} + 2\text{n},$$

producing further neutrinos totalling in excess of 10^{53} erg. The collapse accelerates due to the reduction in internal pressure.

When the density of nuclear matter is reached, the repulsive component of the strong nuclear force brings the collapse of the inner core to a sudden halt. At the boundary of this core, a shock wave is formed, with an energy equal to the kinetic energy of the core at that moment (a few 10^{51} erg). As the shock moves out into the infalling outer core, it photodisintegrates its Fe nuclei too. It is believed that if the mass of the infalling outer core is sufficiently small ($\leq 0.4M_\odot$), the shock may reach the base of the envelope with sufficient energy to lead to a direct explosion. This bounce was once thought to be responsible for the origin of the supernova's energy. In general, this is now not believed to be the case. Once the shock wave reaches the surface it is predicted to produce a short burst of X-rays. This was first witnessed by Soderberg *et al.* (2008) for which an outburst of $L_x \sim 10^{46}$ erg was followed a week later by a faint Type Ib supernova, SN 2008D.

The prompt explosion fails due to photo disintegration and neutrino losses, and within a few milliseconds a dense, neutron-rich core (a proto–neutron star) is produced, accreting matter at a rate of a few tenths of a solar mass per second. If this accretion continued for a second or more, in the absence of another energy source, the proto-neutron star would collapse into a black hole.

Instead, extra energy is supplied by neutrinos which are trapped in the core during the hydrodynamical collapse. These diffuse outwards after the shock is initiated. If the proto-neutron star does not become a black hole, it radiates about 10% of its rest mass as neutrinos over a few seconds, producing a neutron star of radius 10 km. The neutrino emission is the primary output of the event. The detection of around 20 neutrinos following SN 1987A in the LMC by the Kamiokande II and and Irvine–Michigan–Brookhaven (IMB) detectors on 23 Feb 1987 confirms that neutrinos are produced during the death of massive stars (Hirata *et al.* 1987; Bionta *et al.* 1987). The final fate of massive stars is a neutron star if the core mass is below $\sim 2.5M_\odot$ (the Oppenheimer–Volkoff [1939] mass), otherwise the collapse is not stopped by the pressure exerted by the degenerate neutrons and a black hole is formed.

Theoretically, the (still unsolved) challenge of core-collapse supernovae remains the means of converting the collapse of the stellar envelope into an explosion of a few 10^{51} erg in kinetic energy. Current studies are focused upon two- and three-dimensional simulations of the proto-neutron star and neutrino energy deposition in its immediate surroundings. Spherically symmetric simulations with neutrino energy deposition do not lead to explosions. Neutrino energy deposition needs to inflate a bubble of radiation and electron–positron pairs surrounding the neutron star. The outer edge of this bubble becomes the shock wave that ejects the rest of the star, producing the explosion.

Convection may be crucial in efficiently carrying the neutrino energy into the region of the stalled shock. Convection cools the region where the neutrinos are depositing their energy, reducing further losses, and carries the energy to large radii where it can efficiently counter the infalling matter at the shock, which leads to asymmetries. This may offer a physical explanation for the "kicks" of several hundred $km\,s^{-1}$ observed in young pulsars. The challenge of simulations for core-collapse supernova involves ideally three-dimensional calculations in which neutrino physics is considered at sufficiently high spatial resolution, although several groups are currently actively working towards a solution, notably Adam Burrows, Chris Fryer, and Thomas Janka.

As the shock propagates through the stellar envelope, it heats the layers to temperatures higher than in the central burning stages. Explosive nucleosynthesis may occur, modifying the composition from the original quiescent burning stage. This primarily involves Si- and O-burning, with C- and Ne-burning to a lesser degree. The H and He layers are not affected since their densities are too low for explosive burning to occur. The products of explosive nucleosynthesis have been studied by Woosley & Weaver (1995), and involve Fe-peak elements produced by explosive Si-burning.

Heavy elements can be synthesized by neutron capture via the s-process (slow) and r-process (rapid). The former, in which stable isotopes (e.g. ^{208}Pb) are produced on timescales of 10^3 years in low neutron density environments via neutron capture, is primarily synthesized in red giant stars, but the s-process can occur in massive stars during helium burning (Eq. 5.1, Käppeler *et al.* 1994). In contrast, the explosive supernovae (Burbidge *et al.* 1957) represent the likely origin of r-process elements in which the bombardment of nuclei by neutrons from the neutrino driven wind on a timescale of seconds leads to unstable n-rich isotopes which decay to stable n-rich nuclei (e.g. ^{235}U, Woosley *et al.* 1994).

SN 1987A

SN 1987A in the LMC is particularly notable since its immediate progenitor was known to be a *blue* supergiant Sk–69° 202 (B3 Ia, Sonneborn, Altner, & Kirshner 1987) rather than the expected *red* supergiant (Falk & Arnett 1977). Various explanations for the explosion of SN 1987A as a blue supergiant have been proposed, of which the binary merger model is one of the most appealing (see Chapter 6). In addition, SN 1987A possesses a remarkable system of circumstellar rings, which was discovered by Wampler *et al.* (1990) and beautifully resolved by HST, as illustrated in Fig. 5.15. After two decades, material from the SN explosion interacts with the main inner ring, producing a number of bright spots.

The flash of neutrinos from SN 1987A shortly before the optical brightening provided strong evidence that its core had collapsed to a compact object. The inferred energy (10^{53} erg) and decay timescale (4 s) agreed well with models in which a degenerate Fe core collapsed to form a neutron star. However, despite great effort, no firm evidence of a compact object near the center of the supernova has yet been found. If there is a compact object, it must be bolometrically faint, at least a couple of orders of magnitude fainter than the pulsar in the Crab nebula.

Fig. 5.15 Composite BVR image of the circumstellar rings associated with SN 1987A obtained with HST/ACS in Dec. 2006. Dozens of bright spots are seen on the main ring caused by material from the SN explosion interacting with the 0.3 pc radius ring. STScI Press Release 2007-10.

SN 1987A also provides a means of comparing theoretical predictions for an initial 20 $M_\odot$ star with observations of elements produced in quiescent and explosive nucleosynthesis. The early plateau in the light curve of SN 1987A, shown in Fig. 5.16, is thought to originate from the recombination of hydrogen previously ionized in the supernova shock. The length of the plateau is dictated by the envelope mass and explosion energy. Subsequently, the light curve is powered by the radioactive decay of ^{56}Co. γ rays are captured in the ejecta and converted into optical photons, labeled as the "radioactive tail" in the figure, around 100 days after maximum brightness. This phase provides an indication of the amount of ^{56}Ni and ^{56}Co powering the light curve, from which a ^{56}Fe yield of 0.07 $M_\odot$ has been derived for SN 1987A.

The majority of SNe has not been followed photometrically beyond a couple of hundred days, with the exception of SN 1987A from which all very late phase information is based. Dust formation within the ejecta resulted in an increase in the optical decline after ~450 days. A later flattening at 800 days ("freeze-out") occured due to energy release of matter which was ionized in the original explosion but recombined on timescales in excess of the expansion time. Subsequently, a flattening in the light curve originated from isotopes ^{57}Co and ^{44}Ti. Several years after the SN, emission from the circumstellar inner ring exceeded that of the supernova ejecta itself, as the blastwave from the supernova struck the ring, causing a series of "hot spots" on the ring. Since then, many more "hot spots" have appeared, with the radio, optical and X-ray radiation dominated by the impact of the SN debris on its circumstellar environment.

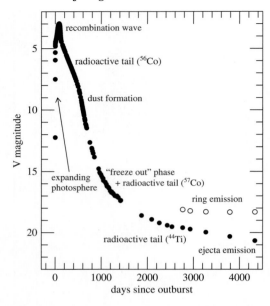

Fig. 5.16 Visual light curve of SN 1987A, in which the various phases are indicated. From Leibundgut & Suntzeff (2003) with kind permission of Springer Science and Business Media.

The amount of dust created in SN 1987A was rather modest in comparison with red giants and asymptotic giant branch stars. However, it seems likely that there was a rapid injection of dust in the Universe during the first 1 Gyr (Todini & Ferrara 2001), before low- and intermediate-mass stars were able to evolve through to such late evolutionary stages. The prime suspects are supernovae from massive stars, although there has been little work on characterizing dust formation in nearby core-collapse SNe.

If core-collapse SNe are indeed important contributors to dust production at high-redshift then we should be measuring dust masses of around 0.1–1 $M_\odot$ in their ejecta. Sugerman *et al.* (2006) claim to have detected mid-infrared excesses consistent with cooling dust in the ejecta of the Type II-P SN 2003gd during the period 499–678 days after outburst. Their radiation transfer model predicts that up to 0.02 $M_\odot$ of dust has formed, although this result is sensitive to clumping. Ongoing studies with Spitzer, plus follow-up with ALMA, sampling cooler dust, may finally resolve this question.

Progenitors of core-collapse supernovae

Aside from SN 1987A, whose progenitor had been spectroscopically identified as an early B supergiant, considerable observational effort has gone into observationally establishing the progenitor stars of nearby (≤10 Mpc) core-collapse SNe. High spatial resolution multi-color Hubble Space Telescope imaging, obtained prior to the SN explosion, has permitted photometric identification of the progenitor in several instances.

These have been successful in supporting RSG progenitors of Type II-P supernovae, such as SN 2003gd in M74 as shown in Fig. 5.17. From comparison between photometric properties of progenitors and evolutionary models, initial masses in the range 8–20 $M_\odot$ are implied. At the time of writing, no firm information on progenitors of less common core-collapse supernovae

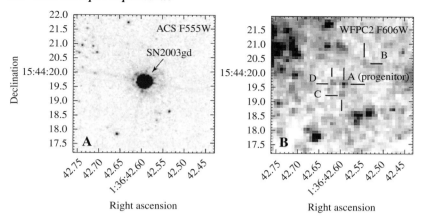

Fig. 5.17 Post- and pre-explosion images of SN 2003gd in M74 from HST ACS and WFPC2. Reprinted from Smartt *et al.* (2004) with permission from AAAS.

are available. Type II-L and IIb SN, possessing a relatively low mass hydrogen envelope, and denser circumstellar media, result either from somewhat more massive single stars (up to perhaps 25 $M_\odot$), or close binaries, with red, yellow or blue supergiant progenitors. Type IIn SNe are rarer still, with a dense circumstellar environment, arising either from a single H-rich star with a very dense wind (possibly a luminous blue variables) or an interacting binary.

If evolution proceeds back to the blue part of the H-R diagram as a Wolf–Rayet star, the resulting supernova is of Type Ib or Ic, depending on whether the star explodes during the WN or WC phase, in which the H-rich envelope has been removed through a stellar wind or mass transfer in a close binary (see Chapter 6). Observationally, SN 2002ap (Type Ic) in M74 so far provides the most stringent constraints upon a potential W-R progenitor, revealing an upper limit of $M_B = -4.2$ mag (Crockett *et al.* 2007). Most, although not all, W-R stars would have been identified from such deep imaging, favoring a lower mass interacting binary in this instance. Overall, the observed frequency of Type Ib/c SN to Type II lies intermediate between theoretical expectations for single, rotating stars and close binaries. Relative to normal Type Ic SN such as SN 2002ap, those associated with GRBs are unusually bright, favoring the core-collapse of a more massive star (e.g. SN1998bw, Galama *et al.* 1998).

Pair-production supernovae

Due to strong mass loss by stellar winds at solar metallicity, remnant core masses are not predicted to exceed $20M_\odot$ (recall Fig. 5.11). He core masses may be significantly more massive in low metallicity environments where winds are weaker (recall Fig. 4.4). If this core were to exceed $\sim 35M_\odot$ an electron–positron pair-instability is predicted to occur. At 2×10^9 K a large part of the energy from gravitational contraction goes into the creation of pairs of electrons and positrons. This process subtracts energy that would otherwise provide pressure support and may trigger a dynamical collapse (Bond, Arnett, & Carr 1984). The central density and temperature quickly increase, starting explosive burning in the carbon–oxygen core.

The energy released results in a complete disruption of the star for initial masses above $\sim 140M_\odot$. As such, pair-production supernovae are amongst the most powerful thermonuclear explosions in the Universe, with total energies ranging from a few 10^{51} erg (140 $M_\odot$) to

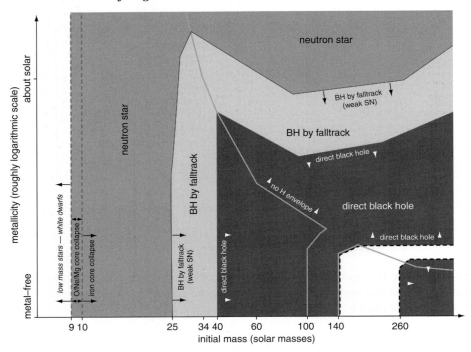

Fig. 5.18 Remnants of massive single stars as a function of initial mass and metallicity. Reproduced from Heger *et al.* (2003), by permission of the AAS.

10^{53} erg (260 $M_\odot$), powered by radioactive ^{56}Ni production, which increases with progenitor mass. The requirements for a high stellar and core mass suggests that such supernovae would only be encountered by single massive stars at very low (primordial?) metallicity. Smith *et al.* (2007) argue that the exceptionally bright Type IIn SN2006gy in NGC 1260 could represent a pair-instability SN. The absolute light curve of SN 2006gy is presented in Fig. 5.19 together with representative light curves for other SN. One signature of pair-production SNe is that explosive burning occurs within an environment in which there are very few excess neutrons, producing a strong "odd–even" effect of elements, i.e. nuclei with odd charge (e.g. Na, Al) are much less abundant than those with even charge (e.g. Ne, Mg). However, such an effect is not observed in the oldest halo stars. Above 260 $M_\odot$ the onset of photodisintegration in the core predicts that most of the star collapses to a black hole (Bond *et al.* 1984).

Neutron stars

Neutron stars represent the final state of the overwhelming majority of massive stars, with remnant masses of $\sim$1.4 $M_\odot$. Neutron stars are most readily identified as pulsars, of which several thousand examples have been discovered since their original identification by Ph.D. student Jocelyn Bell in 1967, which infamously led to the award of the Nobel prize for physics to her advisor Antony Hewish, but not to Bell herself.

From observations of young pulsars, neutron stars are born with rotation rates measured in tens of milliseconds (recall Table 5.2), and strong magnetic fields of $\sim$10^{12} G. Over $\sim$10^7 yr, their rotation periods increase to a few seconds and magnetic fields decay, leaving

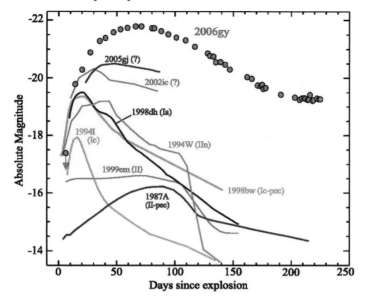

Fig. 5.19 Absolute R-band light curve of the exceptionally bright Type IIn SN 2006gy, together with representative examples of Type II (1999em), Type Ia (1998dh) and Type Ic (1994I) SN, plus several peculiar cases, including SN 1987A and 1998bw. Reproduced from Smith *et al.* (2007) by permission of the AAS.

them effectively dead. Fig. 5.20 presents the period versus period derivate diagram for radio pulsars. Non-pulsing neutron stars may be identified as exceptionally hot thermal sources, since initially their surface temperatures may be as high as 10^8 K. RX J185635–3754 is one such isolated neutron star (Walter, Wolk, & Neuhauser 1996), with a near-thermal spectrum of a 660 000 K blackbody. WFPC2 imaging reveals a distance of ∼120 pc suggesting a mass of 1.7 $M_\odot$ and age of 5×10^5 yr (Lattimer & Prakash 2001; Walter & Lattimer 2002).

Millisecond pulsars can be seen to the lower left of the diagram, and represent systems in which accretion from a companion star resuscitates the pulsar spin, although its magnetic field remains weak. Many cases of millisecond pulsars are now known, including multiple examples in old globular clusters. Figure 5.20 also reveals a rare subset which are highly magnetized, with 10^{14-15} G. Observationally, these are known as anomalous X-ray pulsars (AXPs) and soft gamma-ray repeaters (SGRs). Both classes show evidence for rapid spin-down with respect to normal radio pulsars, and are known collectively as highly magnetic pulsars or magnetars. It was noted early on that such strong fields could be present in neutron stars as a result of flux conservation from the progenitor star (Woltjer 1964). Both undergo occasional bursts of energy, particularly from SGRs, presumably due to a redistribution of the neutron star crust.

To date, three giant flares (∼10^{44} erg) from SGRs have been detected since the 1970s, which were associated with SGR 0526–66 in the LMC, plus SGR 1900 + 14 and SGR 1806–20 in the Milky Way, of which the December 2004 flare of SGR 1806–20 was exceptional. Its fluence was higher than from any extra-solar event in history, and saturated all but a few of the instruments which detected it, even causing an ionispheric disturbance in the upper Earth's atmosphere. Assuming a standard 15 kpc distance to the source (see, however, Bibby

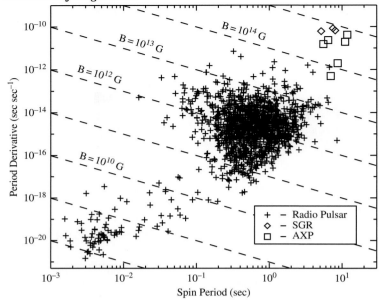

Fig. 5.20 Period versus period derivative for radio pulsars (+ symbols), anomalous X-ray pulsars (AXPs, squares) and soft gamma-ray repeaters (SGRs, diamonds). From Woods & Thompson (2006) with permission of Cambridge University Press.

et al. 2008), the energy released in this hyperflare was $\sim 2 \times 10^{46}$ erg (e.g. Palmer *et al.* 2005). More energy was released by SGR 1806–20 in 0.5 s than the entire solar output over the past 10^5 years, and represents a significant fraction of the total magnetic energy of the neutron star. This suggests we witnessed a catastrophic reconnection event that may have reconfigured the magnetic field of the entire star. At the adopted distance, such a hyperflare could have been detected out to beyond the Virgo cluster, raising the possibility that some apparent short gamma ray bursts may be giant flares from extragalactic magnetars.

Fate of single massive stars

Theoretically, the end state of high mass stars is considered by Heger *et al.* (2003), for which stars of mass above $8M_\odot$ end their life as either a neutron star or a black hole, the latter either formed directly or via fallback, as shown in Fig. 5.18. Historically, it was believed that neutron stars resulted from stars initially between 8 and 25 $M_\odot$ with black holes produced by higher-mass progenitors. There is increasing evidence that very massive stars in metal-rich environments may produce neutron stars instead of black holes, due to extensive continuous/episodic mass-loss during their lifetimes (recall Fig. 5.11). For example, an initial mass of $\geq 40M_\odot$ has been inferred for the magnetar CXO J164710.2–455216 due to its physical association with the young massive cluster Westerlund 1 (e.g. Muno *et al.* 2006).

Pair-production SNe which leave no remnant are indicated in Fig. 5.18 as a white region for metal-poor stars at high initial mass. The corresponding type of supernova explosion is presented in Fig. 5.21, dominated by Type II-P supernovae for intermediate masses, and a Type Ib/c SN at high mass and high metallicity. For decreasing metallicity the direct black hole formation would not produce a normal SN explosion, while pair-production SNe undergo an

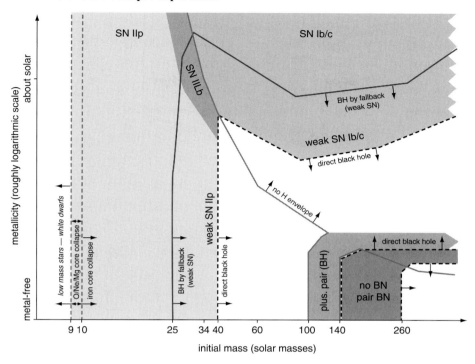

Fig. 5.21 Supernovae types for non-rotating massive single stars as a function of initial mass and metallicity. Reproduced from Heger *et al.* (2003) by permission of the AAS.

ejection prior to core collapse. Non-rotating black holes are named after Karl Schwarzschild, as is the radius of their event horizon, at which point the escape velocity equals the velocity of light. More generally, black holes in nature are likely to rotate, which is described by the Kerr (1963) metric, rather than the Schwarzschild metric.

Gamma ray bursts will be discussed in a cosmological context in Chapter 11. A number of nearby, bright Type Ic SNe have been observationally associated with GRBs, for which SN1998bw = GRB980425 and SN2003dh = GRB030329 serve as prototypes (Galama *et al.* 1998; Hjorth *et al.* 2003). As we shall show, these observations favor the "collapsar" scenario, involving the core collapse of a rotating Wolf–Rayet star to a (Kerr) black hole via an accretion disk, in which the rotational axis provides a means of collimating the jet along the polar direction (MacFadyen & Woosley 1999).

Regarding massive *single* stars, let us consider how likely it is that the iron core is capable of spinning sufficiently rapidly that it cannot collapse directly to a black hole and conserve angular momentum, i.e. the specific angular momentum j of the iron core, M_c, exceeds

$$j \geq \sqrt{6}GM_c/c \approx 2 \times 10^{16}(M_c/2M_\odot) \quad \text{erg s}^{-1}. \tag{5.2}$$

The shear between slowly rotating red supergiant envelopes and the rapidly rotating core tends to slow down the rotation of the latter, i.e. it will be slowed down too much. Equally, those very massive stars which circumvent the red supergiant phase go onto the Wolf–Rayet stage, and so lose mass at a high rate, also causing the core to spin down. Allowance for magnetic

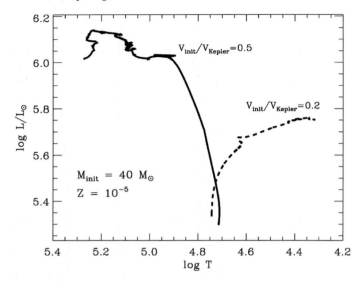

Fig. 5.22 Evolutionary models for $40M_\odot$ low metallicity $Z = 10^{-5}Z_\odot$ models from the Main Sequence to core-carbon exhaustion, initially rotating at $v_e/v_{\rm crit} = 0.2$ (dashed) and 0.5 (solid). From Yoon & Langer (2005).

torques suggest that single massive stars at solar metallicity fall short of the required j by an order of magnitude.

At low metallicity, even very massive stars retain a massive hydrogen envelope, due to the weakness of their stellar winds. Rapid rotation in massive magnetic stars may be capable of avoiding the extended envelope, leading to an almost "chemically homogeneous" evolution. Maeder (1987a) showed that if rotationally induced chemical mixing in massive main-sequence stars occurs faster than chemical gradients from nuclear fusion, the main-sequence star is smoothly transformed into a helium star of the same mass. Since then the physics of rotationally induced mixing has been refined. Nevertheless, chemically homogeneous evolution may occur if rapid rotation is adopted (Maeder & Meynet 2000). In such models, the helium star is born with only a small hydrogen-rich envelope at most, such that magnetic torques can only remove a little angular momentum.

At sufficiently low metallicity, the mechanical mass-loss induced spin-down may be avoided since a combination of observational and theoretical evidence suggest that low metallicity W-R stars may retain sufficient mass and angular momentum in order that an accretion disk can form around the newly formed black hole. Indeed, there is some observational evidence suggesting that Type Ic SN associated with a GRB are preferentially metal-poor with respect to non-GRB Type Ic SNe (Modjaz *et al.* 2008). Alternatively, sufficiently rare channels of binary evolution would represent the main path towards a gamma ray burst in which the progenitor might be spun-up via tidal interactions (Chapter 11). Figure 5.22 shows the evolution of very low metallicity ($10^{-5}Z_\odot$), $40M_\odot$ models from the Main Sequence to core-carbon exhaustion, initially rotating at $v_e = 230$ km s^{-1} and 555 km s^{-1}, respectively. The fast rotator ($v_e/v_{\rm crit} = 0.5$) undergoes chemically homogeneous evolution, and maintains sufficient angular momentum through to the end of its life, as is required for GRB progenitors.

6

Binaries

Most, if not all, stars in the local Universe are formed in multiple systems, such that single stars appear to represent the exception rather than the rule. Detailed studies of Orion indicate a high binary or triple fraction amongst both low- and high-mass stars. Close binaries may have a significant impact upon the evolution of the primary and secondary star, and facilitate the most robust means of establishing binary masses. Constraints upon the winds of early-type binaries may be made from observations of their wind–wind interactions. Dedicated monographs to close binary evolution are provided by Vanbeveren, De Loore, & Van Rensbergen (1998) and Eggleton (2006).

6.1   Massive binary frequency

Observationally, the direct spectroscopically confirmed O-type binary frequency in young open clusters is given by Mermilliod & García (2001) and is presented in Table 6.1. The binary frequency is defined as the relative number of stars in binaries versus singles, such that three stars, of which two are in a binary and one single, implies a binary frequency of 50%. From Table 6.1, the binary frequency may be as high as ∼80% or as low as 14%, suggesting that there may not be a universal frequency.

Table 6.1 *Binary frequency in open clusters according to Mermilliod & García (2001).*

Cluster	O stars	Spect. binaries	Frequency
IC 1805	10	8	0.80
NGC 6231	14	11	0.79
NGC 2244	6	3	0.50
IC 2944	16	7	0.44
NGC 6611	12	5	0.42
Tr 16	20	7	0.35
Cr 228	21	5	0.24
Tr 14	7	1	0.14

Prior to the detection of strong stellar winds from O stars, it was once believed that all W-R stars resulted from close binary evolution (Paczynski 1967). In fact the binary fraction of W-R stars in the Milky Way is ∼40% (van der Hucht 2001), either from spectroscopic or indirect techniques. Due to weaker O star winds in the metal-poor Magellanic Clouds, one would anticipate a higher W-R binary fraction, in which close binary evolution might be anticipated to play a greater role. However, where detailed studies have been carried out, a similar binary fraction to the Milky Way has been obtained.

Nevertheless, the above statistics suffer from several observational biases. Systems with orbital periods in excess of a few weeks will be missing due to poor sampling through the orbit, as are those with small mass ratios, q, the ratio of the secondary to primary mass. Perhaps 50% of all binaries are missing from such statistics. In addition, even supposedly well-studied, apparently single O stars may be binary in nature. A well-known example is HD 93129A, a prototype O2 If star, for which the Fine Guidance Sensor aboard HST has been used to identify a close binary companion (Nelan *et al.* 2004), with a separation of ≤ 0.1 arcsec. An identical approach for W-R stars has been used to identify several binaries, some of which were not previously well established.

6.2   Binary masses

The mass of a star is perhaps its most fundamental property, yet it is generally very difficult to constrain for massive stars to better than a factor of two. The study of binary systems represents the most direct, robust method of measuring stellar masses from Kepler's third law of motion,

$$(M_1 + M_2)P^2 = a^3, \tag{6.1}$$

where M_1 and M_2 are the masses of the primary and secondary in solar masses, P is the orbital period in years and a is the semi-major axis in AU. For the example of two 25 $M_\odot$ O9 dwarfs in a binary system with a circular orbit of period 1 week, the systemic semi-major axis would be 0.25 AU, equivalent to only $\sim 8 R_*(\text{O9 V})$. Considering the two components separately, with respect to the center of mass of the system, $a = a_1 + a_2$,

$$\frac{a_1}{a_2} = \frac{M_2}{M_1}. \tag{6.2}$$

A spectroscopic binary may show lines from either one (SB1) or both (SB2) components of the system. For the more straightforward case of an SB2 system (e.g. BD $+40°$ 4220, Bohannan & Conti 1976), the simultaneous analysis of the radial velocity curves for both stars permits the values of $M_1 \sin^3 i$ and $M_2 \sin^3 i$ to be derived via

$$M_{1,2} \sin^3 i = 1.04 \times 10^{-7}(1 - e^2)^{3/2}(K_1 + K_2)^2 K_{1,2} P, \tag{6.3}$$

where $K_{1,2}$ are the semi-amplitudes of the radial velocity curves (in km s^{-1}), P is the period in days, e the eccentricity ($0 \leq e \leq 1$), and i the inclination of the orbital plane with respect to the sky. This equation provides absolute masses multiplied by a factor $\sin^3 i$, such that the main uncertainty relates to the inclination. If i is known from photometric, visual, or interferometric observations, individual masses follow directly. In the above example of two O9V stars in a 7-day circular orbit, $K_1 = K_2 = 205$ km s^{-1} for an inclination of $i = 90$, i.e. photometric eclipses are seen. Therefore, eclipsing SB2 binaries provide the primary source of direct measurement of stellar masses and radii, from which the mass–luminosity relation for main-sequence stars has been empirically established.

In order to establish the mass–luminosity relation for massive stars, not only does an early-type binary need to be an eclipsing SB2 system, but the components need to be within their

Roche lobes, and young enough not to have suffered any mass exchange via Roche lobe overflow (RLOF). Due to the relative rarity of massive stars and their short main-sequence phase, rather few suitable so-called "detached" binaries exist. Recent compilations of galactic O+O binaries for which relatively accurate mass determinations are available are provided by Gies (2003), and include massive, detached early O-type binaries in the R136 star cluster from Massey, Penny, & Vukovich (2002), for which masses of $\sim 55\ M_\odot$ for two O3 V stars were obtained. These are much lower than one would necessarily expect for O3 V stars from evolutionary models.

The mass–luminosity relation for massive O stars is presented in Fig. 6.1, revealing a dependence of $L_* \propto M_*^{2.5}$ for 10–$25 M_\odot$, and $L_* \propto M_*^{1.75}$ for higher masses, which compares to $L_* \propto M_*^{4.7}$ for solar-type stars. These essentially result from differences in opacities – scattering from electrons dominates in the high interior temperatures of very massive stars since atoms are fully ionized, whilst there is a strong temperature dependence of the opacity in lower-mass stars for which atoms are only partially ionized. One added complication with this approach is that stellar temperatures and bolometric corrections need to be derived or adopted from calibrations. As we have already discussed in Chapter 3, the temperature scale for O stars has recently been revised downward, so errors with previous calibrations will exacerbate uncertainties in masses.

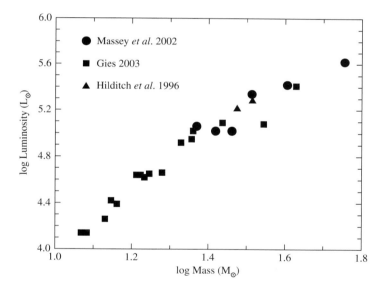

Fig. 6.1 Mass–luminosity relationship for detected OB stars from binary orbits, updated from Gies (2003), Massey, Penny, & Vukovich (2002), and Hilditch, Harries, & Bell (1996).

Let us now turn to the case of an SB1 spectroscopic binary, which will occur if the two stars are very different in luminosity. Obviously, the total mass of the system cannot then be obtained, but the so-called mass function, $f(m)$, provides some information on the systemic mass. The mass function may be written in the form

$$f(m) = \frac{M_2 \sin^3 i}{(1 + M_1/M_2)^2} = 1.04 \times 10^{-7} K_1^3 P (1 - e^2)^{3/2}. \tag{6.4}$$

In Eq. (6.4), if the mass ratio $q = M_1/M_2 \geq 1$, as is usually the case, the minimum mass of M_2 (and also therefore M_1) is $4f(m)$ for an SB1 binary system. In the above example, let us imagine that the O9V primary now has an unseen $10M_\odot$ (black hole) secondary in a 1-week, circular orbit with $i = 90$. In this case, $K_1 = 104 \text{ km s}^{-1}$ and $f(m) = 0.8M_\odot$, so the formal lower limit to $M_2 = 4f(m) = 3.2M_\odot$. We would be able to confirm the black hole nature of the secondary, but not its actual mass.

In addition to O + O or O+B binaries, a number of W-R + OB binaries are known. These have the added complication that one has to rely upon broad emission lines for the W-R components rather than the usual OB photospheric absorption lines. Masses for W-R stars are included in the recent van der Hucht (2001) compilation. In general W-R masses are much lower than for the most massive O stars, with $10 \leq M(W-R)/M_\odot \leq 25$. By way of example, the binary nature of γ Velorum permits a mass determination for both the W-R and O components, with $9.5M_\odot$ for the WC component and a mass ratio of $q = M(W\text{-}R)/M(O) = 0.3$ (De Marco *et al.* 2000).

In a few instances, W-R stars have been found to be extremely massive. These are uniquely late-type H-rich WN subtypes, which are apparently extreme O stars with strong winds at an earlier evolutionary stage than classical H-deficient W-R stars (recall Fig. 2.17). For HD 92740 (WN7ha+O), the mass of the WN component greatly exceeds that of the O star, with $M(W\text{-}R) \sim 55M_\odot$, whilst WR20a (WN6ha+WN6ha) currently sets the record for the highest robust stellar mass, with $M(W\text{-}R){\sim}80M_\odot$ for each WN component, within a 3.7-day orbit, whose radial velocity curve is presented in Fig. 6.2. Yet higher masses no doubt await discovery since the apparent upper limit to stellar masses is ${\sim}150M_\odot$.

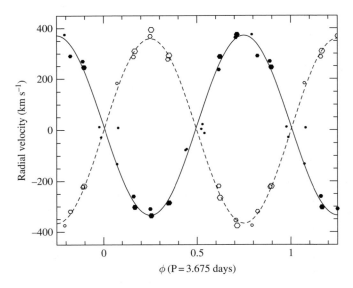

Fig. 6.2 Radial velocity curve of WR20a for an orbital period of 3.675 days. The filled (open) symbols relate to the primary (secondary), whilst the solid (dashed) line represents the best fit orbital solution for a circular orbit, revealing semi-amplitudes of $K_1 \sim 360 \text{ km s}^{-1}$ implying exceptionally high masses of ${\sim}80M_\odot$ for each WN component. From Rauw *et al.* (2004).

6.3 Close binary evolution

The evolution of a binary system depends upon many variables, notably the initial binary period, eccentricity, and mass ratio, and as such is rather more complicated that the single star scenario from Chapter 5. For sufficiently large orbital periods, the components of the binary may evolve independently as if they were single stars.

Within a binary, the mechanical force is zero at the Lagrangian points. The first or "inner Lagrangian point", L_1, always lies between the two components, and the two lobes of the surface it encloses are the "Roche lobes". An example of a detached binary in which neither star fills its Roche lobe is presented in Fig. 6.3(a). As the more massive (primary) star expands to fill its Roche lobe, its outer layers will approach L_1 and the binary becomes an interacting or semi-detached system. Material will pass through L_1 into the potential well of the secondary,

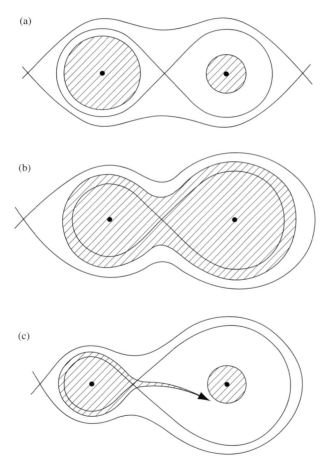

Fig. 6.3 Schematics, showing equipotential surfaces for cases of (a) detached (b) common envelope and (c) semi-detached binaries. The "inner Lagrangian point" L_1 between the stars represents one of the neutral, saddle points in equipotential, and encloses two "Roche lobes". Other Lagrangian points include L_2 and L_3 to the left and right of panel (a), plus L_4 and L_5 which need not concern us. Indeed only L_2 is shown in panels (b) and (c) for simplicity. From Eggleton (2006) with permission of Cambridge University Press.

deflected by the Coriolis force, often forming an accretion disk around the companion, as illustrated in Fig. 6.3(c). This situation is known as Roche lobe overflow (RLOF), with the primary known as the "mass loser" and the secondary the "mass gainer". Well-known low mass binary examples include cataclysmic variables. If both Roche lobes become filled, the system is known as a contact binary, with a common envelope, as shown in Fig. 6.3(b). The qualitative evolutionary scenario of massive close binaries was introduced by Paczynski (1967) and revised by van den Heuvel & Heise (1972). A schematic of this is presented in Fig. 6.4.

If the initial binary period is of the order of a few days the primary will reach its Roche lobe whilst still in the main-sequence core H-burning phase. This is known as "Case A", following the terminology of Kippenhahn & Weigert (1967). If the binary period is in the range between a few days and a few hundreds to a thousand days, the primary will fill its Roche lobe whilst undergoing H-shell burning, known as "Case B". Case B binaries may be further subdivided into early and late phases, where the envelope of the primary (mass loser) is radiative (blue supergiant, Case B_r) or convective (red supergiant, Case B_c) at the beginning of mass transfer. If the period exceeds ~ 1000 days, the separation is greater than the maximum radii of red supergiants, and so the primary may only begin RLOF after core He burning, during the He-shell burning stage, known as "Case C" (see Paczynski 1971). It is much more likely that a star will encounter Case B or C mass transfer than Case A, due to the much larger range of orbital periods sampled. A common envelope will occur instead of Roche lobe overflow if there is an LBV phase, according to Vanbeveren *et al.* (1998). Many early-type binaries provide observational evidence in favor of mass transfer.

In Case A, the primary will continue to lose mass due to RLOF on a nuclear timescale, and cease mass transfer at the end of core H-burning when it contracts. Since the primary retains a significant fraction of its original H-envelope, the star will expand again during the H-shell burning phase, and RLOF will continue during Case B_r. In Case B_r, the primary will undergo RLOF at a high rate, which will cease after most of the H-rich layers of the primary are removed. The primary will be an H-deficient, core He-burning reminiscent of a Wolf–Rayet star that will not fill its Roche lobe. The primary may expand again after core He-burning with a second episode of RLOF. Case A and Case B_r mass transfers are considered to be stable, and the systemic mass and angular momentum are assumed to be *conserved* during mass exchange. High mass transfer rates may lead to a significant expansion of the mass gainer, such that both components may fill their common equipotential surface. Matter would then escape the system, accompanied by a loss of orbital angular momentum, leading to a significant reduction in separation.

In contrast, Case B_c and Case C mass transfer may be dynamically unstable, as a consequence of the fact that the radius of a star with a deep convective envelope shrinks more slowly than the radius of the Roche lobe. The secondary may even expand if the mass-transfer timescale is shorter than the thermal timescale of the outer envelope of the secondary. The secondary then fills its own Roche lobe and the system enters a common-envelope phase (Paczynski 1976) in which the dense core of the primary and companion star orbit around their common center of mass. The orbit decays due to dynamical friction (Bondi & Hoyle 1944) although the details of common-envelope evolution are fairly uncertain.

Situations in which the companion star is degenerate have been proposed in which the compact star spirals in to the center of the primary, producing a so-called Thorne–Żytkow Object (TŻO), named after Thorne & Żytkov (1977). A hypothetical TŻO would have the

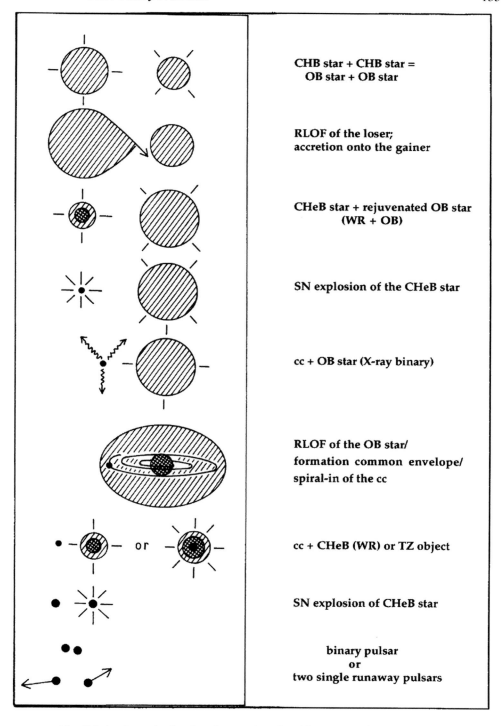

CHB star + CHB star =
OB star + OB star

RLOF of the loser;
accretion onto the gainer

CHeB star + rejuvenated OB star
(WR + OB)

SN explosion of the CHeB star

cc + OB star (X-ray binary)

RLOF of the OB star/
formation common envelope/
spiral-in of the cc

cc + CHeB (WR) or TZ object

SN explosion of CHeB star

binary pulsar
or
two single runaway pulsars

Fig. 6.4 A schematic showing the massive close binary scenario. From Vanbeveren *et al.* (1998).

structure of a red supergiant (Cannon *et al.* 1992), albeit with a peculiar chemical composition. Convection would bring the products of nuclear burning to the surface of TŻOs, namely decay products of rapid-proton (rp)-process, providing enrichment of, e.g., Mo by a factor of 1000 (Biehle 1994). The characteristic lifetime of a TŻO may be as long as 10^5 yr. This may end when the rp-process seed elements are exhausted or the envelope mass falls below a critical mass, leading to a neutrino runaway and possible black hole formation following accretion of material from the envelope onto the neutron star.

In contrast, if both components are compact – neutron stars or black holes – they will lose gravitational wave radiation, causing the components to spiral-in and ultimately merge. If a merger does not occur, both types of Case B mass transfer may lead to an H-deficient core-He burning primary, with the possibility for further RLOF (known as Case BB). During this phase, the primary loses its remaining H-rich envelope. If the secondary is a normal mass gainer, Case BB results in a very large increase in the period. If the companion is compact, Case BB is governed by the spiral-in process, which may result in a hardening of the binary. This has a large effect on whether the system will remain bound after the secondary undergoes an SN explosion, and the possibility for a double compact system. Massive core-He burning stars may lose mass by a stellar wind, such that Case BB RLOF may be inhibited in massive binaries. In this case the binary period is hardly affected, except in low metallicity environments for which the W-R mass-loss rate is reduced. A schematic illustrating the various stages in the evolution of a massive close binary is shown in Fig. 6.4.

Whenever a common envelope forms, under which circumstances does a merger occur versus an ejection of the envelope? Livio & Soker (1988) define an efficiency factor, α_{CE} for which the envelope is ejected if $\alpha_{CE} \leq 0.3$ based on hydrodynamical calculations of the ejection process, with

$$\alpha_{CE} = \frac{2a(M_1 - M_1^c)(M_1 + M_2)}{M_1^c M_2 R_1^0}, \tag{6.5}$$

where M_1 and M_2 are the masses of the primary and secondary, M_1^c is the mass of the H-poor core of the primary, R_1^0 is the radius of the primary at the beginning of the common-envelope phase, and a is the orbital separation of the binary components when the radius of the secondary first exceeds its Roche lobe radius.

The evolution of the mass-losing primary is reasonably well understood, although that of the mass-gaining secondary is less certain. Depending upon the binary separation, the material passing through the Lagrangian point either hits the secondary directly or forms an accretion disk. Accretion leads to an increase in angular momentum, causing the outer layers of the star to spin up. If a Keplerian disk is formed, there is sufficient angular momentum to spin-up the entire star to a velocity which is close to the critical rate. Mass gainers would be expected to exhibit enrichment in nitrogen from the primary and may appear to be rejuvenated stars, i.e. they will look younger than their accompanying mass loser, i.e. a so-called massive "blue straggler".

One of the most appealing scenarios which explains why the progenitor of SN 1987A was a blue supergiant was via close binary evolution. The location of the progenitor Sk–69° 202 in the H-R diagram can be explained through the Case B_c merger of a $16M_\odot$ red supergiant (possessing a $4.5M_\odot$ He core) with a $3M_\odot$ main-sequence star. This is illustrated in Fig. 6.5, and compares the post-merger evolutionary track with that of a $16M_\odot$ single star. The final outcome of the common-envelope phase is a single star of mass $19M_\odot$ possessing a $4.5 M_\odot$ He-core, which spends most of its He core-burning phase as a blue supergiant. The spiral-in time of the merger model ($\sim 10^4$ yr) is much shorter than the remaining evolutionary time of the Case B scenario, so one expects that the merger has been completed before the supernova explosion. Figure 6.5 also illustrates a Case C scenario in which a $16 M_\odot$ star is in the process of merging with a $6 M_\odot$ main-sequence star. At the time of the core-collapse SN, the common-envelope system resembles a blue supergiant, but contains the orbiting cores of the two merged stars.

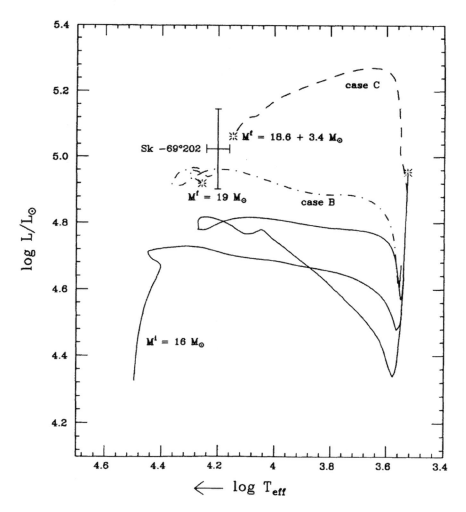

Fig. 6.5 Evolutionary tracks of a $16 M_\odot$ single star (solid) versus the post-merger Case B$_c$ evolution of a $16 M_\odot$ red supergiant with a $3 M_\odot$ main-sequence star (dot-dash), plus a $16 M_\odot$ star in the process of merging with a $6 M_\odot$ main-sequence star (dash). Asterisks mark the locations of the models at the time of the supernovae and the error bars mark the location of Sk–69° 202, the progenitor of SN 1987A. Reproduced from Podsiadlowski, Joss, & Hsu (1992) by permission of the AAS.

The progenitor is expected to be spun-up during the mass accretion or spiral-in phase, leading to rapid rotation, which naturally explains the asymmetric expansion of the ejecta and circumstellar ring structure associated with SN 1987A (recall Fig. 5.15). The equatorial ejection of the circumstellar rings could have occured during the common-envelope phase. The inner ring of SN 1987A is remarkably similar to the ring discovered in HST images of the B supergiant Sher 25 in the young galactic open cluster NGC 3603, as shown in Fig 6.6. The emission-line nebula for Sher 25 has been interpreted as a remnant from a previous red supergiant phase, in this case too, although the stellar spectrum of Sher 25 is remarkably normal with respect to other Milky Way early B supergiants.

Fig. 6.6 HST/WFPC2 composite image (4×4 pc) of the young open cluster NGC 3603 including the blue supergiant Sher 25 in the upper left of center, with a circumstellar ring that resembles the inner ring of SN 1987A (see Brandner *et al.* 1997). STScI Press Release 1999-20.

Podsiadlowski, Joss, & Hsu (1992) consider the range of evolutionary scenarios for systems containing massive progenitors in the range $8M_\odot \leq M \leq 20M_\odot$, assuming that 50% of all massive stellar systems are binaries with periods of less than 100 yr, including the fate of the primary and secondary components. This is shown schematically in Fig. 6.7, and illustrates the importance of binary evolution in the study of core-collapse SNe. Naturally, the probability of individual channels requires a number of simplifying assumptions, regarding the distribution

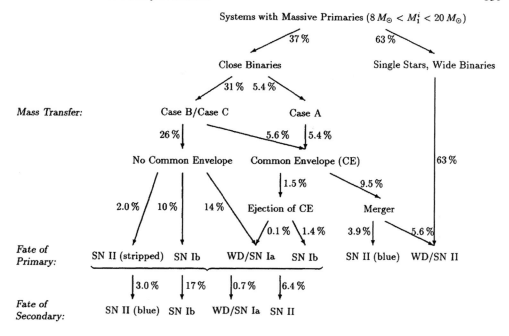

Fig. 6.7 Summary of evolutionary scenarios for systems containing an 8–20$M_\odot$ primary, including probabilities that a system with an initial primary in the chosen mass range passes through the particular evolutionary channel, assuming that 50% of massive stellar systems are binaries with periods of less than 100 yr. Reproduced from Podsiadlowski *et al.* (1992) by permission of the AAS.

of initial binary properties, and that Case A mass transfer always leads to a merger (it is feasible that the primary is transformed into a helium star without passing through a common envelope phase). Similar calculations were later performed for higher-mass close binaries at solar metallicity, as summarized in Table 6.2. For the highest-mass primaries in particular, the remnant might be a neutron star, low or high mass black hole depending upon the treatment of convection (recall Chapter 5), such that the exact dividing lines should be treated with caution.

Runaways and high mass X-ray binaries

The role of a supernova explosion in a massive binary system is generally sufficient to disrupt the binary system. This may lead to the companion becoming a "runaway" star (Blaauw 1961). These are defined as stars with a peculiar space velocity in excess of 30 km s^{-1}, which applies to 10–20% of O-type stars (Gies 1987). Well-known runaway OB stars include HD 30614 (α Cam) and HD 66811 (ζ Pup). The high spatial motion of runaways should lead to a bow shock, as the star with its powerful stellar wind moves through the interstellar medium. A second possibility for the origin of runaways is through dynamical interactions within an open cluster or OB association (Poveda, Ruiz, & Allen 1967). Both scenarios are responsible for the observed properties of runaways.

Following a supernova explosion of the primary, if the system remains bound one would expect to observe enhanced α elements on the surface of the secondary. Such "pollution" has been observed in a few cases (e.g. GRO J1655–40, Israelian *et al.* 1999). So long as

Table 6.2 *Types of remnant resulting from primary components of massive close binaries at solar metallicity, i.e. white dwarf (WD), neutron star (NS), low- or high-mass black hole (LMBH/HMBH) and expected supernova type as a function of primary mass and mass transfer type (Wellstein & Langer 1999).*

Mass ($M_\odot$)	Case A	Case B	Case C	Single
8–13	WD	WD	NS	NS
	–	–	SN Ib	SN II
13–16	WD	NS	NS	NS
	–	SN Ib/c	SN Ib	SN II
16–25	NS	NS	NS	NS
	SN Ib	SN Ib	SN Ib	SN II
25–40	NS	NS	HMBH	HMBH
	SN Ic	SN Ic	SN Ic	SN Ic
> 40	NS/LMHB	NS/LMBH	NS/LMBH	NS/LMBH
			HMBH	or HMBH
	SN Ic	SN Ic	SN Ic	SN Ic

the secondary is a main-sequence star, the accretion of its relatively tenuous wind onto the compact remnant (neutron star or black hole) does not result in a strong X-ray flux. It is only once the secondary OB star evolves to become a supergiant that it will fill its Roche lobe, causing material to flow to the compact remnant via an accretion disk, producing very strong X-rays. Observationally, this situation is known as a high-mass X-ray binary (HMXB).

The X-ray luminosity of such systems are $L_X \sim 10^{38}$ erg s^{-1}, several orders of magnitude greater than that observed from a single OB star or the colliding wind from two OB stars within a binary. Two varieties of HMXB are known; (i) the most massive systems containing OB supergiants, with periods in the range 1.4 (LMC X-4) to 41.5 days (GX 301-2) as illustrated in Fig. 6.8; (ii) Be/X-ray binaries which are often X-ray transients, caused by the periodic increase in X-ray luminosity during the part of the orbit when the compact object accretes from the dense disk associated with the Be star. Indeed, most systems host an X-ray pulsar, identifying the compact object as a neutron star, although some OB supergiant HMXBs contain black holes. In contrast, low-mass X-ray binaries (LMXB) comprise a neutron star or black hole primary, plus a late-type dwarf or white dwarf companion (e.g. GRO J1655–40). The HMXB phase is relatively short (10^4 yr) since the orbit shrinks once RLOF commences, causing the X-ray source to be completely swamped with optically thick material and/or penetrate the mantle of the secondary.

The mass of the OB supergiant and compact source can be accurately measured when the system hosts an X-ray pulsar, if the inclination is known, e.g. it is eclipsing. The most accurate neutron star masses have been derived from binary radio pulsars, which indicate a canonical mass of 1.35 $M_\odot$ in all cases. The theoretically derived minimum mass is 1.1–1.2 $M_\odot$ (Lattimer & Pratash 2001). In contrast, the neutron star in Vela X–1 (HD 77581) is significantly more massive, with 1.87 $M_\odot$. An apparently high neutron star mass of 2.4 $M_\odot$ has been claimed for 4U 1700–37 (HD 153919), although this may be a low-mass black hole (its X-ray source is not an X-ray pulsar). Black hole candidate masses are significantly higher,

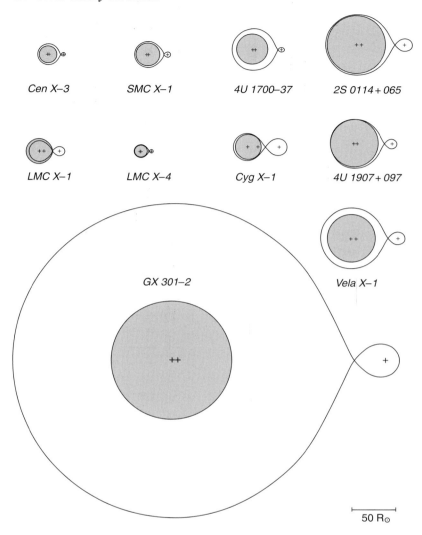

Fig. 6.8 An overview of OB supergiant HMXBs in the Milky Way and Magellanic Clouds, shown to scale. Roche lobe overflow systems (Cen X–3, Cyg X–1, SMC X–1, LMC X–4, LMC X–1) have circular orbits. Wind fed-systems have eccentric orbits, suggesting that orbits circularize with time and increasing mass-transfer rate. From Kaper, van der Meer, & Tijani (2004).

comfortably exceeding $3M_\odot$, as illustrated in Fig. 6.9, suggesting distinct formation mechanisms. In general the OB supergiant components of HMXBs are over-luminous for their mass.

HMXBs are also expected to have received an initial "kick" from the asymmetric supernova explosion, and so exhibit bow shocks as they pass through the interstellar medium. Vela X–1 (HD 77581) is one such case, consisting of an early B supergiant (the original secondary) and an X-ray pulsar (the original primary). An Hα bow shock has been observed in Vela X–1, which is shown in Fig. 6.10, from which it was argued that this system had been ejected from the Vel OB1 association ~2.5 Myr ago.

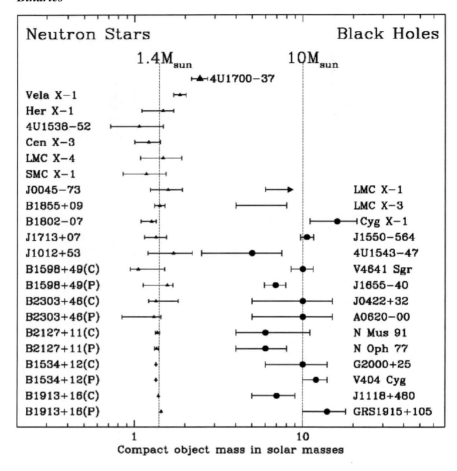

Fig. 6.9 Mass distribution for neutron stars and black holes. From Clark *et al.* (2002).

Progenitor masses to neutron stars are largely based upon theoretical calculations (e.g. Fig. 5.18). Observationally, masses may be inferred from interactions between progenitors and their environments (Nomoto *et al.* 1982), establishing scenarios by which individual X-ray binaries may have formed (Wellstein & Langer 1999) or if the progenitor was a member of a population of coeval stars with well-determined masses. Progenitor masses obtained from the latter, most reliable, method are surprisingly high, in the range $M_{init} \sim 20-50 M_\odot$ for SGR $1900 + 14$, SGR 1806–20, both of which are highly magnetized ($\geq 10^{14}$ G) slowly rotating (5–12 s) pulsars, i.e. magnetars (recall Section 5.6). There is often doubt regarding the physical association between the neutron star and stellar cluster. To date, the most reliable progenitor mass has been obtained for the magnetar CXO J164710.2-455216, for which Muno *et al.* (2006) exploited its proximity to the young massive cluster Westerlund 1 (Fig. 2.7) to infer a progenitor mass $M_{init} \geq 40 M_\odot$ (see also Bibby *et al.* 2008).

Although most HMXBs involve an OB star and compact remnant, one natural, though rare, consequence of close binary evolution would involve the evolution of the OB star through to

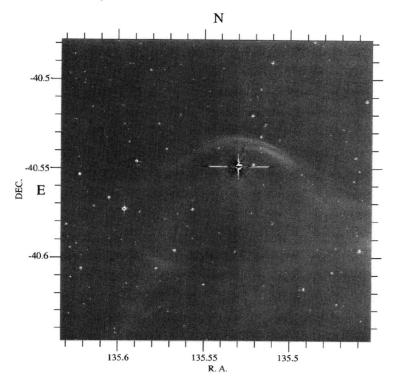

Fig. 6.10 The bow shock observed in Hα around the high mass X-ray binary Vela X–1. From Kaper *et al.* (1997).

the W-R phase, as illustrated schematically in Fig. 6.11. For many years, searches for Wolf–Rayet stars with compact companions proved elusive, until it was discovered that the bright galactic X-ray source Cyg X–3 possessed the near-IR spectrum of a WN star (see Fig. 6.12). Until recently, Cyg X–3, with a period of 4.8 hr and X-ray luminosity of $L_X \sim 10^{38}$ erg s^{-1}, was the only known example of a W-R plus compact companion system. Fortunately, the combination of high spatial resolution X-ray surveys of nearby galaxies plus extragalactic W-R surveys has increased this number to three, including IC 10 X–1 in the northern sky, and NGC 300 X–1 in the south, both of which are probable WN plus black hole systems possessing similar X-ray luminosities to Cyg X–3, albeit with longer periods of ∼35 hr. Such systems are X-ray bright due to an accretion disk around the black hole, fed by the strong W-R wind. Apparently, IC 10 X–1 has a very massive black hole, with 24–33 $M_\odot$ (Prestwich *et al.* 2007). If confirmed, this would establish it as the most massive known stellar-mass black hole.

Once the secondary undergoes a supernova explosion, either the system would disrupt with both compact remnants kicked out at high velocity in opposite directions, or if the system remained bound, double neutron star systems would be formed, which have been observed in a few, albeit rare, cases. PSR 1916 + 13 was the first binary pulsar to be discovered (Hulse & Taylor 1975), earning the discoverers the Nobel prize for physics in 1993. Although only one of the two 1.4$M_\odot$ components can be detected on Earth, this system provided a robust

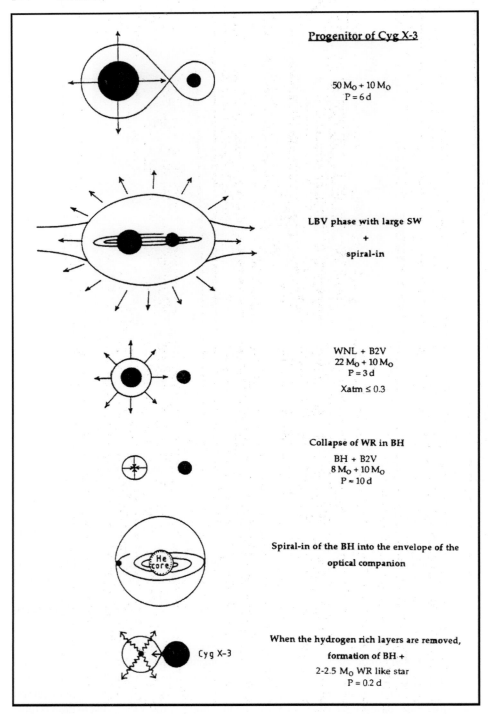

Fig. 6.11 Schematic for the close binary evolution leading to Cyg X–3. From Vanbeveren *et al.* (1998).

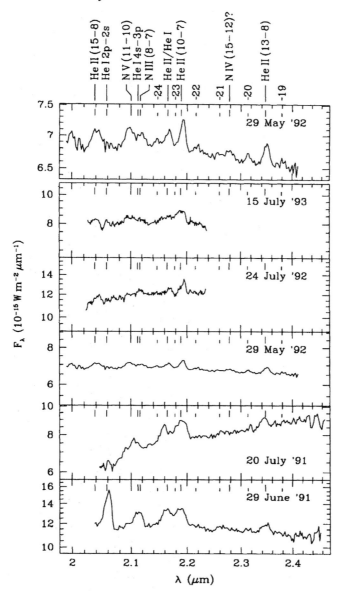

Fig. 6.12 K-band spectroscopy of Cyg X–3 suggesting the presence of a helium (Wolf–Rayet) star. From van Kerkwijk *et al.* (1996).

confirmation of Einstein's General Relativity, through the decrease in orbital period of the 7.75-hr binary, as orbital energy of the system is transformed into gravitational radiation.

Finally, let us consider binary progenitors of gamma ray bursts. The principal requirement is that the specific angular momentum of the iron core at the point of core-collapse is sufficiently large to avoid direct collapse to a black hole. For the single star case, angular momentum is lost either during the red supergiant or Wolf–Rayet phase, except for a rapid rotator at low metallicity undergoing chemically homogeneous evolution. Tidal interactions can cause either component of massive binaries to rotate at the same rate as the binary orbit, spinning it up (or down) as follows. For a star spinning synchronously with its orbit filling a fraction r of its Roche lobe, the ratio of the rotational frequency, ω, to its breakup frequency, ω_{crit}, is

$$\omega/\omega_{crit} = (1 + q)^{1/2} h(q)^{3/2} r^{-3/2}, \tag{6.6}$$

where q is the mass ratio and $h(q)$ is the ratio of the Roche lobe radius to the orbital separation (Podsiadlowski *et al.* 2004). From this the maximum orbital period for which tidal spin-up might provide sufficient angular momentum is a few hours, comparable to the 4.8-hr period for Cyg X–3. From a statistical argument, the number of sufficiently short-period massive binaries is in reasonable agreement with the observed GRB rate.

Alternatively, a rapidly rotating core could be produced during the merger of a black hole and an He core within common envelope evolution, although if the spiral-in occurs due to friction within an H-rich envelope, it is unclear how the merger could occur and still lead to the ejection of the envelope. As we shall see in Chapter 11, nearby GRBs are associated with luminous Type Ic supernovae, which are deficient in both H and He.

6.4   Interacting stellar winds

Two early-type stars within a binary system lead to an interaction of their winds. The wind interaction region involves the compression of their stellar winds, separated by a contact discontinuity. In general, details of the interaction process are investigated by complex hydrodynamics, although an analytical approach following Stevens, Blondin, & Pollock (1992) provides a useful insight into the physics of the colliding winds. The contact discontinuity between the two winds follows from ram pressure equilibrium, involving the density of each wind multiplied by the square of their velocities. One may introduce the wind momentum ratio, R,

$$R = \frac{\dot{M}_1 v_{\infty,1}}{\dot{M}_2 v_{\infty,2}}, \tag{6.7}$$

where the mass loss rates and wind velocities of components i are given by $\dot{M}_i$ and $v_{\infty,i}$, respectively. In the simplest case of $R = 1$, the intersection between the winds occurs in a plane midway between the two stars. In the more likely situation of $R \neq 1$, the contact discontinuity appears as a cone wrapped around the star with the less energetic wind (Rauw 2004). An example for a hydrodynamic simulation for the case of a colliding wind binary with separation 2×10^{13} cm (1.3 AU) and wind momentum ratio $R = \sqrt{5}$ is presented in Fig. 6.13.

If radiative cooling is important, the entire thermal energy produced in the shock is radiated away. If the shock is highly radiative, it collapses and the ionization of the material in the interaction region is set by the radiation of the stars, rather than the shock. This cool material may contribute to the formation of recombination emission lines in the optical spectrum (see Rauw 2004). Indeed, these are several examples in which the high-density material in the post-shock region contribute to the total line emission in early-type binaries (notably WC+O systems), due to the density-squared dependence of the recombination process. If radiative cooling is not important, the shock region is adiabatic. Cooling is more efficient for higher mass-loss rates.

For early-type stars that have reached their terminal velocities of several thousand km s^{-1}, the post-shock plasma temperature is very high ($\geq 10^7$ K), such that the main signature of shock-heated plasma is anticipated in X-rays, especially for the adiabatic case. Since X-rays

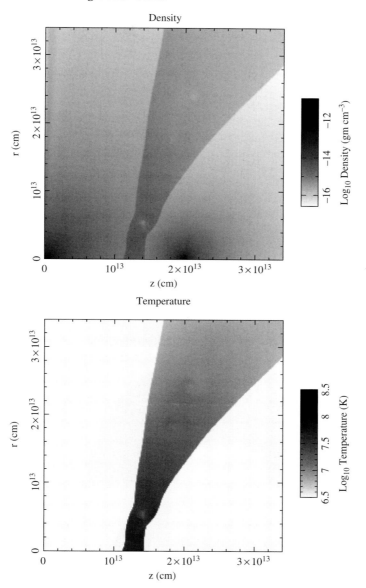

Fig. 6.13 Density (top) and temperature (bottom) snapshots from a hydrodynamic simulation of a colliding wind binary with separation 2×10^{13} cm (1.3 AU) and wind momentum ratio $R = \sqrt{5}$ ($\dot{M}_A = 5 \times 10^{-6} M_\odot$ yr^{-1}, $\dot{M}_B = 1 \times 10^{-6} M_\odot$ yr^{-1} and $v_A = v_B = 2000$ km s^{-1}). From Henley, Stevens, & Pittard (2003).

are also formed in single early-type stars, one expects significantly *harder* X-ray luminosity from massive binaries, as a function of the bolometric luminosity, plus phase-locked modulation. The latter is expected either from the changing opacity along the line-of-sight or changing separation in eccentric binaries. γ Vel (WC8+O) is one such colliding wind binary for which ROSAT X-ray observations revealed substantial phase-locked variability.

The X-ray emission from the shock is absorbed when the opaque wind from the W-R star is in front, but becomes significantly less absorbed at orbital phases when the cavity around the O star crosses the line-of-sight (Willis, Schild, & Stevens 1995).

In close binary systems, it is likely that terminal wind velocities will not be reached, with "radiative braking" relevant, in which the primary wind is decelerated by the radiative momentum flux it encounters as it approaches the companion. Radiative braking alters the bow-shock geometry and reduces the strength of the wind collision. X-ray emission from the shock region may modify the ionization of the wind material, and alter its ability to suffer radiative acceleration. Several 2D hydrodynamical simulations of the close WN+O binary V444 Cyg are presented in Fig. 6.14. The W-R wind may either overwhelm the O star outflow

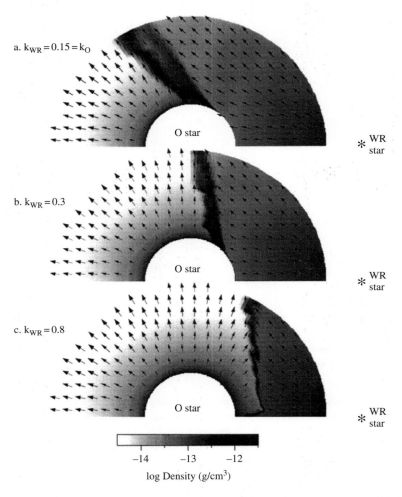

Fig. 6.14 Simulations of the density and velocity structure for the colliding winds in the V444 Cyg (WN+O) system for three CAK K-parameters for the W-R wind: (i) identical CAK K-parameter to the O star; (ii) intermediate value K-parameter; (iii) high K-parameter needed to drive the W-R wind. From Gayley, Owocki, & Cranmer (1997).

and form a shock that wraps around its surface, in the same manner as an O+O collision, or undergo sudden radiative braking in the case of an elevated opacity needed to radiatively drive W-R mass-loss rates, producing a qualitatively different structure.

A subset of early-type stars display a non-thermal (synchrotron) radio emission, in addition to the thermal radio emission produced via free–free emission from the stellar wind. This indicates that a magnetic field must be present in the winds of such stars, and that relativistic electrons exist in the radio emitting region. Shocks associated with a wind collision may act as sites for particle acceleration through the so-called Fermi mechanism (see Rauw 2004). In this case, free electrons would undergo acceleration to relativistic velocities by crossing the shock front. The majority of non-thermal radio emitters amongst W-R systems are known binaries, of which HD 193793 (WC7+O) is the most famous example. HD 193793 is a highly eccentric ($e = 0.84$) system with a 7.9 year period, for which the radio flux is thermal over the majority of the orbit. However, between phases 0.55 and 0.95 (0 corresponds to periastron passage), the radio flux increases dramatically and displays a non-thermal radio index. As a result of the cavity produced by the shock cone in the W-R wind, the non-thermal radio emission can reach us when the O star passes in front of the W-R star.

The effect is illustrated in Fig. 6.15, for the the W-R binary system WR147. Radio emission has been resolved into thermal emission from the WN8 primary to the south, plus non-thermal emission located between the WN star and its early B-type companion to the north. The position of the individual components has been confirmed from HST imaging. The location of the non-thermal component is consistent with the ram pressure balance of the two stellar winds, as expected in a colliding wind system. A sketch of the physical situation is presented in Fig. 6.16. Let us consider the wind momentum ratio, R, of the B star to the WN8 star. It can be shown that

$$r_{OB} = \frac{R^{0.5}}{1 + R^{0.5}} D. \tag{6.8}$$

For WR147, $r_{OB}/D \sim 0.09$, from which a momentum ratio of $R \sim 0.011$ can be obtained. If we use previously published stellar wind properties of $\dot{M} = 2.5 \times 10^{-5} M_\odot$ yr^{-1} and $V_\infty = 950$ km s^{-1} for the WN8 component from Morris *et al.* (2000), the inferred mass-loss rate for a B0.5 star is $\dot{M} = 3 \times 10^{-7} M_\odot$ yr^{-1} for an adopted $V_\infty = 800$ km s^{-1}.

6.5 Dust formation in WC stars

The principal sources of interstellar dust are cool, high mass-losing stars, such as red giants, asymptotic giant branch stars, plus novae and supernovae. Dust *is* observed around some massive stars, particularly LBVs with ejecta nebulae, but aside from their giant eruptions, this may be material that has been swept up by the stellar wind. The intense radiation fields of young, massive stars ought to act as efficient destructors of dust in their local environment.

Allen, Swings, & Harvey (1972) first identified excess IR emission due to heated circumstellar dust in a subset of carbon sequence (WC) Wolf–Rayet stars. In Fig. 6.17 we compare the mid-IR spectrum of two late WC stars: γ Vel (WR11, WC8+O), which does not form dust, and so has the characteristic broad emission lines from its stellar wind, including fine-structure lines of Ne and S; and WR104 (WC9d) which has a very strong dust excess, completely masking its stellar features.

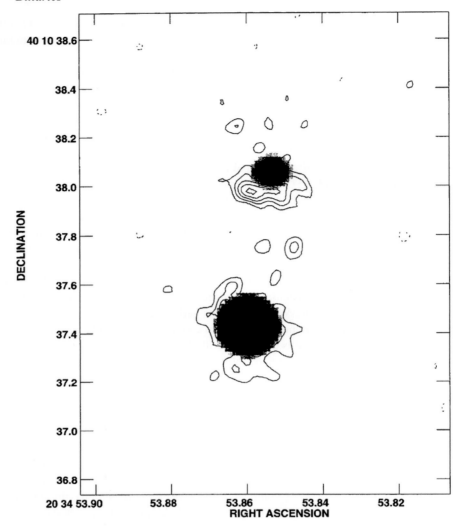

Fig. 6.15 K-band image of WR147 (WN8+B0.5V) superimposed on a 5 GHz (6 cm) radio contour map indicating the WN8 star and thermal radio source in the south and B star plus (colliding wind) non-thermal radio component to the north. From Williams *et al.* (1997).

Some stars form dust persistently, whilst others are episodic, with essentially all dust makers drawn from late WC stars, i.e. WC7 to WC9 subclasses. WC stars are minor contributors to interstellar dust, in general, but the formation and persistence of dust in such a harsh environment remains a puzzle. The problem can briefly be summarized as follows – for the possibility of dust nucleation and subsequent growth, carbon densities need to be very high, and temperatures rather low. Unfortunately, for the case of WC winds, temperatures are very high close to the star where densities are high, and conversely densities are too low far from the star where temperatures are relatively low.

Most WC9 stars are persistent dust makers ("WCd"), of which WR104 shown in Fig. 6.17 is the prototype. A few are episodic makers ("WCed"), and several do not make dust at all.

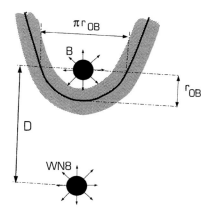

Fig. 6.16 Sketch of WN and B star components of WR147, showing the location of the shocked gas in the wind-interaction region (shaded) on their side of the contact discontinuity (solid). The intersection of the contact discontinuity and the line between the stars is referred to as the stagnation point. From Williams *et al.* (1997).

Indeed, 16/18 persistent dust makers are WC9 stars. A small fraction of WC8 stars form dust persistently or episodically. Only four WC stars with earlier spectral types form dust, though none persistently. HD 193793 (WC7ed+O) is the prototype of the periodic dust-making WC stars. Why are late WC stars preferential dust formers? It is very likely that the ionization from early WC stars is simply too harsh for the possibility of carbon being present in its neutral state.

Observationally, what do we know about the properties of the dust itself? Analysis of the IR spectral energy distribution permits determination of the grain temperature range and emissivity law, ε_λ. Early studies favored graphite grains ($\varepsilon_\lambda \propto \lambda^{-2}$), but Williams, van der Hucht, & Thé (1987) preferred amorphous carbon ($\varepsilon_\lambda \propto \lambda^{-1}$) for HD 193793 during its 1985 formation episode. Attempts to differentiate between these alternatives are hampered by uncertainties in the temperature distribution of the grains. IR spectral features provide much better diagnostics, although most common features are not present in WC stars. For example, the 11.3μm band in carbon-rich AGB stars is not detected in WC stars, nor is the expected 11.5μm graphite resonance. Consequently the favored precursor to grains in WC stars is amorphous (shapeless) carbon.

Williams *et al.* (1987) concluded that the dust grains are coupled to the stellar wind outflow, itself driven by radiation pressure. However, the conditions in the stellar winds at the dust condensation radii did not appear to favor the formation of dust, such that the presence of dust was not readily explained. The problem was how to generate sufficient neutral carbon close to the WC star to form grains along the classical route, $C \rightarrow C_2 \rightarrow C_3 \rightarrow$ monomers $\rightarrow$ clusters $\rightarrow$ grains. Instead, carbon was predominantly ionized in the region where the dust was formed. Indeed, assuming homogeneity and spherical symmetry, carbon remains singly or even doubly ionized due to high electron temperatures of $\sim 10^4$K in the region where dust is forming.

A clue to the explanation for the dust origin resulted from the observation of episodic WC dust makers. HD 193793 is a well-known WC7ed + O4–5 binary in a 7.9-yr, highly eccentric orbit and forms dust solely close to periastron, suggesting that the presence of a main-sequence

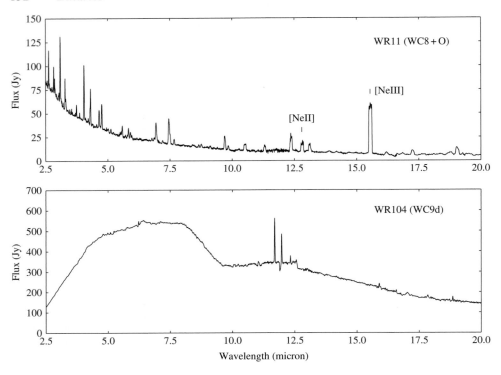

Fig. 6.17 Mid-IR spectral energy distributions of γ Vel (WR11, WC8+O) including strong, broad emission lines from its stellar wind, and WR104 (WC9d) whose mid-IR spectrum is totally dominated by hot dust, completely masking the stellar spectrum.

OB companion is one of the required ingredients for nucleation to occur. HD 193793 is also X-ray and radio bright at periastron, where the power in the colliding winds from each star is at its greatest. Whilst this does not explain how the grains can form, it provides a much more appropriate environment, i.e. high-density, low-temperature, carbon-rich material.

If binarity is responsible for episodic subclass, what is the situation for the persistent dust makers? Either some other process is operating or these are also binaries with (small) circular orbits. Photospheric features from companion OB stars have been observed in some persistent dust makers. Notable amongst these is WR104 (Ve 2-45, WC9d), which is the brightest W-R at mid-IR wavelengths for which conclusive proof of the binary nature of WR104 has been revealed via aperture masking with the Keck I telescope. Remarkably, H- and K-band imaging at epochs taken two months apart revealed that the dust associated with WR104 forms a spatially confined stream that follows a spiral trajectory, a so-called "pinwheel nebula" – see Fig. 6.18. Subsequent observations revealed an orbital period of 243.5 days and inclination of $11°$, based on an Archimedian spiral model, $r = \alpha\theta$, with no significant eccentricity. Other pinwheels have also been identified using this technique, including the five cocoon members of the Quintuplet cluster at the Galactic Center.

It is apparent from observations of HD 193793 and WR104 that binarity appears to play a key role in the formation of dust in WC stars, such that all dust formers may be binaries.

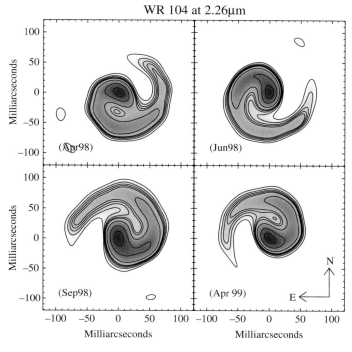

Contours (% of Peak) 1 2 3 4 5 10 30 70

Fig. 6.18 Near-IR images of the pinwheel nebula WR104 (WC9) displaying dust emission within a spiral density wave, presumably formed within the wake of the shock zone trailing behind the O star. From Tuthill, Monnier, & Danchi (1999).

However, not all WC binaries form dust, most likely as a result of their high ionization and/or unsuitable separations. It has been suggested that the presence of hydrogen from the OB companion may provide the necessary chemical seeding in the otherwise H-free WC environment, given the problems faced without hydrogen. However, even this possibility has its difficulties, since chemical mixing between the WC and OB winds is predicted to be unlikely in the immediate vicinity of the shock region. Mixing is thought to occur far away from the stars via Kelvin–Helmholtz instabilities – produced within the interface between the two opposing flows – where the densities are likely to be too low for dust nucleation.

7

Birth of massive stars and star clusters

Massive stars are born in interstellar clouds made up of molecular gas and dusty material. Most of these stars originate from GMCs with typically $\sim 10^5 M_\odot$. Upon collapse, these lead to massive star clusters. Some massive stars are born separate from these massive concentrations of gas and dust in smaller clouds and end up in more compact star clusters. Truly isolated massive stars seem to be rare in our Galaxy. Indeed, de Wit *et al.* (2005) argue that only a few percent of massive stars are born away from clusters. All these massive stars are found highly concentrated towards the Galactic plane where current star formation is still proceeding, albeit at a relatively restrained rate at present.

Consider first an individual massive star. It is formed when gravitational instability overwhelms a cloudlet of gas and dust which then begins a process of collapse. The collapse brings more and more material to the central object in a process of heating and rapid accretion. This phase is very rapid and will only be observed at far-IR wavelengths as the central material aggregates, begins to heat up, and emits radiation or molecular emission at radio wavelengths. Very quickly sufficient material accumulates such that an individual object can be identified. As this material continues to heat from the continuous contraction and accretion of more gas and dust it takes on the characteristic of what is called a "hot core", radiating also now at mid-IR wavelengths.

If the cloudlet is massive enough, the surface of the hot core will be at such a temperature that sufficient Lyman continuum radiation is emitted and an ionized hydrogen region surrounding the object is formed. Once this is observed, the terminology used to describe the situation is that of an ultracompact (or even hypercompact) HII (UCHII, HCHII) region, i.e., now describing the environment rather than the object. One can be assured that once a UCHII region is observed, an O or early B-type star is present. This star may still retain residual natal material in the form of a disk well into and even beyond the UCHII region phase. We have used the term massive young stellar object (MYSO) to describe the star in the presence of such residue from the birth process.

Next, consider the formation processes going on in a GMC, or a smaller molecular cloud, leading to the birth of many massive stars and a large star cluster, or a smaller version with one or only a few OB stars. A working assumption is that the cloud fragments into individual constituents on a timescale which is short compared to all others; thus all stars are "born" together. This fragmentation is inevitable for a collapsing large cloud due to the centrifugal barrier. All the complications connected with single star formation are still present with the addition now of interactions between the individual objects in the newly forming cluster, both gravitational and environmental. The physics involved is indeed very complicated and this important phase of stellar evolution is just beginning to be understood. A comprehensive textbook on the formation of high and low mass stars is provided by Stahler & Palla (2004), while the review article of Zinnecker & Yorke (2007) focuses upon massive star formation.

7.1 Natal precursors of OB stars

Hot cores

The term "hot cores" (or molecular cores) describes the beginnings of the process of the formation of an individual star. The central object heats up as it is optically thick at the wavelengths where the cores radiate in the earliest stages of this ongoing collapse and accretion process, illuminating its surrounding gas and dust. The dust radiates at IR wavelengths and these sources are detected in mid- and far-IR surveys. The cores are visible in the far-IR where the radiation is optically thin. The gas is also kinetically and radiatively energized by the hot core, leading to the appearance of masers. These masers are easily observed as emission features at cm and mm wavelengths and can be considered as tracers and precursors of the very earliest stages of star birth. Masers involve population inversions in molecules such as methanol, water, and the hydroxyl radical, among other species. The inversions arise either from molecular collisions, or radiation, or some combination of them within the star-forming clouds. The maser spatial morphology may be a jet, or a disk, or a ring, but differs from source to source. One would think there would be some connection between these kinds of observations and the nature and evolution state of the natal object, but no physical relationship has yet been established.

Many hot cores are sufficiently luminous and massive that they may end up as O-type stars, and likely represent the period of rapid accretion when most of the stellar mass is established. It is thought that in these phases the natal cloud is still sufficiently thick that any ionized hydrogen region, if present, would not yet be detectable. Other less luminous hot cores might represent massive objects which have not yet accreted enough material to qualify as an O star, but are on their way to do so, or they could be less luminous objects which will end up as less massive B stars. The nomenclature of these earliest phases of massive star birth is somewhat confusing. Most astronomers take the point of view that the term hot core is used before the following UCHII region stage, and not during or after. The terminology leaves much to be desired as once the ionized region is detected that observation is used to describe the status of the stellar phase. Indeed, there is an additional transition phase between hot cores and UCHII regions, namely the hypercompact HII region phase during which accretion is beginning to terminate but has not yet ended. By the UCHII stage accretion is finally complete.

Hypercompact and ultracompact HII regions

Once a massive OB protostar reaches a surface temperature $T_{\mathrm{eff}} \geq 25\,000\,\mathrm{K}$ it is capable of ionizing the high density hydrogen gas in its immediate proximity. This may be observed at radio wavelengths via free–free, bremsstrahlung emission. The pioneering radio survey by Wood & Churchwell (1989) identified many such HII regions within the disk of our Galaxy, revealing a variety of morphologies, with small characteristic sizes of $\leq 0.1\,\mathrm{pc}$ and densities of order $3 \times 10^4\,\mathrm{cm}^{-3}$. Consequently, such regions were coined "ultra-compact" HII regions, to distinguish them from the physically larger, compact HII regions, and lower density classical HII regions, as indicated in Table 7.1. For comparison, the Orion

Table 7.1 *Physical parameters of HII regions from Kurtz (2005), including the emission measure, $\varepsilon = n_e^2 L$, where L is the path length through the nebula.*

Class	r (pc)	ε (pc cm^{-6})	n_e (cm^{-3})
Hypercompact	≤ 0.03	$\geq 10^{10}$	$\geq 10^6$
Ultracompact	≤ 0.1	$\geq 10^7$	$\geq 10^4$
Compact	≤ 0.5	$\geq 10^7$	$\geq 5 \times 10^3$
Classical	~ 10	$\sim 10^2$	$\sim 10^2$
Giant	~ 100	$\sim 10^5$	$\sim 10^1$

Trapezium cluster has a size (0.15 pc) comparable to an UCHII region, but lower density gas of 10^3 cm^{-3}.

The majority of radio surveys of ultracompact HII regions have been carried out at a characteristic frequency of 5 GHz (6 cm). This introduces an observational bias against younger, higher density regions, since the turnover frequency – dividing the optically thin and thick regimes – for a thermal bremsstrahlung spectrum of a classical (spherical, uniform, isothermal) UCHII region peaks at this wavelength. At lower frequencies the flux density falls very steeply $S_\nu \propto \nu^2$, whilst at higher frequencies the flux density declines very slowly, as $S_\nu \propto \nu^{-0.1}$. The optical depth of thermal bremsstrahlung radiation can be approximated as

$$\tau_\nu \approx 0.082 \, \nu^{-2.1} T_e^{-1.35} \varepsilon, \tag{7.1}$$

from Mezger & Henderson (1967), where the emission measure $\varepsilon \propto n_e^2$. Solving for the turnover frequency, ν_t, (at $\tau_\nu = 1$) gives

$$\nu_t \approx \left(0.082 T_e^{-1.35} \varepsilon\right)^{0.48}, \tag{7.2}$$

i.e. the turnover frequency is effectively linear in density, n_e.

A HII region 0.01 pc in size, a factor of 10 times smaller than a typical UCHII region, with a density of $n_e = 10^6$ cm^{-3}, would have an emission measure of $\varepsilon = 10^{10}$ pc cm^{-6} and turnover frequency at 50 GHz (6 mm). At 6 cm, this HII region would be in the optically thick regime, with a flux density perhaps 100 times lower than at 6 mm. Figure 7.1 illustrates the situation. Consequently, small, dense HII regions would have been missed in early radio surveys. An example of such a hypercompact HII region, G75.78 + 0.34, is presented in Fig. 7.2, where 6-cm radio maps indicate the location of the UCHII region, whilst 7-mm maps additionally reveal a hypercompact HII region, spatially coincident with water masers.

The radio spectrum of G75.78 + 0.34 actually has an index of +1.4, instead of the nominal +2, indicating that there is a density gradient in the HII region, with limits on the turnover frequency, ν_t, implying either an emission measure in excess of $\varepsilon \sim 2 \times 10^{11}$ pc cm^{-6}, or an electron density $n_e \geq 6 \times 10^6$ cm^{-3}. A number of hypercompact HII regions are now known, with densities approaching that of hot molecular cores, and sizes of ≤ 0.03 pc (6000 AU). This characteristic size scale approaches that for single/binary star formation, rather than a star cluster, which is expected for ultracompact HII regions.

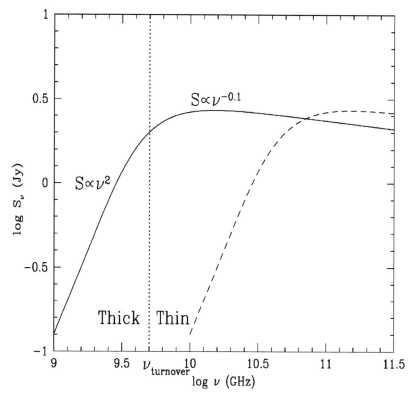

Fig. 7.1 Radio spectrum of a thermal bremsstrahlung emission from a spherical, uniform, isothermal ultracompact HII region (solid) with turnover frequency (dotted line) at 5 GHz (6 cm) dividing the optically thin high-frequency regime from the optically thick low-frequency regime. The dashed line shows a HII region with an emission measure, ε, 100 times higher, resulting in a turnover frequency at 50 GHz (6 mm). The abscissa unit is in GHz. From Kurtz (2005).

Evidence that HCHII regions are still *accreting* material includes broad radio recombination line profiles in some instances, indicating substantial large-scale velocity structure, most likely involving both infall and outflow. In contrast, there is generally no evidence of outflows being actively driven from UCHII regions, implying that accretion has ceased at this stage. Individual UCHII regions are known to exist within GHII regions (e.g. G49.49–0.37 within W51A). In general, the detection process of ultracompact HII regions has generally favored relatively isolated sources along the Galactic plane, away from the source confusion present in large assemblages. Most exciting stars within UCHII regions would be considered to be MYSOs, but a few (e.g. G29.96–0.02) may be "naked" if they have no evidence of circumstantial material.

Stellar content of UCHII regions
Circumstellar extinction is high towards UCHII regions, with typically $A_V \sim 50$ mag, since the ionized hydrogen is surrounded by the natal dust cocoon. The well-known

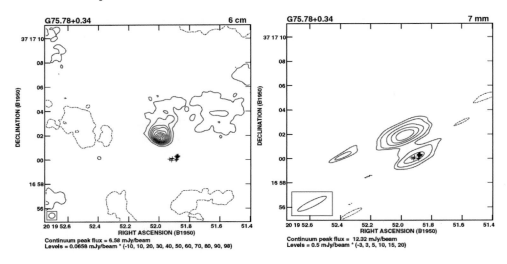

Fig. 7.2 Continuum images of G75.78 + 0.34 at 6 cm, revealing an ultracompact HII region in the center, with crosses marking water masers, plus at 7 mm additionally revealing a hypercompact HII region spatially coincident with the maser emission. From Kurtz (2005).

UCHII region G5.89–0.39 represents an excellent example. The ionizing star is not seen in either H-band or K-band imaging, but does appear at longer wavelengths (e.g. 4.5 μm) due to the lower dust absorption (see Section 8.1).

Of the small fraction of UCHII regions that can be seen at 2 μm, the dominant spectral signature is nebular Brγ emission, with He I and H$_2$ also usually seen. In principal, nebular He I and Brγ emission can be used as indirect probes of the ionizing star, but mid-IR fine structure metal lines offer a more sensitive probe. The mid-IR spectrum of the UCHII region G29.96 – 0.02 from the Infrared Space Observatory (ISO) is presented in Fig. 7.3, revealing these nebular lines superimposed upon a dust continuum. Dust bands from polycyclic

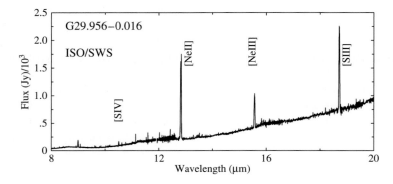

Fig. 7.3 Mid-IR ISO/SWS spectrum of G29.96–0.02, revealing nebular emission lines superimposed on a hot dust continuum, together with dust bands and molecular ices. From Crowther (2005).

aromatic hydrocarbons (PAHs) and silicates are also observed in the mid-IR. Direct spectro-scopic classification of the ionizing star(s) would of course be preferable, but this is possible only in unusual low line of sight and central extinction cases, such as G29.96 – 0.02 which resembles an early O star, from a comparison with template stars (e.g. Fig. 2.15).

UCHII regions are believed to contain a cluster of stars, including at least one O or early B star. However, high foreground and internal extinction together with the need for high spatial resolution prevents direct observations of their stellar content in most cases. Conse-quently, one relies upon nebular properties to infer stellar parameters of the ionizing star(s) of individual UCHII regions.

The most common technique is to use radio continuum fluxes to derive the number of Lyman continuum ionizing fluxes, $N(\mathrm{LyC})$. From Table 3.1 it is apparent that the hottest, most massive main-sequence star in a cluster would likely dominate the Lyman continuum flux. The Lyman continuum flux, $N(\mathrm{LyC})'$ required to maintain ionization of a homogeneous, spherical, dust-free nebula according to Kurtz, Churchwell, & Wood (1994) is

$$N(\mathrm{LyC})' \geq 8.04 \times 10^{46}\, T_e^{-0.85}\, U^3, \tag{7.3}$$

where T_e is the electron temperature (K) and U is the excitation parameter (pc cm^{-2}),

$$U = r n_e^{2/3} = 4.553 \left(\frac{\nu^{0.1}\, T_e^{0.35}\, S_\nu\, d^2}{a(\nu, T_e)} \right)^{1/3} \tag{7.4}$$

for frequency ν (GHz), integrated flux density S_ν (Jy), distance d (kpc), and $a(\nu, T_e)$ is a factor of order unity, tabulated by Mezger & Henderson (1967).

Complications arise since a significant fraction of the ionizing photons may be absorbed by dust, and re-radiated at far-IR wavelengths. If a fraction $1 - \zeta$ of the ionizing photon flux is absorbed by dust instead of gas, the actual Lyman continuum flux $N(\mathrm{LyC}) = N(\mathrm{LyC})'/\zeta$ after correction for dust absorption.

Alternatively, analysis of mid-IR metal fine-structure lines (recall Fig. 7.3) provides an indirect tracer of the extreme UV energy distribution of the O stars within the UCHII region. Common examples are listed in Table 7.2 together with required ionization energies and threshold wavelengths. A few are available within the N-band from ground-based telescopes, e.g. [Ne II] 12.8 μm, [S IV] 10.5 μm, although the majority require space-based facilities such as ISO or Spitzer. High ionization lines such as [O IV] 25.89 μm are believed to be produced by Wolf–Rayet stars or in shocks, and are not anticipated from normal main sequence OB stars.

In recent years, significant progress has been made in the development of atmospheric mod-els for OB stars (Chapter 3). A comparison between the threshold wavelengths for selected fine-structure lines in Table 7.2 with predicted ionizing flux distributions, such as Fig. 3.4, demonstrates their diagnostic power regarding stellar temperatures. For example, UCHII regions exhibiting strong [Ne III] 15.5 μm emission indicates a substantial photon flux short-ward of 303 Å, unique to hot early-type O stars amongst main-sequence stars. Unfortunately, predictions from non-LTE atmospheric models disagree with one another at wavelengths significantly below the Lyman edge at 912 Å. Nevertheless, attempts have been made to de-rive effective temperatures of OB stars by using non-LTE, extended, line-blanketed model atmospheres together with photoionization codes.

G29.96–0.02 serves as a very important illustration of the current limitations of the nebular analysis technique. Morisset *et al.* (2002) studied the [Ne II–III] and [S III–IV] fine-structure lines of G29.96–0.02, deriving a temperature of $T_{\mathrm{eff}} \sim 35\,000$ K which corresponds to an O8

Table 7.2 *Threshold photon energies χ (13.60 eV for hydrogen) and wavelengths λ_{limit} (911.8 Å for hydrogen) for selected ions possessing mid-IR fine-structure nebular emission lines, relevant for ionizing flux distributions for O and W-R stars.*

χ (eV)	λ_{limit} (Å)	Ion	Spect. line(s) (μm)
15.76	786.7	$Ar^0 \longrightarrow Ar^+$	6.98 [Ar II]
21.56	574.9	$Ne^0 \longrightarrow Ne^+$	12.81 [Ne II]
23.34	531.3	$S^+ \longrightarrow S^{2+}$	18.71, 33.48 [S III]
27.63	448.7	$Ar^+ \longrightarrow Ar^{2+}$	8.99 [Ar III]
29.60	418.8	$N^+ \longrightarrow N^{2+}$	57.33 [N III]
34.83	356.4	$S^{2+} \longrightarrow S^{3+}$	10.51 [S IV]
35.12	353.0	$O^+ \longrightarrow O^{2+}$	51.81, 88.35 [O III]
40.96	302.7	$Ne^+ \longrightarrow Ne^{2+}$	15.55, 36.01 [Ne III]
54.93	225.7	$O^{2+} \longrightarrow O^{3+}$	25.89 [O IV]

dwarf (recall Table 3.1). However, this inverse problem is not without difficulties, since the dominant ionizing star of G29.96 − 0.02 has been *directly* observed in the near-IR, revealing an O5V star with an effective temperature of $T_{eff} \sim 41\,000$ K (Hanson, Puls, & Repolust 2005b). This problem represents a serious discrepancy, and needs to be resolved in order to progress our understanding of UCHII regions from mid-IR spectroscopy. Clumping and geometry in HII regions no doubt play important roles. In addition, the ISO observation samples the integrated nebular emission from G29.96 − 0.02 which may comprise multiple O stars, biasing the nebular results towards a later O subtype than that spectroscopically observed by Hanson *et al.* Whatever the cause, we cannot hope to understand more complex cases, such as the integrated mid-IR spectra from starburst galaxies, until we have resolved such apparently simple cases.

Do UCHII regions host star clusters?

The radio continuum and mid-IR fine structure techniques discussed above in principle provide information on the most massive star in a UCHII region. Once the effective temperature of the exciting star has been evaluated by either of these methods, its luminosity and mass can be estimated by consulting stellar models. However, do UCHII regions host individual massive stars or star clusters? The presence of a cluster of stars within an UCHII region, as opposed to a single hot, luminous star, is observationally challenging to assess on the basis of their compact nature ($\sim$0.1 pc), large distance (typically $\sim$5 kpc) and high visual extinction ($A_V \sim 30$ mag).

In principle, the ratio of the far-IR to radio flux of an UCHII region should distinguish between a cluster and a single star. The former is sensitive to the re-radiated UV flux from all the massive stars within a UCHII region, while the latter is dominated by the the extreme UV radiation from the hottest star(s). One may generally assume that dust absorbs essentially all of the stellar radiation. In reality, if a fraction ζ of the extreme UV photon flux is photoionizing the gas, the integrated far-IR luminosity, L_{FIR} is

$$L_{FIR} = (1 - f_{esc})\zeta L_{EUV} + f_{FUV}L_{FUV}, \qquad (7.5)$$

where L_{EUV} is the EUV luminosity emitted shortward of the 912 Å Lyman edge, L_{FUV} is the far UV (FUV) luminosity, f_{esc} is the fraction of stellar ionizing photons that escape and f_{FUV} is the fraction of stellar FUV photons absorbed by dust. Photoionized gas recombines through far-UV lines – including resonant scattering of Lyα – plus a nebular continuum, while ζL_{EUV} contributes to emission from hot dust in the mid-IR. Typically $f_{esc} = 0$ is a reasonable assumption for UCHII regions, although some of the FUV photons may escape from the HII region if there are holes in the dust cocoon, $f_{FUV} < 1$.

Observationally, L_{FIR} may be calculated from the measured Infrared Astronomical Satellite (IRAS) fluxes, as we shall show for IR luminous galaxies in Eq. (9.13). However, for Milky Way star-forming regions the IRAS far-IR spatial resolution of $\sim$2 arcmin corresponds to a spatial scale of 3 pc at a distance of 5 kpc. Therefore, IRAS luminosities inferred for many UCHII regions will include contributions from other nearby star-forming complexes or hot cores. This effect is further compounded by the arcsec resolution sensitivity of radio surveys.

During the 1990s the Midcourse Space Experiment (MSX) Galactic plane survey was carried out at mid-IR wavelengths at 18 arcsec spatial resolution using the Spirit III instrument. More recently, Spitzer has conducted a mid-IR survey of the inner Milky Way through the Legacy Galactic plane survey GLIMPSE (Benjamin *et al.* 2003) with the IRAC instrument, providing a spatial resolution of $\sim$2 arcsec. To illustrate the improvement, Fig. 7.4 compares MSX with Spitzer observations of the UCHII region G10.30−0.15.

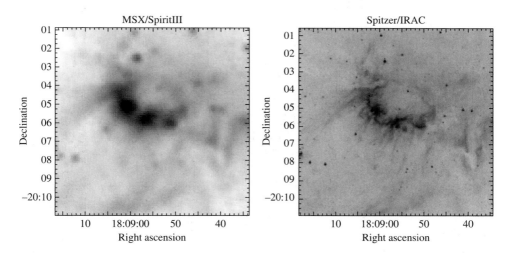

Fig. 7.4 Comparison between 8 μm 10 × 10 arcmin images of the UCHII region G10.30–0.15 obtained with MSX (left) and the Spitzer GLIMPSE survey (right). North is up and east is to the left in both, logarithmic scale, images. From Crowther (2005).

However, reliable IR luminosities from UCHII regions require high spatial resolution imaging far-IR wavelengths, where the dust emission peaks. The MIPS instrument aboard Spitzer provides a spatial resolution of 18 arcsec at 70 μm, a major improvement over IRAS. Surveys are underway to establish the cluster versus single-star nature of UCHII regions with Spitzer MIPS. In addition, a far-IR all-sky survey commenced in 2006 with AKARI (Murakami *et al.* 2007).

In order to derive the integrated IR luminosity from an individual far-IR flux density measurement, either empirical calibrations or dust modeling needs to be carried out. Radiative transfer modeling of the dust surrounding massive protostars was first carried out in a seminal study by Wolfire & Cassinelli (1987). Recently, clumpy 2D and 3D radiative transfer models have been developed by Barbara Whitney and colleagues. At shorter wavelengths, the presence of an accretion disk would also affect the shape of the spectral energy distribution.

The density and temperature distribution from a typical clumped dust radiative transfer model is presented in Fig. 7.5, together with high and low extinction sight-lines indicated, corresponding to $A_V = 401$ and 13 mag, respectively. Predicted spectral energy distributions for these two viewing angles are presented in Fig. 7.6, illustrating that wide differences in mid-IR colors and strengths of silicate features may result solely from different sight-lines. Indeed, mid-IR spectroscopy of several UCHII regions, normalized at $100\,\mu$m, compares well to different viewing angles for a single clumpy model. In all cases, the flux peaks around $130\,\mu$m.

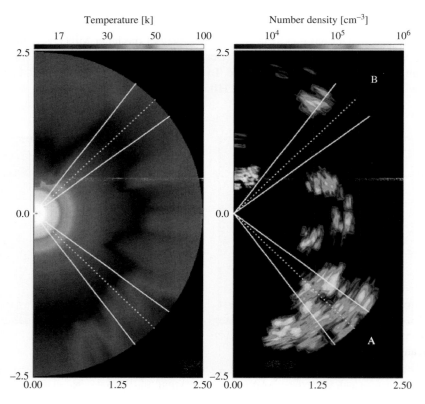

Fig. 7.5 Density and temperature distributions for a clumpy radiative transfer model, with the spatial scale in parsecs, and high (A) and low (B) column density sightlines indicated. From Indebetouw *et al.* (2006).

Finally, the advent of mid-IR instruments on 8–10 m ground-based telescopes has provided an additional method of establishing whether UCHII regions host star clusters. The diffraction limit of such telescopes in the N-band is 0.3–0.4 arcsec, corresponding to a physical scale of 0.01 pc. Consequently, mid-IR ground-based imaging permits the dust and gas distributions

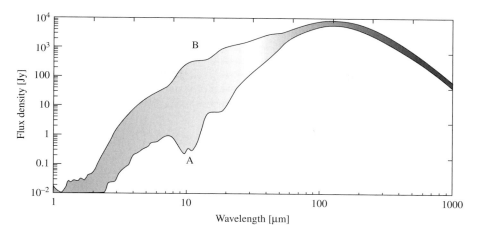

Fig. 7.6 Spectral energy distributions of clumpy radiative transfer models, illustrating the differences due to sight line, including the high (A) and low (B) extinction viewing angle from the previous figure. From Indebetouw *et al.* (2006).

of UCHII regions to be spatially resolved. Indeed, nebular fine-structure maps of [S IV] 10.5 μm and [Ne II] 12.8 μm in the UCHII region G70.29 + 1.60 (K3–50A) indicates multiple massive stars – one mid- and two late-type O stars – according to Okamoto *et al.* (2003). Such techniques provide great promise for the near future.

7.2   The initial mass function

Stellar mass function

The stellar IMF is one of the most fundamental quantities in the research areas related to star formation, stellar populations, and galaxy evolution. The concept of the IMF was introduced by Salpeter (1955) in a study of the average star-formation rate in the solar neighborhood. (He called this function the "original" mass function.) Mathematically, the IMF is a probability distribution of the stellar mass from the lowest to the highest mass. Observationally, this function describes the relative number of stars we count in mass bins of unit size. Therefore we must correct the raw observed number counts for evolutionary and dynamical effects, something particularly relevant for massive, short-lived stars with significant mass loss.

We can define the number of stars formed in the mass interval dM_* and time interval dt as

$$\phi(M_*)\psi(t)dM_*dt. \tag{7.6}$$

$\psi(t)$ is the total mass of stars formed per unit time, also called the star-formation rate. Usually it has units $M_\odot$ yr^{-1} kpc^{-2}. $\phi(M_*)$ is the mass distribution, or IMF. Equation (7.6) defines the rate of stellar birth per unit time, mass, and area. We implicitly assume a dependence on area although this is not explicitly spelled out in Eq. (7.6). A fundamental assumption in this equation is that the time- and mass-dependence of the star formation can be separated and that $\phi(M_*)$ is time-independent, and vice versa that $\psi(t)$ has no mass-dependence. The justification for this simplification is the absence of observational counter-evidence.

Since $\phi(M_*)$ is a probability distribution, its absolute number is meaningless and can be defined arbitrarily. The usual convention is the normalization

$$\int_0^\infty M_*\phi(M_*)\mathrm{d}M_* = 1. \tag{7.7}$$

$\phi(M_*)$ constrains only the *relative* proportion of stars. For instance, we can use this function to calculate the number of G-type stars expected for every O star if we can specify the mass intervals occupied by G and O stars.

Later we will see that observations of *massive* stars suggest an IMF following a power law:

$$\phi(M_*) \propto M_*^{-\alpha}, \tag{7.8}$$

where α is called the slope of the IMF over a *linear* mass interval. Sometimes the number of stars is counted per *logarithmic* mass interval. For clarity we call this function $\xi(M_*)$. The relation between $\phi(M_*)$ and $\xi(M_*)$ is

$$\phi(M_*) = \log e \, M_*^{-1}\xi(M_*). \tag{7.9}$$

If $\xi(M_*)$ is parameterized in terms of a power law, the common notation is

$$\xi(M_*) \propto M_*^{-\gamma}, \tag{7.10}$$

with $\gamma = \alpha - 1$. Note that the constant of proportionality in Eqs. (7.8) and (7.10) differs by a factor of $\log e$. Finally, and to add to the confusion, Eq. (7.10) can be redefined via a power-law exponent parameter of opposite sign, Γ, with $\Gamma = -\gamma$. While any of the three power-law exponents α, γ, and Γ can be found in the literature, most researchers studying starburst galaxies prefer to define the power-law IMF with Eq. (7.8). Throughout this volume we shall endeavor to follow this custom and count stellar numbers on *linear* mass intervals, with the power-law exponent of the IMF called α.

The IMF slope is defined over a specific mass interval. Originally, Salpeter (1955) proposed $\alpha = 2.35$ for the mass range $0.4 \leq M_* \leq 10 M_\odot$. In the present context, the most significant results for massive stars were by Massey, Johnson, & Degioia-Eastwood (1995a) for Galactic OB associations revealing $\alpha = 2.1 \pm 0.1$ for stars more massive than $7 \, M_\odot$ and Massey *et al.* (1995b) for Magellanic Cloud OB associations, obtaining $\alpha = 2.3 \pm 0.3$. Consequently, Massey and collaborators provided a direct extension of Salpeter's work to higher mass. Strictly, the single power-law exponent between solar-type stars and massive stars should be known as a Salpeter–Massey IMF. Scalo (1998) reviews the massive star IMF.

Subsequent revisions at lower masses have included studies of the Orion Trapesium cluster (e.g. Muench, Lada, & Lada 2000), permitting Kroupa (2001) to propose three α intervals, $\alpha = 2.3 \pm 0.7$ for $0.5 \leq M_*/M_\odot$, $\alpha = 1.3 \pm 0.5$ for $0.08 \leq M_*/M_\odot < 0.5$ (low-mass M dwarfs) and $\alpha = 0.3 \pm 0.7$ for $0.01 \leq M_*/M_\odot < 0.08$ (brown dwarfs). This is commonly known as a Kroupa IMF.

The observed form of the IMF closely mimics the core or clump mass function within which star formation occurs (Testi & Sargent 1998), providing an estimate of the star formation efficiency of $\sim$30%, i.e. only 1/3 of the mass of a collapsing clump finds its way into stars. The characteristic mass scale for stars is believed to be related to the Jeans mass, M_Jeans:

$$M_\mathrm{Jeans} = \left(\frac{5R_g T}{2G\mu}\right)^{3/2}\left(\frac{4\pi\rho}{3}\right)^{-1/2}, \tag{7.11}$$

where ρ is the gas density, T the temperature, R_g is the gas constant and μ is the mean molecular weight of the gas. The Jeans mass, introduced by the physicist James Jeans, is the critical mass for gravitational collapse of a gas cloud, and from Eq. (7.11) depends critically upon the gas density and temperature, favoring dense, cold molecular clouds. Fragmentation of the gas cloud is believed to continue until $\sim 0.1 M_\odot$, close to the boundary between stars and brown dwarfs. In primordial gas the temperature is believed to be an order of magnitude higher than in local molecular clouds since metals efficiently cool gas clouds, such that the characteristic mass of the first, Population III, stars differs dramatically from that in the present Universe, as we shall see in Section 11.1.

Cluster mass function

Individual star clusters span a wide range of masses from perhaps $10 M_\odot$ in Taurus to in excess of $10^6 M_\odot$ in starburst galaxies. The number of massive stars formed depends critically on the form of the cluster mass function. If all clusters were modest affairs, the form of the IMF would prohibit the formation of massive stars on statistical grounds. Indeed, massive stars are favored in more massive clusters, according to Larson (2003), with the most massive star, M_{up}, related to the cluster mass M_{clu} as follows:

$$M_{up} = 1.2 M_{clu}^{0.45}. \tag{7.12}$$

This provides a reasonable match to the most massive stars within a wide range of nearby clusters, which is shown in Table 7.3.

Table 7.3 *Comparison between cluster mass and most massive star within these clusters for young Milky Way and LMC clusters (adapted from Weidner & Kroupa 2006).*

Cluster	$M_{clu}(M_\odot)$	$M_{up}(M_\odot)$	Age (Myr)
Tau-Aur	25 ± 15	2.2 ± 0.2	1–2
ρ Oph	100 ± 50	8 ± 1	0.1–1
NGC 2264	355 ± 50	25 ± 4	3.1
Orion (M42)	2200 ± 300	45 ± 5	< 1
NGC 6611 (M16)	$(2 \pm 1) \times 10^4$	85 ± 15	1.3
R136	5×10^4	145 ± 10	1–2

Within the most massive clusters the upper stellar mass limit approaches $150 M_\odot$, but is this limit a statistical or physical one? As recently as the 1980s, R136a, the central region of the cluster R136, was believed to be a single star with a mass of several thousand solar masses. Speckle interferometry first resolved R136a into multiple components, which has subsequently been improved upon with HST (e.g. Massey & Hunter 1998). Observationally, care has to be taken to consider clusters sufficiently young that stellar evolution will not have caused the most massive stars to undergo supernova explosions.

R136 is sufficiently young and massive that it represents the best cluster to distinguish between a statistical and a physical upper limit to the mass of stars. Massey & Hunter (1998) obtained an IMF with $\alpha = 2.35$ for stars in the mass range $3 \leq M_* \leq 120 M_\odot$, i.e. 2000 stars with a total mass of $M_{clu} = 1.7 \times 10^4 M_\odot$ for R136a. Using a Kroupa IMF with a different slope of $\alpha = 1.3$ between 0.1–0.5 $M_\odot$ would imply a total cluster mass of $M_{clu} = 5 \times 10^4 M_\odot$, distributed amongst 85 000 stars in the mass range 0.1–100$M_\odot$. If there is no physical upper mass limit, one would expect 10 stars with masses above 150$M_\odot$, where none are observed in R136.

An analysis for the Arches star cluster (see Section 7.4) came to similar conclusions, in spite of a more metal-rich environment and apparently flatter stellar mass function than R136a. Kim *et al.* (2006) estimated $\alpha = 1.9$ for the Arches cluster for $1.3 < M_* < 50M_\odot$, and presented numerical calculations from which an initial IMF of $\alpha = 2.0$–2.1 is inferred. Depending upon the exact IMF assumed, the expected number of stars above 150 $M_\odot$ is ∼18–33 stars, as indicated in Fig. 7.7, where again none are observed. Of course, results for R136 and the Arches cluster only hold providing the most massive stars have not yet undergone supernova explosions. This *should* be the case since these clusters are believed to be 1–2 Myr old. Hence, there appears to be a physical upper limit to the mass of stars in the local Universe. These rules do not necessarily apply in the early Universe, as we shall see in Chapter 11.

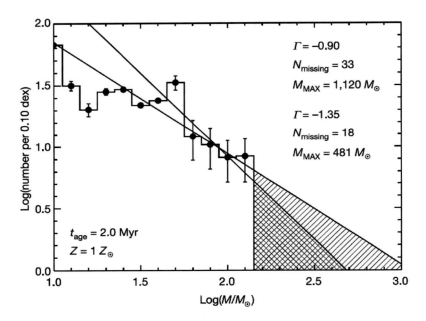

Fig. 7.7 Stellar mass function in the Arches cluster where $\Gamma = 1 - \alpha$. From Figer (2005).

In spite of the apparent upper limit to stellar masses, the total number of massive stars formed is the sum of stellar IMFs contributed by the sum of star clusters. Consequently, the cluster mass function, $\zeta(M_{clu})$, needs to be considered. This can be described by the following power law:

$$\zeta(M_{clu}) \propto M_{clu}^{-\beta}, \qquad (7.13)$$

where $dN_{clu} = \zeta(M_{clu})dM_{clu}$ is the number of clusters in a mass interval $[M_{clu}, M_{clu} + dM_{clu}]$ and M_{clu} is the cluster (rather than cloud) mass. According to Hunter *et al.* (2003) and references therein, observational evidence suggests $2 \le \beta \le 2.4$ for clusters with masses between 50 and 1000 $M_\odot$ in the solar neighborhood, intermediate mass clusters of $10^3 \le M_{clu}/M_\odot \le 10^4$ in the Magellanic Clouds, and high mass clusters of $10^4 \le M_{clu}/M_\odot \le 10^6$ in the Antennae galaxies. The upper cluster mass limit depends upon the environment, with gas rich mergers favoring a higher limit than quiescent star formation.

7.3 Formation of high-mass stars

Most massive stars in our Galaxy are born in massive stellar clusters which are formed from gravitational collapse of small regions within GMCs, such as the Orion nebula (Trapezium) cluster within the Orion GMC. Some are found in less luminous clusters derived from smaller molecular clouds and a few percent appear to be born more or less isolated from other stellar birth activity (de Wit *et al.* 2005). Surveys of GMCs find them to be strongly confined to the Galactic plane; in other spiral galaxies, GMCs are aligned with, and define, the spiral arms, but the exact numbers and orientations of the arms in our Galaxy is not well shown. This may be because in our Galaxy distances of spiral arm features such as these are determined by kinematic measurements, which are dependent on the *assumptions* of a simple radial dependence of orbital velocity (the rotation "law") in the absence of peculiar (random) velocities of the clouds. Roughly one third of the GMCs in our Galaxy show no evidence of any star formation; the rest are either associated with known stellar clusters, or star birth features, or other signposts of stellar activity such as masers. The third of GMCs without current star activity may merely represent the time before such action begins; alternatively, but seemingly less likely, they are GMCs that will disrupt first and never provide stars. Turbulence and magnetic processes likely play a role here, but details remain to be shown.

We have seen that massive stars form predominantly within massive clusters produced during the collapse of GMCs and apparently follow a universal IMF. However, observational details are scarce due to the high visual extinction inherent to their natal cocoon during the formation process. In the low-mass case, the overall pre-main sequence evolution is fairly well established involving accretion onto a young protostar via a circumstellar disk, within an envelope of infalling dust and gas, plus jets and molecular momentum-driven outflows produced by the protostar, leading to a mature phase in which the protostar enters the zero-age main sequence (Shu, Adams, & Lizano 1987). As we shall show, theories of low-mass star formation do not readily "scale-up" for the high-mass case. Nevertheless, individual stars with masses of order $100 M_\odot$ are observed, so a variety of mechanisms have been proposed to explain their origin.

Accretion rates for regions of low-mass star formation are predicted to scale with the gas temperature, $T^{3/2}$, such that low accretion rates of 10^{-6} to $10^{-5} M_\odot$ yr^{-1} result for isothermal gas of temperature 10–20 K within GMCs. At such rates, the formation timescales (≥ 1Myr) would represent a significant fraction of the main-sequence lifetime of the highest mass stars. Observational evidence suggests much shorter formation timescales of ~ 0.1 Myr for high-mass stars (McKee & Tan 2003). In addition, the Kelvin–Helmholtz contraction timescale is

$$t_{KH} = \frac{GM_*^2}{R_* L_*},$$

where M_*, R_*, and L_* are the protostar mass, radius, and luminosity, respectively. This is sufficiently short for high-mass stars ($t_{KH} \sim 3 \times 10^4$ yr for a 30 $M_\odot$ star) that they begin burning hydrogen before accretion from their surrounding envelope ceases, for which the free-fall time is

$$t_{ff} = \frac{3\pi}{\sqrt{32G\bar{\rho}}},$$

where $\bar{\rho}$ is the average initial density of the cloud at its Jeans limit, i.e. $t_{ff} \gg t_{KH}$ ($t_{ff} \ll t_{KH}$ for low-mass stars). Consequently, high-mass stars move along the hydrogen-burning main sequence whilst accretion continues, hidden within their molecular clouds. The high luminosity protostars produce an extreme UV environment which would rapidly photoevaporate an accretion disk. Accretion disks and massive outflows may not be universal in the highest mass cases. Circumstellar disks have been inferred for some less massive early B stars (Blum *et al.* 2004). UV photons will also photoionize the cold gas in the vicinity of hot stars to 10^4K, evacuating their immediate environments. A final challenge is that hot, luminous stars possess radiatively driven winds (recall Chapter 4) which would potentially repel the (spherically) infalling material once the mass of the protostar exceeds $\sim$10 $M_\odot$ (Yorke & Krügel 1977).

These issues have motivated extreme solutions, such as coalescence of low-mass protostars within extreme dense star clusters in order to achieve a rapid build up of the final stellar mass (Bonnell, Bate, & Zinnecker 1998). Although this scenario is not presently favored, as the required high stellar densities of 10^8 pc^{-3} have never been observed, two competing theoretical models are presently favored for the formation of massive stars, termed the competitive accretion and turbulent core models. It should be stressed that details of star formation in general remain poorly understood, of which high-mass star formation represents the most challenging case.

Competitive accretion versus turbulent core models

In the competitive (Bondi–Hoyle) accretion scenario, massive stars form by accreting gas through gravitational collapse that was initially not bound to the star. Protostars compete for the mass available within the cluster. Simulations based upon 3D smoothed particle hydrodynamics (SPH) predict high accretion rates within the center of the cluster where gas densities are high and velocity dispersions low, as illustrated in Fig. 7.8 (Bonnell, Clarke, & Bate 2006). Indeed, massive stars form preferentially within the centers of stellar clusters – known as primordial mass segregation – where the high densities and pressures lead to a small thermal Jeans mass.

This scenario has the advantage that it both explains why O and B stars form almost exclusively in rich clusters (de Wit *et al.* 2005) and predicts a spectrum of stellar masses (from initially equal mass particles) which closely mimics the observed form of the initial mass function (e.g. Bonnell *et al.* 2006). A large fraction of close, massive binaries naturally follows, also in broad agreement with observations.

One major deficit of these SPH simulations is that solely hydrodynamics and gravity have been considered to date, i.e. once material comes within a specific distance of a protostar it is assumed to be accreted. It has long been established that radiation pressure hinders, or perhaps completely halts, accretion from protostars more massive than $\sim$10$M_\odot$. In the above picture, stellar mergers might occur in drag-induced binary systems. Alternatively, Yorke & Sonnhalter (2002) present 2D simulations for the hydrodynamic collapse of a rotating

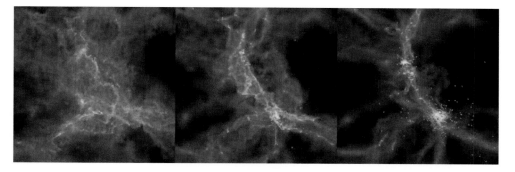

Fig. 7.8 The evolution of a $1000M_\odot$ cluster competitive accretion simulation, at times of 0.9, 1.3, and 1.7 t_{ff}, from left to right, respectively. The column density is plotted between 0.0075 and 150 g cm^{-2}. From Bonnell *et al.* (2006).

molecular cloud clump in the presence of a circumstellar disk. They conclude that massive stars of $\geq 20M_\odot$ may be formed via accretion through a disk as a result of the non-isotropic distribution of radiative flux, in which the radiative force is strong at the poles and weak at the equator. Observations of some high-mass star forming regions reveal evidence for massive molecular outflows with properties in common with the low-mass case (e.g. Beuther *et al.* 2002). Subsequently, Krumholz, McKee, & Klein (2005) demonstrate how massive outflows provide an optically thin cavity through which radiation may freely escape, effectively circumventing the radiation pressure limit to protostellar accretion. Rather contrary to intuition, outflows which drive material away from the protostar may increase the mass of the final star.

In contrast to the competitive accretion scenario, the turbulent core model – an extension of the classic low-mass paradigm – proposes that most of the mass that goes into a star was initially gravitationally bound to it. High accretion rates (of order $10^{-3}M_\odot$ yr^{-1}) follow from supersonic turbulent motions and high pressures within clusters, providing the necessary rapid formation timescales (McKee & Tan 2003). A snapshot of one such simulation is presented in Fig. 7.9.

One deficiency with this general scenario is that massive stars might be expected to preferentially form in relative isolation, rather than within the centers of dense stellar systems. Indeed, the form of the IMF is not directly predicted by this model but results from the core mass function, which has a similar form (Testi & Sargent 1998). Massive star formation in isolation would seem to imply no episodes of cloud fragmentation, presumably requiring either small, low-mass molecular clouds or a very small initial net angular momentum of the cloud. Perhaps a few of the low-mass tail of the molecular cloud mass distribution are responsible for the small fraction of isolated massive stars.

Both scenarios possess appealing aspects, notably the formation of high-mass stars within the center of massive clusters for the competitive accretion scenario, and rapid formation via high protostellar accretion rates in the turbulent core case. However, stellar feedback and magnetic fields are neglected. Consequently, while both remain credible scenarios for the moment, it is hoped that future observations will permit one case to be unambiguously favored over the other.

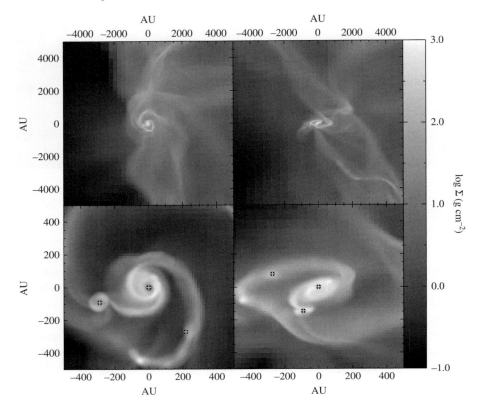

Fig. 7.9 Column density on scales of 10 000 and 1 000 AU (top and bottom) in two projections (left and right) for a simulation of a spherical, initially centrally condensed core of mass 100 $M_\odot$ and 0.1 pc radius core after 2.7×10^4 yr (0.5 mean density t_{ff} times). A triple system with masses 1.8, 8.3, and 0.05 $M_\odot$ has formed in the center. Reproduced from Krumholz, Klein, & McKee (2007) by permission of the AAS.

7.4   Massive stellar clusters

The majority of young open clusters in the Milky Way are fairly modest affairs, low density and low mass. The Pleiades star cluster (M45), for example, contains a few hundred stars including a few intermediate mass B stars and has an age of $\sim$100 Myr. Within the Milky Way, massive open clusters are rare, although several are visible to the naked eye.

To northern hemisphere observers, the Trapezium cluster within the Orion nebula (M42) and NGC 6611 (M16) within the Eagle nebula are the brightest young massive clusters (Table 7.3). In the southern sky, Trumpler 14 and 16 within the Carina nebula (NGC 3372) and NGC 3603, are spectacular examples of open clusters. It is apparent that massive clusters are located within larger star forming complexes, typically OB associations or Giant HII regions, which will be discussed in Chapter 9.

Newly born clusters

We will take the phrase "newly born clusters" to represent those that contain at least some stars still buried in their natal dust clouds or having substantial material remaining in their natal disks (i.e. containing MYSOs). It has been found that these clusters typically contain stars with a range of evolving environmental properties and it may be possible to list them in order of age. When discussing cluster ages, it is common to assume that *all* stars are born together at the *same time*. Examples of newly born clusters are the radio selected GHII regions W49A, W31, and W43 and the optically visible cluster M17; we will argue below that this is also their age order. Note again the nomenclature used for three of these refers to an environmental property, the radio free–free emission by which these clusters were discovered (M17 is also a strong radio source).

The lifetime of the newly born cluster phase is thought to be less than 10^6 years; while the total O star phase is something like 10^7 years. The lifetime of the most luminous radio emission phase is a *few* 10^6 years, less than the O-star lifetime (because the most luminous stars in the cluster are already beginning to die out). Recall that in our Galaxy the GHII phase is the most readily observed through its radio emission, given that the Galactic dust is transparent to this wavelength while optical and even NIR wavelengths are relatively opaque in the Galactic plane. GHII regions can be found throughout the Galaxy, even well beyond the center, but optically visible ones are only observable relatively nearby. Near-IR wavelengths can probe intermediate distances. Recent near-IR imaging surveys of radio selected GHII regions have been undertaken by Blum and colleagues (Blum, Damineli, & Conti 1999, 2001). In most of them, a cluster or clusters are found embedded in the center of the radio source. In a few GHII regions, no cluster is seen at the K-band, but these objects are typically the most distant or located beyond the Galactic Center such that the intervening line of sight dust is likely absorbing the stellar near-IR.

W49A Figure 7.10 presents a composite JHKs image of the W49A region (G43.2 + 0.0), one of the brightest Galactic giant HII regions – W49B is an unrelated SNR outside of the field of view. While the image is centered on the main radio source, there is *no* obvious cluster at that position and most of the stellar images are foreground and *not* related to it. Careful examination of the (false) color image does reveal a number of very red sources, scattered towards the center of this field, and probably related to W49A. In Fig. 7.10 several resolved sources are labeled with an explanation as follows: W49A South is an individual HII region in which several stellar sources are seen buried within the "fuzzy" resolved object (the HII region). Similar remarks can be made about objects CC, S, Q, and O3. The stellar sources are faint in the K-band, so IR spectra have not yet been published. The "Welch ring" represents the location of a cluster of radio detected hypercompact and UCHII regions which are arranged in an oval morphology. Object F is sufficiently bright that a K band spectrum has been obtained (Conti & Blum 2002), revealing an MYSO (star J2 is probably similar). *None* of the other radio sources in the Welch ring are visible in K but most appear as point sources at MIR wavelengths. However, the western sources of the ring are not visible even at 20 μm, leading one to suppose they are buried within pervasive (cluster?) dust which absorbs their radiation. One has the impression that this cluster is in the very youngest stages of star birth, in which only sources F and J2 are visible in the K band. This might be an effect of clumpiness or orientation (cf. Fig. 7.6).

Careful examination of Fig. 7.10 reveals a faint cluster of stars to the northeast of the O3 HII region; this is not so obvious in the black and white reproduction. The over-arching view of W49A, then, is that it is a close association of a number of individual HII regions, of which the Welch ring is the strongest in the radio regime. Conti & Blum (2002) have used

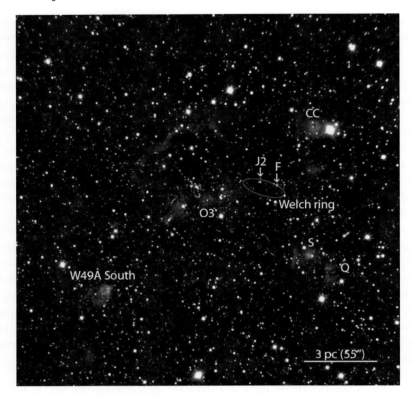

Fig. 7.10 Composite JHKs-band imaging of the W49A region (5×5 arcmin), in which several radio continuum sources from De Pree, Mehringer, & Goss (1997) are indicated, including two UCHII regions (J2 and F) in the Welch "ring" (Welch *et al.* 1987). The central cluster lies to the northeast of the O3 HII region. Reproduced from Alves & Homeier (2003) by permission of the AAS.

the radio properties of the hypercompact and UCHII regions in this image to investigate their evolution. They suggest that in a plot of radio luminosity vs. physical size one finds that the larger the HII region, the more likely the central exciting object is to be seen in the K band. One expects that as an UCHII region evolves, the star will provide more or less the same (Lyman) luminosity but that the cocoon will expand and the dust will thin out. Thus a HII region will get larger and the central object will become more visible at K. *If* all massive stars in W49A were born at the same time, then the size of the HII region will indicate its age. In W49A the HII region sizes are different with no dependence on the luminosity but with some connection to location – those to the "outskirts" being more visible. Other effects may be at play: It could be that initially higher density regions take longer to "clear out", or clumpy/orientation effects dominate, or perhaps stars in W49A, which is a physically large association, do *not* all form at quite the same time. In any case, in contrast to the next newly born clusters we will examine, the radio sources in W49A are mostly invisible in K.

W31   Figure 7.11 presents K-band imaging of a star cluster (G10.2–0.3) within the W31 star forming region. This cluster is readily visible in the K band but is not obvious at

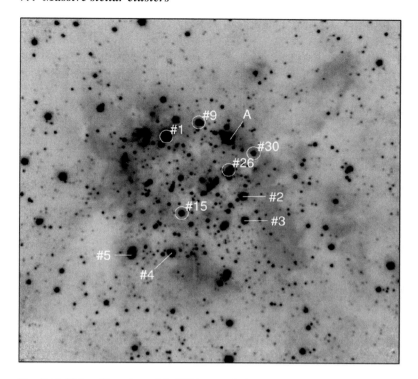

Fig. 7.11 K-band imaging of the W31 central cluster (G10.2 – 0.3), in which a number of O stars (2, 3, 4, 5), massive YSOs (1, 9, 15, 26, and 30) and a foreground star (A) are identified. Reproduced from Blum, Damineli, & Conti (2001) by permission of the AAS.

all in the H (or shorter wavelength optical) bands. Given that hot stars have similar intrinsic JHK magnitudes, this illustrates the impact of dust extinction on detection of massive stellar clusters in the Galactic plane. K-band spectra have been published for W31 1 and four other members of W31; the latter show O-type spectra which can be classified. W31 1 has evidence of surrounding natal material (e.g. Br series emission, [Fe II] emission, K-band excess) and is probably an MYSO B-type star. A JHK color–color plot of W31 reveals some stars, including the known O types, along the normal reddening line, and a number of other sources with K band excesses. These color criteria typically indicate MYSOs with natal material still present in each of them. In contrast to W49A, already some O-type stars in this cluster are free of their natal dust, while others remain partially embedded but still visible in K. Thus W31 must be older than (at least some parts of) W49A on the basis of its stellar constituents. Ghosh *et al.* (1989) have carried out a radio survey of W31 and discovered a number of UCHII regions. Several of these are coincident with K band point sources which are MYSOs, but several others have no stellar counterparts. This cluster is under further investigation.

M17 Hanson, Howarth, & Conti (1997) obtained JHK photometry and K-band spectroscopy of the very nearby massive star cluster M17. Parts of this object are seen optically but other portions are revealed only with NIR imaging. The JHK color–color plot indicates that many stars lie along the normal reddening line, with some others having K-band excesses. The line of sight extinction to this cluster is lower than others discussed in this section so a large number of K-band spectra have been obtained. Many O stars are identified at this wavelength from their absorption lines while a few other objects have K-band excesses and spectra indicative of an MYSO nature (e.g. CO emission features, Brγ and He I emission, etc.). Optical spectra of a few of the MYSOs reveal underlying late-O and early-B type stellar features (which were otherwise lost in the K band due to dilution of the stellar continuum by the dust emission). In addition, the optical spectra included classical "shell" lines in the Balmer series which have been already known in Be stars with disks. As such, this cluster is more evolved than W49A and W31.

All these massive clusters are younger than 1 Myr, but it is not yet possible to assign specific numbers as the modeling of the stellar environment is not yet complete and, even more importantly, the observational data are lacking. What is needed is a complete description of the spectral energy distributions of these clusters with 3D spatial dependence along with knowledge of the gas and dust content. Fuller knowledge of the properties of the exciting stars would be another parameter which would help age date the clusters.

Young massive clusters

Within the disk of the Milky Way, the highest mass young clusters are typically located within giant HII regions, either optically visible or selected from radio surveys. Classic examples are Trumplers 14 and 16 within the Carina nebula, and the centrally condensed cluster NGC 3603 with a core stellar density of $10^5 M_\odot \, \mathrm{pc}^{-3}$. Their ages are only a few million years, since they host early O-type main sequence stars and hydrogen-rich WN stars with masses of order $50–100 M_\odot$, whilst their total stellar masses are somewhat inferior to $10^4 M_\odot$.

Another example is the central cluster (G30.8–0.2) within the W43 giant HII region. Figure 7.12 presents K-band imaging of the star cluster. Near-IR spectra of the three brightest stars reveal the most luminous one to be a WN7 star (some broad emission lines indicative of a strong wind) while the other two are normal O-type stars. An examination of the JHK color–color plot indicates that most cluster members are along the reddening line, while only a few have K band excesses. The general impression is that while some MYSOs are present, their fractional numbers are relatively low compared to, say, W31. The W-R nature of W31#1 suggests a larger age to the cluster, but it is more likely a very luminous H-burning star that has strong stellar winds throughout its evolutionary history, analogous to those seen in similar clusters within Carina OB1 (Section 2.4), with a similar age of a few Myr.

Near-IR surveys during the 1990s revealed three previously unknown young massive clusters close to the Galactic Center, named the Arches, Quintuplet and Galactic Center clusters, with somewhat higher masses of order $10^4 M_\odot$ and ages in the range 2–5 Myr, as shown in Table 7.4. It should be emphasized that masses and radii are calculated in different ways for these clusters. For example, the stellar mass of R136 is $2 \times 10^4 M_\odot$ including solely stars above $2.8 M_\odot$, but a factor of three higher if the measured IMF if extrapolated to $0.1 M_\odot$.

R136 aside, the Arches cluster is notably the youngest, and perhaps the most centrally concentrated (see Fig. 7.13), with a stellar mass of $5 \times 10^3 M_\odot$ above 6 $M_\odot$, or a total stellar mass of $1.2 \times 10^4 M_\odot$ extrapolating the IMF to 0.1 $M_\odot$. Large numbers of early O stars and H-rich WN stars are observed in the Arches cluster from near-IR spectroscopy. The location of this cluster, only a few tens of pc from the Galactic Center means that it will be disrupted

Fig. 7.12 Near-IR imaging of the W43 central cluster (G30.8–0.2) in the K-band, in which three stars No. 1 (WN7), 2 (early O), 3 (early O) are identified. Several separate regions make up the W43 radio complex, not all at the same distance. Reproduced from Blum, Damineli, & Conti (1999) by permission of the AAS.

Table 7.4 *Comparison of young, massive clusters from Hunter et al. (1995), Figer et al. (1999), and Stolte et al. (2006) and references therein.*

Cluster	M_{clu} ($M_\odot$)	Size (pc)	r_{core} (pc)	ρ_{core} ($M_\odot\,pc^{-3}$)	Age (Myr)	MF α
Orion (M42)	$\geq 10^3$	3	0.2	10^4	≤ 1	2.2 ± 0.1
NGC 3603	$\geq 7 \times 10^3$	4.4	0.2	10^5	1–3	1.9 ± 0.1
Quintuplet	$\geq 10^4$	–	0.5	10^4	3–5	–
Arches	10^4	1	0.2	10^5	2–3	1.9 ± 0.15
R136a	5×10^4	4.7	~ 0.1	10^5	1–5	2.3 ± 0.1
Antennae	$10^4 - 10^6$	1–10	–	10^3	1–20	–

over a timescale of order 10^7 yr, due to the intense tidal field, according to N-body simulations by Portegies-Zwart *et al.* (2001).

The Quintuplet and Galactic Center clusters are somewhat older systems, as witnessed by the presence of ~ 20 Wolf–Rayet stars and 1–2 RSGs within each. Indeed, the five cocoon

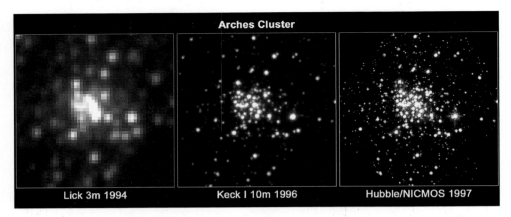

Fig. 7.13 Near-IR images of the Galactic Center Arches cluster obtained with increasing spatial resolution from left to right (see Figer *et al.* 1999). STScI Press Release STScI-2005-05.

stars after which the Quintuplet cluster was named have been revealed to be dusty "pinwheel" WC binary systems. In addition, the Quintuplet cluster hosts two LBVs, including the "Pistol star" which possesses a prominent ejecta nebula and is apparently amongst the most luminous stars presently known (Figer *et al.* 1998).

Beyond the Galactic Center region itself, further high mass young stellar clusters have been identified via near-IR imaging surveys and follow-up spectroscopy. Notably, Westerlund 1 (Wd1), discovered optically by Bengt Westerlund in the 1960s (recall Fig. 2.7) possesses a veritable zoo of massive stars, which includes 24 Wolf–Rayet stars, three RSGs, six yellow hypergiants, an LBV and a B[e] supergiant. The age of Wd1 is $4 - 5$ Myr, such that it is a more massive counterpart to the Quintuplet cluster. It is located close to the galactic bar, with an expected metallicity somewhat in excess of the Solar Neighborhood and a mass potentially as high as $10^5 M_\odot$. As such it is the most massive young cluster yet known in the Milky Way. Ongoing infrared surveys will no doubt identify further high mass young clusters (e.g. Figer *et al.* 2006). Wd1 is *not* a radio GHII region, nor a particularly strong radio source.

In external galaxies, there are numerous young, massive, compact clusters known including R136a at the heart of the 30 Doradus region of the LMC for which a mass of $5 \times 10^4 M_\odot$ has been reported. R136a will be discussed in detail within Chapter 9. Further afield, young high mass (10^5 to $10^6 M_\odot$) compact clusters have been identified in starburst and spiral galaxies, albeit spatially unresolved due to their large distances, collectively known as "super star clusters" (SSCs) or more generally young massive clusters (YMCs). Nearby, well studied examples include NGC 1569-A, NGC 1705-1 and M82-F, whilst large numbers have been observed within the merging Antennae galaxies NGC 4038/9 (Whitmore *et al.* 1999), with typical properties listed in Table 7.4.

Dynamical interaction between stars in young, dense clusters leads to primordial mass segregation, in which the most massive stars quickly sink towards the cluster core. Alternatively, clusters may form in a mass segregated state via the competitive accretion process outlined above. In principle, the core density may become so high ($\geq 10^8$ stars pc^{-3}) that stellar collisions occur at a sufficiently early stage that a runaway collision process occurs from which hypothetical intermediate-mass black holes (IMBHs) may result, with masses of order $10^3 M_\odot$. The density could be considerably different if the encounters occur between protostars with large accretion disks, and still lower if between binary protostar systems with a third star or protostar. Both binary systems and protostars have much larger cross sections than dwarfs for interactions. Regarding IMBHs, a number of X-ray luminous candidates exist, often labeled ultraluminous X-ray sources (ULXs), including cluster MGG-11 in M82 for which Portegies-Zwart *et al.* (2004) claim a lower mass of 350 $M_\odot$, although this interpretation remains controversial.

For spatially unresolved star clusters we do not have the luxury of being able to measure the mass function (and hence the mass) directly, and so have to resort to indirect mass indicators. Let us assume that the cluster is in virial equilibrium, i.e. a spherically symmetric, gravitationally bound system of equal mass stars and an isotropic velocity distribution. Dynamical masses, M_{dyn} may be estimated using

$$M_{\mathrm{dyn}} = \eta \frac{\sigma^2 r_{\mathrm{eff}}}{G} \qquad (7.14)$$

where η is a dimensionless parameter which depends upon the adopted cluster profile, r_{eff} is the effective radius, and σ is the velocity dispersion along the line-of-sight. η is typically adopted to be 9.75 (see, however, Bastian *et al.* 2006), whilst r_{eff} may be measured from high-spatial resolution HST imaging for sufficiently nearby ($\leq 10-20$ Mpc) clusters, and σ may be measured by comparison of high-resolution spectroscopy of cluster red supergiants with individual templates, for which $\sigma \sim 9$ km s^{-1} due to macro-turbulence within their atmospheres. Therefore, providing the mass of the cluster is not too small ($\sigma > 10$ km s^{-1}), one should be able to extract its velocity dispersion via a cross-correlation technique.

The majority of dynamical mass estimates agree with a Kroupa IMF for the observed light-to-mass ratio at their deduced ages, as illustrated in Fig. 7.14. For a few, relatively young cases, peculiar mass functions are obtained, e.g. top-heavy (i.e. deficient in low mass stars) for cluster F in M82 (Smith & Gallagher 2001). This apparent lack of long-lived low mass stars suggests it will not remain bound over sufficient time to become an old globular cluster. However, N-body simulations by Goodwin & Bastian (2006) indicate that such young (≤ 50 Myr) clusters are undergoing violent relaxation, as a result of rapid gas removal due to ionizing radiation, stellar winds, and later supernova explosions, such that dynamical mass estimates are in error, and it may *not* be necessary to resort to non-standard initial mass functions in such cases. For star formation efficiencies of $\leq 30\%$ clusters are predicted to become completely unbound (Fig. 7.14). We shall return to the topic of young massive extragalactic clusters in Chapter 9.

Regarding the mortality of star clusters, Oort (1958) noted a lack of clusters older than a few Gyr in the Solar Neighborhood, for which masses rarely exceed $10^3 M_\odot$. Theory predicts that the survival time of clusters depends on their initial mass, in the sense that low mass clusters dissolve quicker than massive clusters (Spitzer 1958). *N*-body simulations allowing for the rapid gas removal suggest that many clusters may suffer so-called "infant mortality" on time-scales of a few tens of Myr although it is not yet established whether this effect is cluster mass dependent (Whitmore, Chandar, & Fall 2007). Large numbers of stars would be dispersed into interstellar space, potentially explaining the presence of large numbers of B stars within the diffuse interstellar medium in external galaxies. Indeed, Kroupa, Aarseth, & Hurley (2001)

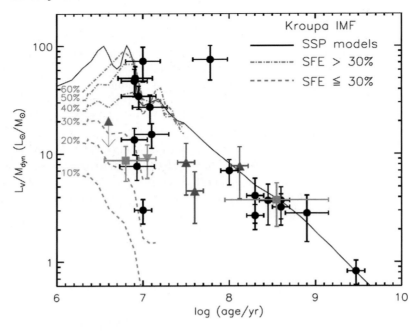

Fig. 7.14 Light-to-mass (V-band) ratios for a number of young extragalactic star clusters for which dynamical masses have been derived (Bastian *et al.* 2006 and references therein). The clusters older than a few 10^7 yr are well fitted by solar metallicity simple stellar population (SSP) models for a Kroupa IMF (solid line). Some of the youngest clusters show significant deviations, apparently due to early, rapid gas removal, from which star formation efficiencies (SFE) may be estimated (dashed lines for SFE $\leq$30%; dash-dotted lines for SFE > 30%). From Goodwin & Bastian (2006).

demonstrate how the Orion nebula cluster may be a precursor of the Pleiades through rapid gas removal followed by the loss of two thirds of the initial stellar content within 100 Myr.

For those clusters surviving for longer timescales, the dissolution time, t_{dis}, depends upon the initial mass, M_i, as

$$t_{\text{dis}} = t_4 \left(\frac{M_i}{10^4 M_\odot} \right)^{0.62}$$

where $t_4 \sim 6.9$ Gyr, based on N-body simulations accounting for stellar evolution and dynamical encounters of stars (Baumgardt & Makino 2003). Empirically, the same exponent *is* observed in the Solar Neighborhood, SMC, M51, and M33, although t_4 differs dramatically from galaxy to galaxy, 0.04 Gyr in M51 and 8 Gyr elsewhere (Boutloukos & Lamers 2003). For the Solar Neighborhood, $t_4 = 1.3 \pm 0.5$ Gyr for clusters with $10^2 \leq M/M_\odot \leq 10^4$ for which the effect dominating cluster distribution timescales appears to be via encounters with giant molecular clouds (Lamers & Gieles 2006).

Overall, observations point to a sequence of massive star formation progressing from the gravitational collapse of a small part of a giant molecular cloud leading to a hot core, involving rapid mass accretion, heating up dust which radiates in the far infrared. Once a massive, visually obscured protostar is capable of ionizing its immediate surroundings, radio observations can detect the free–free bremsstrahlung radiation from its environment. This first involves a hypercompact HII region at which stage mass accretion is close to ending, and later an ultracompact HII region, at which stage accretion has finally terminated. Within the Milky Way, such regions likely host star clusters, typically comparable in content to the Orion nebula cluster (ONC) within larger star forming regions, i.e. OB associations. Star clusters are heavily biased towards low-mass stars, since collapsing molecular clouds fragment into sub-solar mass clumps, which subsequently undergo accretion, for which both competitive accretion and turbulent core models remain credible scenarios. More massive young clusters are known within the Galaxy, albeit in very small numbers. Further afield, in the extreme environments of interacting or merging galaxies (e.g. Antennae NGC 4038/9) large numbers of massive compact clusters are known.

8

The interstellar environment

Several textbooks discuss the interstellar medium in depth, notably *The Physics of the Interstellar Medium* (Dyson & Williams 1997) and *The Physics and Chemistry of the Interstellar Medium* (Tielens 2005). This chapter focuses on aspects relevant to massive stars, namely the properties of interstellar dust, ionized nebulae surrounding individual O stars (giant HII regions are discussed later on), wind blown bubbles and ejecta nebulae around LBVs and W-R stars.

Diffuse gas in the ISM may be in a neutral or ionized form, of which 90% is in the form of hydrogen, either in an atomic, molecular, or ionized state. Cold ($\sim$100 K), atomic hydrogen can be traced via the 21 cm (1420 MHz) hyperfine line, first predicted by van de Hulst (Bakker & van de Hulst 1945) and observed by Ewen & Purcell (1951). This provided the key means of mapping out the structure of the Milky Way and external galaxies. The Lyman series of neutral hydrogen can also be observed in the UV against a suitably hot background source.

Most of the cold ($\sim$10 K) molecular ISM is in the form of H_2. This molecule, however, does not emit at radio wavelengths. Since H_2 is well correlated with carbon monoxide (CO), the CO emission at 1.3 and 2.6 mm is used as a proxy for H_2. From CO radio emission, the distribution of molecular clouds can be observed, which span a huge range of mass in the Milky Way, the most massive and densest of which are relevant for massive star formation. In addition, lines of molecular hydrogen from the Lyman–Werner series can also be observed in the far-UV against a hot background source.

Finally, warm neutral ($\sim$8000 K) or hot (10^5-10^6 K) ionized gas may be observed as UV/optical absorption lines towards suitable background sources. Warm ionized, low density gas is observed in HII regions, most readily at Hα in the visible or via radio recombination lines, e.g. H166α (21 cm), H110α (6 cm), H76α (2 cm).

8.1   Interstellar dust

Observational evidence in favor of dust was inferred from the obscuration or "extinction" of starlight. "Dark" clouds are well known within regions of star formation in the Milky Way, plus dark lanes within external spiral galaxies viewed along their disks. A review article on this topic is given by Draine (2003).

The properties of dust grains are such that their transparency increases at longer wavelengths. Dust absorption is high in the optical and especially ultraviolet regions, and low in the near-IR and especially radio regions. A significant fraction of starlight is absorbed by dust and re-radiated in the infrared.

With the advent of the first space-based UV satellites, the form of the UV/optical interstellar extinction law was first investigated. Extinction is most reliably derived using the pair method – the spectrophotometric comparison of two stars of the same spectral class, for which one star has negligible dust whilst the second is heavily reddened. Comparison of the two spectra permits one to determine the extinction $A_\lambda = 2.5 \log(F_\lambda^0/F_\lambda)$ as a function of wavelength λ, where F_λ is the observed flux and F_λ^0 is the flux in the absence of extinction. This method has been used to measure extinction curves for many sight lines, in many cases from the far-UV to the near-IR.

Seaton (1979) characterized the Galactic extinction law using representative early-type stars with varying extinction characteristics observed with the IUE satellite using the pair method. The average UV/optical extinction law for the Milky Way is presented in Fig. 8.1. Recall from Chapter 2 that the average total-to-selective extinction of the diffuse ISM is given by $R_V = A_V/E(B-V) = 3.1$ where A_V is the absorption in the V-band and $E(B-V)$ is the color excess (Cardelli, Clayton, & Mathis 1989). Individual sight lines vary from $R_V = 2.1$ to 5.6. Two characteristics are apparent. The extinction law may be parameterized by a power law with slope $A_\lambda \propto \lambda^{-\beta}$ where $\beta = 0.9$ for the Milky Way. The second feature is the strong, broad "bump" at $\lambda 2175$ Å. The precise origin of this feature remains uncertain, however it is widely attributed to carbonaceous grains (Draine 2003).

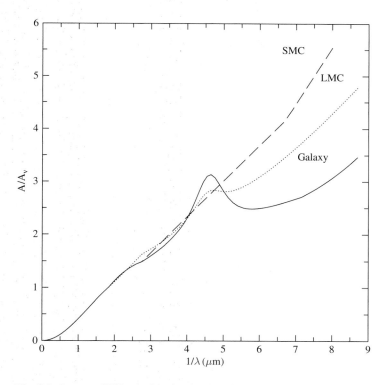

Fig. 8.1 Average UV/optical extinction law for the Milky Way according to Seaton (1979) (solid), together with the LMC extinction law of Howarth (1983) (dotted), and the SMC extinction law of Bouchet *et al.* (1985) (dashed).

IUE has also been used to obtain stellar extinction laws for stars within the Large and Small Magellanic Clouds. These are also reproduced in Fig. 8.1, and indicate steeper slopes than in the Milky Way with $\beta = 1.0$ for the LMC and $\beta = 1.2$ for the SMC, plus a weakening of the $\lambda 2175$ feature as one shifts from the Milky Way to the LMC and SMC. Extinction laws at shorter wavelengths have been studied with FUSE, and support extrapolations of these power laws to the Lyman edge.

The Milky Way and Magellanic Cloud extinction laws described above were obtained for individual stars. In contrast, observations of starburst galaxies at a median distance of 60 Mpc comprise a mixture of stars and gas, for which Calzetti, Kinney, & Storchi-Bergmann (1994) demonstrated differences in extinction properties, in the sense that lower extinction was indicated for the stars than the interstellar gas. This extragalactic starburst extinction law, with an attenuation slope of $\beta = -0.95$, also lacks the 2175Å feature, as shown in Fig. 8.2. This law has been widely applied to nearby starbursts and high-redshift star-forming galaxies. The physical scales sampled by ground-based observations of high-redshift star-forming galaxies are reminiscent of the large aperture IUE observations used to construct the starburst extinction law. The choice of extinction law may have a significant effect upon the resulting stellar population. In the case of a spatially resolved, nearby star cluster, one of the stellar based extinction laws is most appropriate.

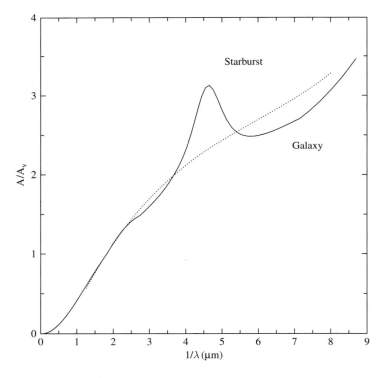

Fig. 8.2 Average UV/optical extinction law for the Milky Way according to Seaton (1979, solid), together with the Calzetti, Kinney, & Storchi-Bergmann (1994) starburst obscuration law (dotted).

Since many early-type stars in the Milky Way are visibly obscured, the IR extinction law has been investigated by Rieke & Lebofsky (1985). Their results support a general power law decrease in extinction at longer wavelength, with $A_\lambda \propto \lambda^{-\beta}$ with $\beta \sim 1.6$, except that other discrete features are present, including the 9.7 μm silicate feature which is analogous to the UV "bump". In Table 8.1 we compare the UV, optical, and IR extinction for a standard Galactic $R = 3.1$ extinction law with $E(B - V) = 1.0$ mag. Typically, an extinction of $A_V = 30$ mag is estimated towards the Galactic Center. From the table, the K-band extinction towards the Galactic Center is only 3.3 mag, while the M-band suffers merely 0.7 mag extinction. Hence studies of the Galactic Center have focused upon IR and longer wavelengths.

Table 8.1 *A comparison of UV, optical, and IR interstellar extinction for a standard $R = 3.1$ Galactic extinction law of Cardelli et al. (1989) and Rieke & Lebofsky (1985) with $E(B - V) = 1.0$ mag. Indebetouw et al. (2005) provide updates to the IR extinction law for the Spitzer IRAC filters.*

Filter	$\lambda(\mu m)$	A_λ	Filter	$\lambda(\mu m)$	A_λ
	0.125	10.20	I	0.90	1.50
	0.180	7.81	J	1.21	0.87
	0.220	9.77	H	1.65	0.54
U	0.365	4.83	K	2.18	0.35
B	0.440	4.11	L	3.55	0.18
V	0.555	3.10	M	4.79	0.07
R	0.700	2.33	N	10.47	0.16

Dust extinction within HII regions may be determined from the observed hydrogen line flux ratios, e.g. Hα/Hβ, since case B recombination theory predicts an intrinsic ratio which is fairly independent of electron density and temperature, i.e.

$$\frac{I(\text{H}\alpha)}{I(\text{H}\beta)} = 2.86. \tag{8.1}$$

If the observed nebular fluxes are $F(\text{H}\alpha)$ and $F(\text{H}\beta)$, the extinction can be deduced from $E(B - V) \approx 0.77c(\text{H}\beta)$, where

$$\frac{F(\text{H}\alpha)}{F(\text{H}\beta)} = \frac{I(\text{H}\alpha)}{I(\text{H}\beta)} 10^{-c(\text{H}\beta)[x(\text{H}\alpha) - x(\text{H}\beta)]} \tag{8.2}$$

with $x(\text{H}\alpha) - x(\text{H}\beta) = -0.346$ for a standard $R_V = 3.1$ extinction law. Similar relations hold for other extinction laws or pairs of hydrogen recombination lines (Osterbrock & Ferland 2006).

In addition to interstellar dust, absorption by interstellar atomic and molecular gas also affects the observed stellar spectrum. In the optical, interstellar CaII H and K and NaI D lines[1] are well known, whilst numerous interstellar lines from ionized metals are seen in the far-UV. The strongest UV interstellar lines are due to neutral hydrogen in the Lyman series

---

[1] The nomenclature "H", "K", and "D" goes back to Joseph von Fraunhofer (1787–1826). Fraunhofer interpreted the dark lines in the solar spectrum as due to absorption by gases in the outer portions of the Sun and in the Earth's atmosphere. Fraunhofer lines are designated by letters, starting with "A" for the Telluric oxygen line at 7594 Å, and ending with the CaII H and K lines at 3968 and 3934 Å, respectively.

(e.g. Lyα at 1215Å) plus molecular Lyman–Werner hydrogen transitions in the λ912–1120Å FUSE spectral range. There is a reasonably well established relationship between the column density of hydrogen (neutral and molecular) and interstellar dust extinction for the Milky Way, namely

$$N(\mathrm{HI})/E(B-V) = 4.8 \times 10^{21} \, \mathrm{cm}^{-2}\mathrm{mag}^{-1}, \tag{8.3}$$

from Bohlin, Savage, & Drake (1978). The dust-to-gas mass ratio is typically assumed to be $\sim$0.01. Similar relationships hold for sight lines towards stars in the Magellanic Clouds, except that the reddening per H atom is a factor of 5 (18) times *lower* than for the Milky Way for the LMC (SMC), respectively (Koornneef 1982; Fitzpatrick 1985).

8.2   Ionized hydrogen regions

Hot, luminous stars affect their circumstellar environment in a variety of ways, most notably by ionizing the surrounding gas. Stellar photons with energies higher than 13.6 eV, the value necessary to ionize a hydrogen atom, are absorbed by interstellar atoms and ions within a distance of many tens of parsecs from the star. Subsequent re-emission gives rise to a rich nebular emission spectrum. Regions where we observe this spectrum are called *HII regions*, and the brightest ones are referred to as giant HII regions (Chapter 9). The brightness of the emission lines makes HII regions detectable at distances from the observer where the stars themselves are hard to see. This is a result of the "packing" of the *stellar* Lyman continuum photons into a few bright *nebular* emission lines. Therefore HII regions have become important diagnostics for the properties of the ionizing stars themselves.

Orion nebula

One of the most striking constellations high up in the mid-summer sky (or the mid-winter sky, for observers in the northern hemisphere) is Orion, the hunter in Greek mythology. Those who can enjoy light-pollution free skies away from metropolitan areas find a relatively bright patch of light near the three Orion belt stars – the Great Orion Nebula. Known since ancient times, it carries the designation M42 in C. Messier's list of 110 nebular objects first published in 1774. The Messier Catalog includes a variety of astronomical objects, such as planetary nebulae, star clusters, gaseous nebulae, supernova remnants, and galaxies. Their distinguishing property is the diffuse, nebular appearance when viewed in a small telescope.

M42 turned out to be what its original name suggested: a gaseous nebula. In 1864, W. Huggins observed M42 and found its spectrum to show many bright lines, indicating that it is a gas cloud emitting light in the same way a neon or mercury vapor lamp does. A modern spectrum of M42 is reproduced in Fig. 8.3. Strong nebular emission lines of hydrogen, helium, nitrogen, oxygen, and other elements are superposed on a faint continuum. This spectrum is typical for an HII region.

M42 is the second-closest HII region in the sky. (The title for the closest region goes to the Gum nebula which is less well studied due to its location in the southern hemisphere.) The Orion region is host to many newly formed – and forming – stars. Among them are several O-type stars. The O stars are not exceptional by galactic standards; their spectral types are O7 and later. At the center of the nebula is a cluster of four stars called the Trapezium. The brightest star in the Trapezium, known as Θ^1 Orionis C, is an O7 V star with $T_{\mathrm{eff}} \approx 39\,000$ K. It is the source of most of the UV radiation which causes the nebula to glow. The Trapezium

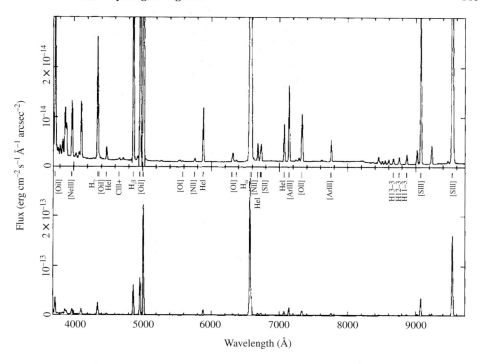

Fig. 8.3 Optical spectrum of the Orion nebula from 3800 to 9400 Å. The scales of the upper and lower sections emphasize weak and strong emission lines, respectively. Reproduced from Baldwin *et al.* (1991) by permission of the AAS.

stars and many other stars in the region formed out of a surrounding cloud of dust and gas only a few million years ago.

The Orion nebula would be quite inconspicuous, were it not for its small distance from us of about 500 pc. Astronomers have discovered numerous HII regions both in our Galaxy and in external galaxies. Many of them are much larger (in terms of physical, not angular size) than the Orion nebula. A famous example is the Tarantula nebula in the Large Magellanic Cloud, which will be discussed in detail in Section 9.2. This *giant HII region* is 100 times more distant than the Orion nebula, yet it appears about equally bright and large to the naked eye. Therefore the energy input into the Tarantula nebula (or 30 Doradus, as it is known to astronomers) must be about 10 000 times larger than that powering the Orion nebula. As we will see, this reasoning is roughly correct: 30 Doradus hosts many more stars, and some of them are among the most luminous and massive known.

We mention here that the Orion nebula is somewhat exceptional in that its surface brightness is a factor of several times higher than expected from its ionizing star cluster. On the average, extended HII regions with similar stellar content should have the same surface brightness, that is, their flux measured in a certain wavelength interval and within a fixed angular area, e.g. 1 arcsec^2, should be similar. The different *apparent* brightness is simply the result of different distances: a more distant HII region appears fainter because it subtends a smaller solid angle than a closer region. Since *surface* brightness is defined as a brightness per unit surface (in the terminology of stellar atmospheres it is an *intensity*), it does not depend on distance. (If an object is moved to a ten times larger distance, its flux will be 100 times smaller but at the same time the solid angle it subtends will be 100 times smaller as well. Therefore the flux per unit solid angle remains the same.) Why then does the Orion nebula behave differently? It is because we neglected the effects of dust grains which can absorb, scatter, and reflect light. Dust is commonly found in HII regions, but the Orion nebula is peculiar in its dust geometry since most of the dust is located behind M42 as seen from Earth. The dust acts like a giant mirror reflecting light back towards us, thereby increasing the surface brightness by a factor of several.

Determination of O star content

Within an HII region, the number of hydrogen ionizing photons emitted per second, $N(\text{LyC})$, is derived from the reddening corrected hydrogen line luminosity using

$$N(\text{LyC}) = \frac{\alpha_B(H^0)\lambda}{\alpha_\lambda^{\text{eff}} hc} L, \tag{8.4}$$

where $\alpha_B(H^0)$ is the case B hydrogen recombination coefficient, $\alpha_\lambda^{\text{eff}}$ is the effective recombination coefficient for the line with wavelength λ, and L is the line luminosity. As extensively discussed by Osterbrock & Ferland (2006), this assumes that the HII region is optically thick to Lyman continuum photons and free of dust (case B ionization-bounded).

If the escape fraction of ionizing photons is f_{esc}, for canonical values of the electron temperature and density, $T_e = 10^4$ K and $n_e = 10^2$ cm^{-3}, respectively,

$$N(\text{LyC}) = 2.10 \times 10^{12} L(H\beta)/(1 - f_{\text{esc}}),$$

where $L(H\beta)$ is in erg s^{-1} and $N(\text{LyC})$ is per second. Similar relations hold for other hydrogen recombination lines, scaled according to their case B recombination coefficients.

Typically, this ionizing luminosity is expressed in terms of the number of "equivalent" stars of a given subtype, normally O7 V,

$$N(\text{LyC}) = N_{O7\,V} N(\text{LyC})^{O7\,V} \tag{8.5}$$

where $N_{O7\,V}$ is the number of O7 V stars present in the region and $N(\text{LyC})^{O7\,V}$ is the Lyman continuum luminosity produced by a typical O7 V star. Historically, a typical O7 V star has been considered to possess an ionizing flux of 10^{49} photon s^{-1}, although $10^{48.86}$ photon s^{-1} was suggested by Panagia (1973). Recent calibrations for Galactic O stars (recall Table 3.1), suggest a somewhat lower ionizing flux of $N(\text{LyC}) = 10^{48.8}$. $N(\text{LyC}) = 10^{49}$ photon s^{-1} *is* probably a reasonable value for O7V stars at low metallicity, since they are somewhat hotter than Galactic counterparts.

For a larger stellar complex, such as a starburst galaxy, it is possible to estimate the actual number of O stars by taking the (present-day) mass function into account, which relates to the equivalent number of O7 V stars via

$$\eta_0(t) = N_{O7\,V}/N_O \tag{8.6}$$

for which Fig. 8.4 provides the time evolution of $\eta_0(t)$ using a standard Salpeter IMF for a variety of metallicities. This stays fairly constant for the first 2 Myr of a burst, and then decreases due to the evolution of the O star populations and subsequent disappearance of the most massive stars. At larger times, the number of O stars drops to zero, such that $\eta_0(t)$ becomes meaningless.

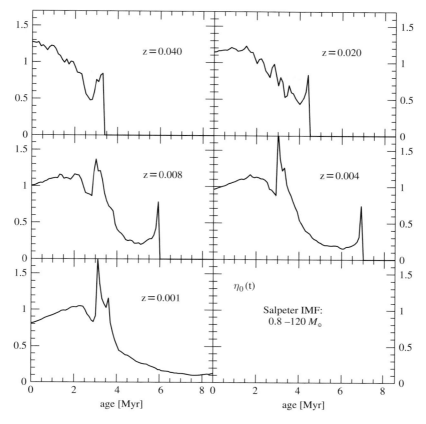

Fig. 8.4 Evolution of $\eta_0(t)$, defined by Eq. (8.6), with age for a variety of metallicities and a high-mass Salpeter IMF. Reproduced from Schaerer & Vacca (1998) by permission of the AAS.

8.3   Wind blown bubbles

A typical massive O star emits about one third of its total luminosity at wavelengths below 912 Å in the Lyman continuum. If this O star were observed not in isolation, but as part of a stellar population having an age of a few Myr and following a standard Salpeter-type IMF, the fractional emission of this population below 912 Å would be about 20%. The output of highly energetic photons with energies above 13.6 eV, the amount required to ionize hydrogen, shapes the thermal balance of the surrounding ISM and gives rise to the spectacular HII region spectra and the associated phenomena.

Hot, massive stars have stellar winds with momentum fluxes $\dot{M}v_\infty$ of about 30% of the radiative momentum L_{Bol}/c. The mechanical luminosity of the wind, $1/2\dot{M}v_\infty^2$, is only a small fraction (few percent) of L_{Bol}. Therefore the mechanical luminosity of an O star, while not negligible, is not dominant in energizing the ambient ISM. This changes 5 to 10 Myr after the star was born. First, O stars evolve into W-R stars, whose powerful winds have mechanical luminosities reaching 30% and higher of the stellar luminosity. Second, supernovae with their prodigious, instantaneous mechanical energy release appear at that epoch.

The mechanical energy provided by one supernova event is about 10^{51} erg. This is comparable to the wind energy of a massive O star integrated over its lifetime of 10^7 yr. Recall from Table 4.2 that typical O star mass-loss rates are $\dot{M} = 10^{-6} - 10^{-5}\,M_\odot\,\mathrm{yr}^{-1}$ with wind velocities of $v_\infty = 2000$–3000 km s^{-1}. Finally, the stellar ionizing thermal luminosity decreases with time since the overall evolutionary trend is towards lower effective temperature. Therefore the relative proportion of the mechanical and thermal luminosity increases with time, with the two quantities being comparable at about 10 Myr. After that epoch, the powering of the interstellar environment becomes almost totally determined by the non-thermal energy input. These processes are the topic of this section.

The mathematical description and physical interpretation of the interaction between a stellar wind and the surrounding interstellar gas were given by Castor, McCray, & Weaver (1975) and Weaver *et al.* (1977). Initially the stellar wind is in a brief phase of *free expansion* that lasts until a sufficiently large mass of interstellar gas has been swept up to stop the expansion. The duration of this phase is a few hundred years, at which time a bubble of radius 0.1 to 1 pc has formed. This is too short to be of observational significance so that this phase is of academic interest only. The resulting bubble has a distinct three-zone structure: the expanding wind, the shocked wind close to the interaction zone with the ambient ISM, and the shocked ISM itself. Both the shocked wind and the shocked ISM are at temperatures above 10^7 K. Radiative cooling is negligible at such high temperatures because very few cooling mechanisms (such as photon emission in spectral lines) exist in this energy regime. Therefore radiative energy losses are small and the shocked zones will expand adiabatically. This second phase is therefore called the *adiabatic phase*. The adiabatic phase will come to an end when radiative cooling in the swept-up ISM is no longer negligible. We can estimate the epoch for this to become important by calculating the timescale for radiative cooling:

$$t \approx 1.7 \times 10^3 \sqrt{\frac{L_\mathrm{W}}{n_0}}. \tag{8.7}$$

Here t is the age in years, L_W is the wind luminosity in 10^{36} erg s^{-1}, and n_0 is the ISM density in cm^{-3}. Inserting typical numbers, we find that the adiabatic phase of the swept-up ISM lasts only a few thousand years before radiative cooling becomes important. At that time the bubble has a radius of order 1 pc. Similar to the phase of free wind expansion, the probability of observing the adiabatic phase is negligibly small.

As soon as the shell of swept-up interstellar material has reached a temperature somewhat below 10^6 K, radiative cooling becomes even more important, and the shell rapidly cools down to 10^4 K. The bubble has reached the *radiative phase*, also called the "snow-plow" phase, which is long enough and which produces a bubble large enough to be detectable by observations. A schematic of the velocity, density, and temperature structures of the bubble is in Figs. 8.5 and 8.6. The two figures are for so-called energy and momentum conserving bubbles. Since the former is the astrophysically more relevant case, we will first discuss an energy conserving bubble. The term "energy conserving" indicates that radiative losses of the *hot compressed wind* are negligible so that the bubble is driven by the gas pressure of the hot interior. This should not be confused with the definition of the radiative phase itself, which is tied to the properties of the *thin, swept-up ISM shell*. This shell is radiative, both for an energy and a momentum conserving bubble.

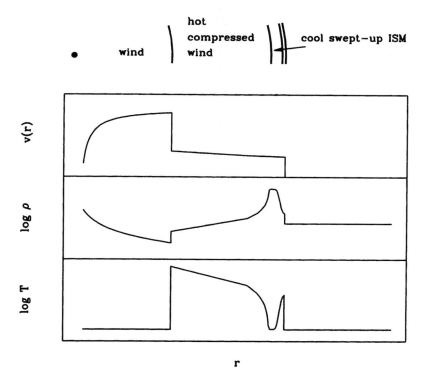

Fig. 8.5 Schematic of an energy conserving bubble. The top panel indicates the geometry of the star, the wind, the swept-up shell, and the surrounding ISM. The three curves show the run of velocity (upper), density (middle), and temperature (lower). The horizontal axis is not to scale. Reproduced from Lamers & Cassinelli (1999) with permission of Cambridge University Press.

The energy conserving bubble in Fig. 8.5 is driven by *gas pressure*. Its four main components are the cool expanding wind, the hot shocked wind, the cool swept-up ISM, and the ambient ISM. The density and temperature determine if and how a particular structure is observable. For the material to be observable in X-rays, it must be hot and have reasonably high emission measure (i.e., density). Figure 8.5 suggests that the hot wind zone has the most favorable conditions for the detection of X-rays. We expect to see an extended X-ray emitting sphere at soft X-rays. In contrast, the optical emission will preferentially come from the swept-up ISM because the intersection between the two conditions of large emission measure and temperature $\approx 10^4$ K is met in that region. Observationally we expect a thin ring of emission around the central star, possibly with some weaker emission inside. The boundary between the cold ($\sim 10^4$ K) shell and the ISM is an isothermal shock. The high compression rate leads to a very thin structure. The hot shocked wind region is still too hot for radiative cooling and continues to expand adiabatically. The temperature gradient at the boundary between the hot wind region and the cold shell is so large that thermal conduction by electrons transfers heat to the shell. As a result, the inner surface of the shell evaporates and material flows into the hot wind region, increasing the mass and lowering the temperature in the hot region. Notice that the density and the mass in the shell are orders of magnitude higher than in the shocked wind region.

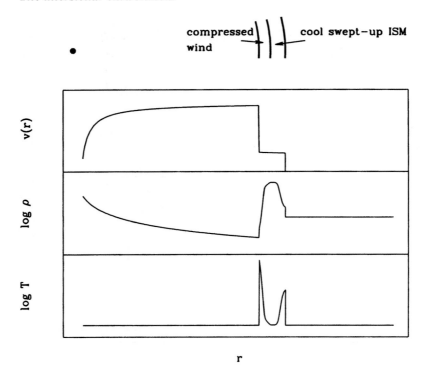

Fig. 8.6 Schematic of a momentum conserving bubble. The individual panels show the same items as in Fig. 8.5. Reproduced from Lamers & Cassinelli (1999) with permission of Cambridge University Press.

The evolution of the structure can be described by solving the equations of momentum and energy. The most interesting quantities are the radius of the bubble $R(t)$, its expansion velocity $v(t)$, and the thickness of the swept-up shell $d_S(t)$. Weaver *et al.* (1977) found:

$$R(t) = 28 \left(\frac{L_W}{n_0} \right)^{1/5} t^{3/5}, \tag{8.8}$$

$$v(t) = 17 \left(\frac{L_W}{n_0} \right)^{1/5} t^{-2/5}, \tag{8.9}$$

$$d_S(t) = 3 \times 10^{-4} T_S \mu^{-1} \left(\frac{L_W}{n_0} \right)^{-1/5} t^{7/5}. \tag{8.10}$$

$R(t)$ is in pc, L_W in 10^{36} erg s^{-1}, n_0 in cm^{-3}, t in Myr, $v(t)$ in km s^{-1}, $d_S(t)$ in pc, and T_S in K. μ and T_S are the mean molecular weight and the temperature of the shell, respectively. A typical bubble has a size of tens of pc, expands at velocities of tens of km s^{-1}, and has a shell thickness of a few pc. The thickness of the shell depends strongly on its ionization state. It shrinks from a few pc to a fraction of a pc if it recombines from ionized to neutral, with a corresponding drop of T_S from 10^4 K to 10^2 K. This numerical example assumes typical O-star parameters. Corresponding bubbles develop around W-R stars, the differences being somewhat larger bubble sizes and higher expansion velocities. L_W of W-R stars is 1 to 2 orders of magnitude larger but since $R(t) \propto L_W^{1/5}$, the bubble has nearly the same size. The case of a W-R bubble is somewhat more complicated because W-R stars are surrounded by and interact with circumstellar material from their progenitors. The predicted morphologies of the wind-blown bubbles are in reasonable agreement with actual observations.

After a few 10^6 yr the mass-losing star will evolve into a RSG if its initial mass was less than $\sim 30\,M_\odot$. The wind velocity drops drastically in this stage. The bubble can continue its expansion until it stalls by the external pressure of the ISM. However, the RSG will explode as a type II supernova and repressurize the bubble. If the star has an initial mass above $30\,M_\odot$, it will likely form a supernova immediately following the hot-star phase.

If radiative cooling in the shocked wind region becomes dominant, the interior bubble rapidly cools down and ceases driving the swept-up shell by its gas pressure. Since the wind is expanding, the shell is driven by dynamic pressure. We call this case "momentum conserving". The resulting bubble geometry is in Fig. 8.6. Most wind-blown bubbles and outflows powered by stars and starbursts are "energy conserving", i.e., their interiors are so hot that radiative losses are small. However, cooling is not entirely negligible, for instance if the mass of the interior bubble is increased by cold exterior material flowing into the hot shocked wind zone. This process is called mass loading and results in some mass deposition at the boundary layer due to inhomogeneities and instabilities. Figure 8.6 suggests that the main observational difference between an energy and a momentum conserving bubble is the size of the X-ray emitting region. Since most of the compressed wind zone is not at coronal temperatures, the X-ray emission is significantly reduced. The hot compressed wind is confined to a thin shell close to the swept-up ISM.

The physics of the interaction between supernovae and the ISM is analogous to the stellar wind case. Review articles are those of Woltjer (1972), Chevalier (1991), and Shull (1993). The main difference between the two processes is the timescale of the energy release. The stellar-wind phase lasts for several Myr and provides a constant supply of mechanical luminosity during that period, while a supernova event occurs essentially instantaneously. Since the total luminosities associated with a stellar-wind-blown bubble and a supernova bubble are similar, supernova bubbles have properties similar to those described above.

In Fig. 8.7 we reproduce an optical image of the wind-blown bubble NGC 7635 surrounding the O6.5 IIIf star BD $+60°2522$. The star and the bubble are embedded in the HII region S162. Both the bubble and the HII region are thought to be ionized by BD $+60°2522$. The morphology and the kinematics of the wind-blown bubble are roughly consistent with the theoretical picture of an energy conserving bubble as described above. The shell is composed of interstellar material swept up by the powerful stellar wind of the mass-losing O star.

Very few examples of wind-blown bubbles around individual O stars as striking as that in Fig. 8.7 are known. The ISM is dusty, inhomogeneous, and has density gradients. As a result, idealized spherical bubbles may rarely form, and if they do, they are difficult to detect. More importantly, massive stars do not form in isolation but in dense clusters. The combined effect of multiple, overlapping stellar winds leads to the creation of so-called "superbubbles" whose physical scales are an order of magnitude larger than those of single wind-blown bubbles. Such superbubbles will be discussed within the context of starburst phenomena in Section 10.4.

Fig. 8.7 Narrow-band [OIII] + Hα + [SII] image of the stellar-wind-blown bubble NGC 7635 and the surrounding nebulosity S162. NGC 7635 is the sphere in the image center; it has a diameter of ∼3′, corresponding to about 2 pc. The central star BD + 60°2522 is the brightest, overexposed star inside the bubble. Reproduced from Moore *et al.* (2002) by permission of AAS.

8.4   Ejecta nebulae around LBVs and W-R stars

Luminous massive stars are often surrounded by circumstellar material whose morphology resembles that of, e.g., nebulosities around AGB stars. The phenomenon is particular prevalent among LBVs and W-R stars. These stars are known for their irregular eruptions and strong winds, and the ejected gas can be linked to the extended atmospheres of these stars. In this respect, ejecta nebula are fundamentally different from stellar-wind blown bubbles: the optical emission of the former is due to stellar material whereas in the latter case the emitting matter is predominately interstellar. A recent review of the subject was given by Chu (2003).

Spectacular ejecta nebulae have been found around LBVs. The famous Homunculus nebula around η Car is a bipolar nebula which was ejected during the giant eruption of η Car in 1843 (Fig. 2.4). As of today, most LBVs are known to be surrounded by ejected nebulosities, indeed Bohannan (1997) proposed that the presence of an ejecta nebula may be a defining characteristic of LBVs. In Table 8.2 we summarize the properties of several well studied examples.

Table 8.2 *Properties of eight well-studied LBV ejecta nebulae. From Nota &*
Clampin (1997).

LBV	Distance (kpc)	Size (pc)	v_{exp} (km s^{-1})	Dyn. timescale (yr)
P Cygni	1.8	0.2	140	10^2–10^3
η Carinae	2.6	0.2	600	150
HD 168625	2.2	0.13×0.17	40	10^3
HR Carinae	5	1.0	~100	5×10^3
AG Carinae	6.1	1.1×0.96	70	10^4
He 3-519	8.0	1.14	60	10^4
R127	51	1.9×2.2	28	4×10^4
S119	51	1.9×2.1	25	5×10^4

Fig. 8.8 Composite multi-band image of AG Carinae highlighting filamentary structures in
the ejecta nebula (Nota, private communication).

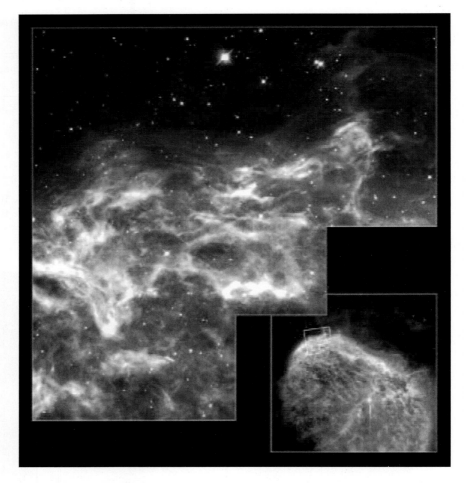

Fig. 8.9 HST narrowband (large) and ground-based (small, lower right) image of NGC 6888, the nebula surrounding the W-R star HD 192163 (WR 136). The HST image is $150''$ (corresponding to about 1 pc) on a side. Reproduced from Moore, Hester, & Scowen (2000) by permission of the AAS.

The sample in Table 8.2 includes LBVs both in the Galaxy and in the LMC. (The latter are at a distance of 51 kpc.) The nebulae have sizes of order 1 pc and expansion velocities of tens to hundreds of km s^{-1}. Assuming constant outflow velocities, the derived dynamic timescales are of order 10^2–10^4 yr. This is consistent with the recurrence timescale of LBV outbursts.

A detailed high-resolution image of AG Car is reproduced in Fig. 8.8. The AG Car nebulosity is a good proxy for the morphology of most LBV ejecta nebulae. The shell-like structure shows enhanced brightness aligned with a preferred axis of symmetry. Nota *et al.* (1995) interpret the observed morphologies as due to a density contrast in the LBV vicinity. This contrast could result from a circumstellar disk or a prior RSG wind. Alternatively, an undetected binary might be responsible. The chemical composition and kinematics of the ejecta nebula provide valuable clues for the evolutionary connections between LBVs and their predecessors. Chemical abundances derived from emission nebulae are often less model dependent than those determined for the star itself whose atmosphere is subject to significant non-LTE conditions. Depending on the age of the nebula, the observed abundances are tracers of the predecessor surface abundances and allow one to infer the footprints of envelope convection and/or rotationally induced mixing (Lamers *et al.* 2001). The observed abundance patterns, together with the observed kinematics, favor a blue supergiant predecessor for LBVs with ejecta nebulae.

About half of all known Galactic W-R stars are surrounded by circumstellar nebulosities. One of the most prominent examples is NGC 6888 and its central star HD 192163. Its close proximity of 1.5 kpc makes it an ideal target for observations at high spatial resolution. The HST image in Fig. 8.9 shows intricate detail in the outer shell. The cavity inside the shell was probably formed by the interaction between the current W-R wind and the slower wind of the W-R progenitor. The nature of the progenitor is still under debate. Previously, the progenitor was assumed to be an RSG based on the abundance pattern, morphology, and kinematics. Helium and nitrogen are both found to be overabundant, and oxygen is underabundant. This pattern is roughly consistent with the stellar evolution prediction for RSGs. More recent models, however, suggest that rotating blue stars may produce a similar abundance pattern. In this case, WR 136 may have never evolved to the red side of the H-R diagram but rather become a W-R star immediately following the blue supergiant phase. The slow predecessor wind inferred from the shell-like structure of NGC 6888 would have been produced by an LBV progenitor.

The WN8 star WR 124 is surrounded by the irregularly shaped nebulosity M1-67 (Fig. 8.10). Since both the star and the nebula have the same radial velocity of about 200 km s^{-1}, there

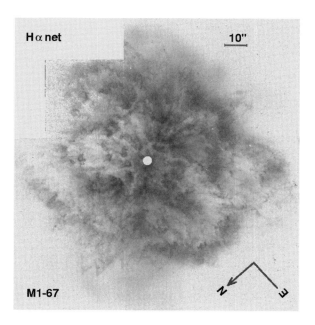

Fig. 8.10 HST WFPC2/Hα image of M1-67. The physical scale is 3×3 pc in which the saturated central star WR 124 has been replaced by a white disk. Reproduced from Grosdidier *et al.* (1998) by permission of the AAS.

is little doubt of a physical association between the star and the nebula. The estimated age of the nebula is $\sim 10^4$ yr. M1-67 is an example of a WR ejecta nebula that underwent only slight mixing with the surrounding ISM. This provides the opportunity of an uncontaminated measurement of the W-R and W-R progenitor abundances. The determined N and O abundances are higher and lower by factors of several than the standard ISM abundances, respectively. These values are not extreme and make it likely that the nebula was ejected during a prior LBV phase. An LBV origin of the nebular helps understand its kinematics as well. The velocities are small, hinting at a progenitor star with a slow wind, like those observed in LBVs.

Interferometric imaging of the WC star WR 104 has revealed highly axi-symmetric curved plumes of hot dust at distances of tens of AU from the central star (recall Fig. 6.18). The dust is streaming away from WR 104 on a linear spiral trajectory. Observations taken at multiple epochs indicate a rotation period of about 220 days. Most likely, an unseen binary component is responsible for the morphology. The colliding winds of the two components carry unequal momenta, and dust will form on the interface of the winds and be subsequently carried out in the flow.

9

From giant HII regions to HII galaxies

Giant HII regions are extensive regions of ionized hydrogen (and other elements) powered by hot stars. Their high concentration of very massive stars often makes them appear as the optically most luminous structures in galaxies. Nearby giant HII regions allow us to study star formation and evolution in great detail and help us understand unresolved giant HII regions in distant galaxies.

9.1 Giant HII regions: definition and structural parameters

In this section we will concentrate on the properties of giant HII regions, such as the famous 30 Doradus nebula in the LMC. The focus of this book is on hot, massive stars, and we are interested in HII regions because their properties allow us to learn about the stars powering them. As opposed to smaller, Orion-like HII regions, giant HII regions are sufficiently rich in O stars that the entire upper end of the mass spectrum is sampled and an unbiased view of the hot-star population can be obtained.

Giant HII regions are among the most conspicuous objects in nearby late-type galaxies, in particular when observed in narrow emission lines, such as Hα. Systematic studies of nearby giant HII regions such as that by Kennicutt (1984) have established their fundamental properties. These HII regions typically have diameters of 100 pc or larger, densities of a few particles per cm^3, and ionized gas masses of order 10^5 $M_\odot$. In Fig. 9.1 we show a well-known example, NGC 604 in the Triangle galaxy M33. This giant HII region is powered by hundreds of O plus a few W-R stars with ages of a few Myr. It is only with the advent of HST that individual stars in the very center of extragalactic HII regions can be studied by means of spectroscopy.

In Table 9.1 (adopted from Kennicutt 1984) some properties of nearby extragalactic HII regions are given. The extragalactic regions are contrasted with several Galactic counterparts. Since the objects in this table are Hα-selected, there is a bias against the most luminous galactic HII regions, which are heavily reddened and become prominent only in IR and radio surveys. The host galaxies of the objects in Table 9.1 include the LMC, NGC 6822, M31, M33, M81, M82, M101, and others. Even for this narrow range of galaxy types, the giant HII regions span several orders of magnitude in parameter space. As a group, they form a rather heterogeneous sample with several characteristic and largely independent parameters such as morphology, density, and luminosity. Their common property is the large number of ionizing photons as derived from the Hα recombination flux. Since each recombination must be balanced by an ionization, we can relate the number of hydrogen ionizing photons N(LyC) to the Hα recombination luminosity L(Hα) using Eq. (8.4). More generally, *any*

197

Fig. 9.1 HST emission-line plus continuum composite of the giant HII region NGC 604 in the neighboring spiral galaxy M33, located at a distance of 0.9 Mpc. More than 200 O and W-R stars ionize the surrounding gas and shape it with their powerful winds. The field size is about 400 pc. STScI Press Release 1996-27.

strong recombination line is a measure of $N(\mathrm{LyC})$. Since $N(\mathrm{LyC})$ can be predicted from stellar model atmospheres and is a unique function of T_{eff} (see Table 3.1 in Section 3.2), we can infer the number of ionizing stars from the recombination fluxes. The O star numbers listed in Table 9.1 were derived in this way, assuming an O7 V star (with a Lyman continuum flux of $\sim 10^{49}$ photon s^{-1}) is typical for the ionizing population as a whole. In Section 9.3 we will refine this approach by accounting for the actual mass distribution of the O stars.

Giant HII regions are ionized by an entire cluster of O stars, in some cases counting hundreds or even thousands of stars. A convenient working definition is to require a giant HII region to be ionized by at least 10^{50} photon s^{-1}, the output of a single O3 star. This is equivalent to about three to five mid-O stars, and roughly ten times the Orion nebula. In our Galaxy, the Carina nebula, NGC 3603, and W49 qualify as giant HII regions (recall Chapter 7). While there is a continuous transition from normal Orion-like HII regions to those with ionizing photon budgets above 10^{50} photons s^{-1}, it is important to realize that giant HII regions are not just scaled-up M42s. Along the same lines, would we consider a 100 $M_\odot$ star a superposition of 100 solar-type stars? Certainly not. The parameter space occupied by giant HII regions provides the necessary conditions for a whole new set of astrophysical phenomena which would otherwise not occur in less luminous regions.

Table 9.1 *Integrated properties of nearby giant HII regions. The first three entries are Galactic regions, all other entries are extragalactic. Adapted from Kennicutt (1984), for an assumed O7V Lyman continuum flux of 10^{49} s^{-1}.*

Region	Diameter (pc)	$L(H\alpha)$ (erg s^{-1})	n_e (cm^{-3})	N(O7 V)
Carina	200:	6×10^{38}	13	45
NGC 3603	100:	1.5×10^{39}	25	110
W49	150:	2×10^{39}	15	160
NGC 6822B	200	2×10^{38}	1.7	15
NGC 6822A	160	4×10^{38}	5	30
NGC 6822C	130	7.5×10^{38}	8	55
SMC N19	220	2.2×10^{38}	2.3	15
SMC N66	220	6×10^{38}	4	45
M31A	250	5×10^{38}	3	35
M31B	240	5×10^{38}	3	35
M31C	190	4×10^{38}	4	30
M33C (NGC 592)	360	3×10^{38}	1	20
M33B (NGC 595)	400	2.3×10^{39}	3	160
M33A (NGC 604)	400	4.5×10^{39}	4	320
M81A	450	1.7×10^{39}	2	125
30 Doradus	370	1.5×10^{40}	6	1100
NGC 2366A	560	1.5×10^{40}	4	1100
NGC 2403A	600	1.4×10^{40}	4	1000
M82A	450:	4×10^{40}	16	2800
M82B	300:	3×10^{40}	15	2200
M101A (NGC 5471)	800	5×10^{40}	4	3500
M101B (NGC 5461)	1000:	7×10^{40}	3.5	5000
M101C (NGC 5455)	750	2.5×10^{40}	3	1800

A distinguishing property of giant HII regions is the existence of highly supersonic motions, as measured from the width of the nebular emission lines. When observed at sufficiently high spectral resolution, e.g., with an echelle spectrograph or with a Fabry–Perot interferometer, the lines are resolved with widths of more than 10–20 km s^{-1}. This is much larger than the thermal broadening for a temperature of 10 000 K, and it is higher than the sound speed for this temperature. Such velocities can indicate large-scale macroscopic motion in giant HII regions. Since supersonic velocities cannot be generated by stellar radiation, the kinematic properties in giant HII regions must be fundamentally different from those of normal HII regions like the Orion nebula, which do *not* have supersonic velocities. Moreover, supersonic flows will rapidly dissipate, so that their energy must be replenished continuously. Therefore we need to identify an additional powering source in giant HII regions which is otherwise not important in Orion-type nebulae.

One hypothesis is that gravity could be responsible for the broadening. Pressure-supported systems like elliptical galaxies or globular clusters are self-gravitating systems in which the spectral-line broadening results from the velocity dispersion of individual stars in the gravitational well of the gas–star system. Applied to giant HII regions, we could interpret the supersonic gas motions as due to the gravitational potential of the massive ionizing stars. As we have seen, giant HII regions are characterized by a larger number of massive ionizing stars than normal HII regions. Therefore it is reasonable to suspect that gravity could be the cause of the supersonic motions.

There is an alternative explanation. Hot, luminous stars have powerful winds, starting on the main sequence and increasing in strength when the supergiant and W-R phases are reached. These winds, with velocities of thousands of km s^{-1}, can stir up the gas in the HII region and induce highly supersonic motions. How would we be able to distinguish between the two interpretations, ordered gravitational motion vs. random wind-induced large-scale turbulence? Careful studies of the nebular line profiles have revealed that the profile shapes are not consistent with gravitation-induced motion. If gravity were the dominant broadening mechanism, the lines should have Gaussian profiles with width of order

$$\sigma \approx (G M_{\text{tot}}/R)^{1/2}, \tag{9.1}$$

where G is the gravitation constant, M_{tot} is the total mass of the HII region (including both stars and gas), and R is a characteristic size (for instance the half-mass radius). Instead of observing such a profile, actual data indicate wings at much higher velocities. Such wings are expected from shells triggered by stellar winds and favor non-gravitational forces as the cause of the supersonic motions in giant HII regions (Melnick, Tenorio-Tagle, & Terlevich 1999). This makes sense: giant HII regions harbor the most massive stars, which have the highest mass-loss rates. Figure 9.1 supports this view: the central ionizing star cluster in NGC 604 has evacuated its immediate surroundings and blown a ring-shaped cavity with a radius of tens of pc. The entire HII region exhibits filaments of ionized gas which are likely the result of mechanical energy input by winds.

Sometimes the existence of supersonic gas velocities is taken as the defining property of a giant HII region (Terlevich & Melnick 1981). An observational determination of the intrinsic gas velocity is often quite challenging since very high spectral resolution is necessary. Therefore we will rather rely on an equivalent definition which is commonly used: *giant HII regions must be powered by at least 10^{50} photons s^{-1}, corresponding to one O3 star or ten O7 stars.* This criterion defines giant HII regions in a manner analogous to a definition via the line width: they are much more luminous than normal Orion-like regions.

The existence of such large and luminous HII regions raises several important issues related to the formation of massive stars. (i) What is the spatial morphology of the underlying stellar population? Is there a single very rich cluster or are there many "normal" OB associations in close physical proximity? (ii) If there is substructure, how is star formation triggered and how does it propagate over regions larger than 100 pc? (iii) What are the initial conditions in the interstellar medium to produce giant HII regions and how do they depend on the properties of the host galaxy? (iv) What is the shape of the IMF of the clusters that produce such nebulae? We will revisit these questions in various sections in this and the following chapter.

9.2   30 Doradus – the Rosetta Stone

The most famous and best studied giant HII region is located in our close neighbor galaxy, the LMC. Photographs of the LMC, such as the one reproduced in Fig. 9.2, show a spectacular nebulosity to the north-east of the main LMC disk. The nebulosity was originally thought to be a star and called 30 Doradus by seventeenth century astronomers, meaning

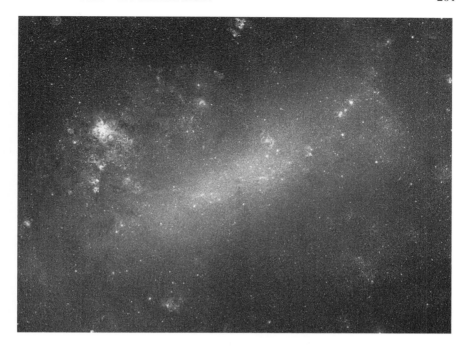

Fig. 9.2 Wide-angle photograph of the Large Magellanic Cloud. The brightest nebulosity near the upper left of the stellar bar is 30 Doradus. Copyright Anglo-Australian Observatory/David Malin Images.

that the object is star number 30 in the constellation of Dorado, where the LMC is located. The true nature of 30 Doradus (or 30 Dor, as it is commonly abbreviated) was recognized by Abbé Lacaille in 1751. Lacaille undertook a journey to the Cape of Good Hope in 1751–1752 to observe and document the southern stars. One of his reports, published in 1755, contains a catalog of 42 nebulae, nebulous star clusters, and nebulous stars. 30 Doradus is listed as entry 2 in the group of nebulae, thereby establishing the non-stellar nature of this object.

30 Dor is often called the Tarantula nebula. Figure 9.3 makes it clear why this is quite an appropriate name. Images taken in the light of strong nebular emission lines reveal thin tendrils of glowing gas which have been likened to the legs of a spider. The central region of the nebula is the body of the spider. It is an extremely dense clump of gas and very hot stars. These hot stars in 30 Dor are responsible for powering the nebula and give it its spectacular appearance.

The greater 30 Doradus region and its distinct substructures are known under a variety of designations, all of which refer to well-defined entities and astrophysical phenomena. Table 9.2 gives a summary, listing both the physical sizes and the apparent diameters as seen from 50 kpc, the distance of the LMC. The 30 Doradus region comprises all stars and nebulosities visible on deep images and which are thought to be physically related to the star-formation event in the region. The 30 Dor nebula in Fig. 9.3 extends over about 1/5 of the diameter of the 30 Dor region and defines the location of the bulk of the ionized gas. Although the 30 Dor nebula is often referred to as NGC 2070 (entry 2070 in the New General Catalog of non-stellar objects), most astronomers refer to the ionizing *star cluster* in the

Fig. 9.3 Emission-line image of 30 Doradus, the Tarantula nebula. Its powering source is R136, a cluster composed of thousands of hot blue stars with masses up to 100 $M_\odot$. Copyright Anglo-Australian Observatory/David Malin Images.

Table 9.2 *Characteristic sizes in the greater 30 Doradus region. From Walborn (1991).*

	Linear	Angular
LMC	5 kpc	5°
30 Doradus region	1 kpc	1°
30 Doradus nebula	200 pc	15′
30 Doradus cluster	40 pc	3′
R136	2.5 pc	10″
R136a	0.25 pc	1″

center of 30 Dor when using the designation NGC 2070. This cluster has a total diameter of about 40 pc and contains at least a few hundred thousand stars. The center of NGC 2070 is defined by a very dense cluster of young, massive stars, called R136 (from a list of stars compiled at Radcliffe Observatory in South Africa). R136 hosts a whole population of O3 stars, some of the earliest, hottest, and most massive stars known in the Universe (Massey & Hunter 1998).

Finally, Table 9.2 lists "R136a" as the smallest distinct substructure other than individual stars. R136a gained notoriety in the early 1980s when this object was suggested to be a single supermassive star. At that time astronomers relied on ground-based images to determine its size and concluded it is a single stellar object. Since its luminosity and temperature by far exceeded that of any other star, it had to be much more massive than the most massive star known. Masses larger than a thousand solar masses were discussed. Such masses would pose challenges to theories of the formation and stability of stars. Today we know that R136a is a small cluster hosting several $\sim$100 $M_\odot$ stars (still very massive by any standard — but not unique and outstanding) whose compactness mimics a single stellar source at the distance of the LMC. While the debate of R136a as a supermassive star is little else than a historic curiosity, it serves as a reminder for the interpretation of more distant stellar systems: spatially resolved imagery and/or high-quality spectral synthesis are required when the morphology of compact starburst regions outside the Local Group of galaxies is investigated.

After this historical perspective we move on to place 30 Dor into astrophysical and cosmological context. First, multi-wavelength observations from X-rays to the IR will be discussed. We will use the data to identify distinct stellar evolutionary phases, which in turn can be related to the star-formation history of the region. This will help us focus on several key properties of the stars and the gas and their interrelation.

The most extensive and detailed data sets exist at optical wavelengths. High-resolution broadband imagery from space reveals distinct stellar populations whose evolutionary states and spatial locations are correlated. This is evident from the schematic in Fig. 9.4 (from Grebel & Chu 2000), which provides a glimpse of 30 Dor's past, present, and future. The four quadrants show the same optical HST image, but each image highlighting different stellar generations. The upper left quadrant emphasizes Hodge 301, a 20 Myr old star cluster representing the oldest generation in the image. About 20 Myr ago, this cluster may have resembled the center of R136 as it appears today. (Hodge 301 is, however, less massive.) Most likely, the stellar winds and supernovae originating in Hodge 301 have triggered subsequent star formation and may have contributed to the formation of the massive stars currently observed in the center of NGC 2070.

The present generation of stars is labeled in the upper right and lower left quadrants. The former shows stars classified spectroscopically as OB supergiants, and the latter identifies O main-sequence and W-R stars. These populations are found towards the center and have a typical age of a few Myr. The masses of these stars reach up to $\sim$100 $M_\odot$, placing them into the top left portion of the H-R diagram. As we explained in Chapter 2, the relation between mass, age, and spectral type is highly degenerate in this part of the H-R diagram. It is not straightforward to establish evolutionary sequences simply from spectral types. For instance, stars classified as "W-R" may actually be younger and less evolved than some of the OB supergiants. Detailed atmospheric analyses are required for definitive answers.

The effects of the present-day star formation can be clearly seen in the network of filaments and shells originating from the center of the image. The shells have velocities of tens to hundreds of km s^{-1} and are thought to result from the interaction between stellar winds and supernova ejecta. This is suggested by the spatial correlation between the shells and the X-ray emitting gas, which is heated by shocks associated with supernova remnants. The total kinetic energy in the shells is about half that of the entire 30 Dor complex, and it exceeds the gravitational binding energy of the region by a large factor. Eventually, the gas will leave the gravitational well of the stellar cluster.

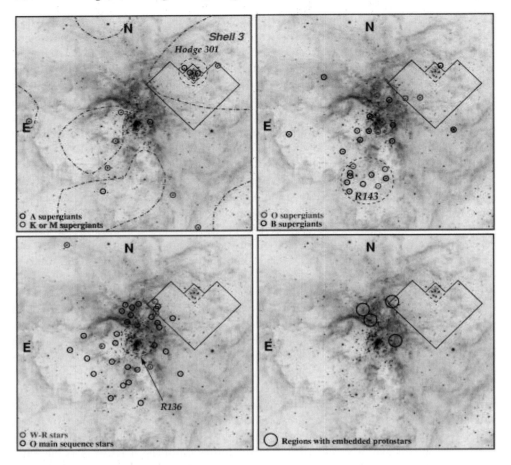

Fig. 9.4 Stellar generations in 30 Dor. Top left: oldest generation (20 Myr); top right: B (10 Myr) and O (3 Myr) supergiants; bottom left: O main-sequence and W-R stars (5 Myr); bottom right: protostars. Reproduced from Grebel & Chu (2000) by permission of the AAS.

Some nebular shells are the nursery for the next stellar generation. The fourth (lower right) quadrant of Fig. 9.4 indicates the location of regions with dust-embedded protostars. They are located on and along the giant shells surrounding R136 and are likely the result of triggered star formation by R136 itself. Their precise nature has still eluded astronomers but they may well evolve into a future R136.

The large dust obscuration at optical wavelengths precludes detailed studies of the newly forming protostars. Observations in the near-IR between 1 and 2.5 μm reduce the dust attenuation by up to a factor of 10. Imaging at these wavelengths has discovered many new sources, including multiple systems, clusters, and nebular structures (Walborn *et al.* 1999). The newly formed stellar generation in the periphery of R136 turns out to host groups of massive, early-type stars embedded in nebulosity. Sometimes, jet-like structures are found. The most spectacular and brightest IR source resides at the top of a massive dust pillar several parsecs from and oriented directly toward R136. In general, the jet-like structures consist of detached but aligned extended IR sources. It has been suggested that the IR sources could define the impact points of a highly collimated, bipolar jet on the surrounding dark clouds. These outflows from young stars in 30 Dor were the first such detections outside our own Galaxy. Their morphologies are strikingly similar to those seen in the nearby Orion nebula. These results make 30 Dor a prime laboratory to investigate star-formation processes in giant extragalactic HII regions, which would otherwise become unobservable in more distant objects. The observed phenomena emphasize the complex "microcosmos" of 30 Dor where we witness multiple star-formation events, each triggering and terminating the other.

Ultraviolet observations of 30 Dor provide insight into the properties of the hottest, most massive stars which are not hidden by dust obscuration. The observational emphasis is on spectroscopy rather than imaging: since the stellar *continuum* of hot stars is not too different in the optical and the UV as both wavelength regimes are close to the Rayleigh–Jeans tail of the spectrum, the gain in information from the optical to the UV does not always justify the observational expense of UV imaging. The stellar *line spectrum*, however, is radically different in the UV. It is only in the UV that we can measure the velocities of the powerful winds shaping the gas in 30 Dor. The central star cluster, when observed through a large aperture in the UV, such as that of the IUE satellite, shows a characteristic spectrum of a young population with P Cygni profiles of NV λ1240, SiIV λ1400, and CIV λ1550 (Vacca *et al.* 1995). Remarkably, similar spectra are observed in galaxies nearby and at cosmological distances, suggesting that 30 Dor is in some (but not all!) aspects a scaled-down version of galaxy-wide starbursts when dilution effects due to the older galaxy host population are taken into account.

Spectra of individual stars in R136 have helped establish the fundamental stellar parameters from sophisticated atmospheric analyses. R136 contains the largest concentration of hot, massive stars anywhere in the sky for which individual UV spectra can be collected. It turns out that many stars have the same very early spectral type of O3. Correspondingly, they occupy only a relatively narrow range in T_{eff} around 45 000 K. This may sound implausible but it is a simple consequence of stellar evolution. At the very top of the observed H-R diagram, stars evolve almost vertically in the first 3 Myr of their lives (Fig. 9.5). Although the interior of two 100 $M_\odot$ stars with ages 1 and 3 Myr are fundamentally different because of nuclear burning, their surface temperatures are almost the same, and so are their absolute visual (but not bolometric!) magnitudes. The stellar mass would decrease by almost a factor of 2 over this age range due to strong stellar winds with mass-loss rates of about 10^{-5} $M_\odot$ yr^{-1}. However, determining masses of single O stars even to within a factor of 2 is by far non-trivial, and large systematic uncertainties exist. In retrospect, this behavior explains why R136a was suggested to be a supermassive star in the 1980s. The alternative hypothesis of the object being a dense star cluster of almost identical massive stars was considered but rejected as too contrived. At that time, the properties of hot, massive stars were not well enough understood to expect the degeneracy between T_{eff}, M_{V}, mass, and spectral type.

The final stages of stellar evolution, when stars have their most powerful winds as W-R stars and/or explode as supernovae, can be traced with the shock-heated gas seen in X-rays (Wang 1999). The 30 Dor nebula was the first giant HII region for which strong diffuse X-ray emission was first detected by the Einstein Observatory and subsequently studied in great detail with the Chandra X-ray Observatory (Townsley *et al.* 2006a, b). The thermal soft X-ray emission in the $\sim$1 keV range is consistent with the presence of hot gas of a few times 10^6 K.

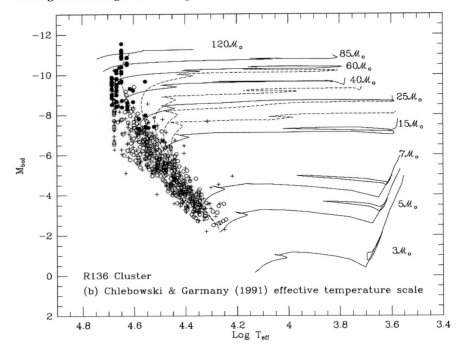

Fig. 9.5 H-R diagram for R136. Filled circles: stars with spectral types; open circles: stars with only photometry; plus signs: stars with uncertain reddening. The evolutionary tracks are from Schaerer *et al.* (1993). The dashed lines are isochrones at 2 Myr intervals. Reproduced from Massey & Hunter (1998) by permission of the AAS.

X-rays and optical line emission are spatially anti-correlated: the hot gas is located in the optical cavities delineated by the shells and filaments, which in turn are associated with dense molecular clouds in the center of the nebula. The cavity around R136 itself is particularly noteworthy. The kinematics of the HII gas indicate an outflow which is expanding much more rapidly in the east than in the west. The expansion is most likely driven by the over-pressured X-ray emitting gas flowing out from the center to the halo of 30 Doradus.

Stellar winds and energetic photons from massive stars and supernova explosions are responsible for the heating of the gas and the spatial morphology. The evolution of the cold, warm, and hot gas in 30 Dor is clearly driven by the energy release from massive stars. The total mass-loss rate from all stars combined is of order 10^{-3} $M_\odot$ yr^{-1}. With a typical terminal wind velocity of a few 10^3 km s^{-1}, the central cluster R136 alone has a stellar wind luminosity of 3×10^{39} erg s^{-1}. Integrated over its lifetime of several Myr, the cluster will release about 10^{53} erg of mechanical energy. A substantial fraction of this energy should be contained in the hot gas, the precise amount depending on the importance of the radiative cooling of the hot gas[1] and on the amount of the ambient ISM entrained in the shocked

---

[1] Radiative cooling acts like a loss-term by removing energy from the shock-heated gas.

material. We will return to these processes in Section 10.4 when we discuss outflows on a much grander scale, driven by powerful starbursts and extending over galaxy-wide sizes.

We conclude the review of the Rosetta Stone 30 Doradus by placing it into context with other high-mass star-formation regions. How do the *stars* in R136 compare with those in other regions of high-mass star formation? In Table 9.3 we provide a comparison of the *stellar* light output from several such regions, arranged in order of increasing age. The five examples in this table have comparable sizes of a few pc. Columns 3–6 list: L_{1500}, the cluster luminosity at 1500 Å, M_B^{clu}, the absolute B magnitude, M_B^{gal}, the total absolute B magnitude of the host galaxy, and M_B^{3Myr}, the B magnitude the cluster would have if observed at an age of 3 Myr. (Clusters fade with age and older clusters would be brighter if observed at an earlier epoch.) R136 ranks at the bottom of the list. A typical ionizing O star has an L_{1500} of a few times 10^{35} erg s^{-1} Å^{-1}. The total UV luminosity of R136 is produced by a few hundred such stars, consistent with empirical star counts.[2] NGC 1741-B1, a superposition of several massive star clusters in the starburst galaxy NGC 1741, is two orders of magnitude more luminous than R136. The two clusters NGC 1569-A and NGC 1705-A are outstandingly bright. What makes them particularly extraordinary is their brightness relative to the host galaxy (cols. 4 and 5): they provide a significant fraction of the blue and ultraviolet light of the galaxy and become even more impressive if fading due to age effects is taken into account.

Table 9.3 *Comparison of R136 with other sites of massive star formation. From Leitherer (1998).*

Region	t (Myr)	L_{1500} (erg s^{-1} Å^{-1})	M_B^{clu}	M_B^{gal}	M_B^{3Myr}
R136	3	6×10^{37}	-11.3	-17.9	-11.3
NGC 4214-1	5	2×10^{38}	-13.0	-18.8	-13.5
NGC 1741-B1	5	6×10^{39}	-16.7	-20.3	-17.2
NGC 1569-A	10	3×10^{38}	-14.1	-16.2	-15.1
NGC 1705-A	15	6×10^{38}	-14.0	-17.0	-15.5

A ranking of 30 Dor in terms of *nebular* properties is done in Table 9.4. Quantities listed are the diameter, the Hα luminosity, the corresponding number of ionizing photons, N(LyC), the total number of ionizing stars, and the mass of ionized gas. The O-star numbers in this table should not be confused with those in Table 9.1. Here, we have counted all O stars between spectral types O3 and O9, whereas in Table 9.1 only O5 V stars are included in the census. The local comparison HII region is again the Orion nebula. 30 Dor exceeds Orion by three orders of magnitude in all parameters listed. Column 4 of this table gives the range of values observed for giant extragalactic HII regions. In terms of distance, most of them are within, or close to, the Local Group of galaxies. 30 Dor is a fairly typical giant HII region. Note that

---

[2] The number of stars of *all* masses is several orders of magnitude higher since there is only about one O star for every 10 000 solar-type stars. The UV luminosity of solar-type stars is too low to make any contribution to the combined L_{UV} of the population.

Table 9.4 *Range of physical parameters in Orion, 30 Dor, giant HII regions (GHII), and the starburst galaxy NGC 7714. From Kennicutt (1991).*

	Orion	30 Dor	GHII	NGC 7714
Diameter (pc)	10	400	100–1000	600
$\log L(\text{H}\alpha)$ (erg s^{-1})	37	40.2	38–41	42
$\log N(\text{LyC})$ (photon s^{-1})	49	52	50–53	54
Ionizing O stars	6	1000	10–10 000	> 10 000
$\log M(\text{HII})$ ($M_\odot$)	2–3	5.9	3–7	7

the comparison in Table 9.4 is more appropriate for 30 Dor than that for R136 in Table 9.3. This is because 30 Dor has the right age for being observed as a bright HII region.

The entries in the last column are for the nuclear starburst galaxy NGC 7714. This galaxy is one of the prototypes of the starburst class, which will be discussed in Chapter 10. NGC 7714 surpasses 30 Dor by one to two orders of magnitude for all listed quantities.

9.3   Stellar population diagnostics

Only the closest giant HII regions can be studied in as much detail as 30 Dor. Ground-based observations of even relatively nearby objects such as NGC 604 in M33 fail to resolve the ionizing star clusters into individual stars. Adaptive optics techniques can mitigate the problem, but the analysis of the stellar content from the H-R diagram becomes increasingly prohibitive for more and more distant giant HII regions. The subject of this section are the *diagnostic methods* that allow astronomers to investigate the properties of massive stellar populations which are unresolved into individual stars. The stellar populations in giant HII regions and galaxies inferred with these techniques are the subject of Section 9.4.

Giant HII regions are powered by young star clusters. Images taken with filters encompassing bright nebular lines such as [OIII] λ5007 or Hβ are often dominated by light from the gas. By carefully choosing wavebands where nebular lines are weak, we can obtain images of the underlying stellar clusters. This is illustrated in Fig. 9.6, which compares two sets of HST images of two giant HII regions in the nearby galaxies NGC 2366[3] and NGC 4214. One set includes the nebular emission lines, whereas the other does not. The pure continuum images nicely unveil the ionizing stellar clusters. The two clusters in Fig. 9.6 are about 4 Mpc away, too distant for obtaining spectra of individual cluster members. Therefore we need to rely on an analysis of the integrated light.

Star clusters, even those as massive as NGC 2070, form very quickly. The star-formation process lasts for a few million years, which is short in comparison with the timescale of stellar evolution, except for extremely massive stars. Therefore asking the question "What is the star-formation rate of this star cluster?" makes little sense because very few stars are currently forming. Rather, the relevant parameters of the cluster and the HII region are, e.g., the chemical composition, the IMF, the mass, and the age.

---

[3] The HII region itself has the designation NGC 2363 and should not be confused with its galaxy host, the HII galaxy NGC 2366.

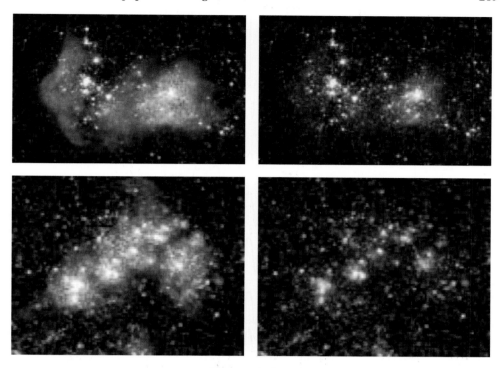

Fig. 9.6 Continuum plus emission line (left) and continuum (right) color composites taken with HST's WFPC2. The images show two giant HII regions in NGC 2366 (top) and NGC 4214 (bottom). Images courtesy J. Maíz-Apellániz.

Measuring the cluster mass and determining the IMF are intimately related. Massive stars are overwhelmingly luminous for a typical population mix, precluding almost always the direct detection of the light from low-mass stars. Yet, it is the low-mass population where most of the stellar mass is concentrated and whose mass fraction determines the star-formation efficiency. The most direct technique for obtaining the starburst mass relies on gravity, not on radiation. Velocity dispersion measurements using resolved faint absorption features from red supergiants give the most reliable cluster masses if the theoretical assumptions on the dynamical state of the cluster are met. RSGs are often the only stellar species in a young stellar population whose spectral lines are detectable and sufficiently narrow to allow measurements of the velocity dispersion. Derived values are of order 10 km s^{-1}. This technique has now been applied to a handful of clusters with masses of $10^4 \, M_\odot$ or higher (e.g., Larsen, Brodie, & Hunter 2004).

Here we will concentrate on techniques to determine cluster ages. The single stellar populations in clusters are the reason for the tight relation between spectral morphology and cluster age. The *optical* continuum light from a single stellar population is almost always emitted by stars close to the main-sequence turn-off because these are the brightest stars.[4]

---

[4] Depending on the cluster age, the near-IR and UV spectral regions can also be dominated by few luminous post-main-sequence stars, such as RSGs or post-AGB stars.

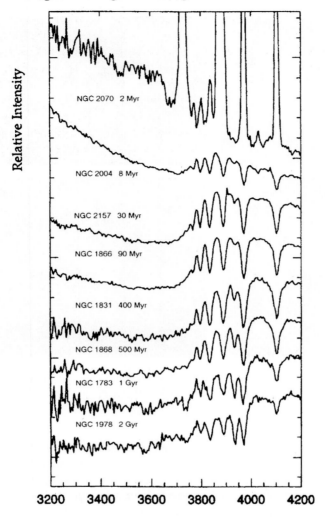

Fig. 9.7 Spectra of young and intermediate-age LMC clusters. The cluster age increases from top to bottom. From Bica, Alloin, & Schmitt (1994).

Consequently, the integrated cluster spectrum can often be compared to a single stellar spectral type. This can be seen in Fig. 9.7, which is a montage of integrated spectra of eight LMC star clusters with ages ranging from a few Myr to more than 1 Gyr. At the youngest age (a few Myr), we observe a typical HII region spectrum with strong nebular emission lines. With increasing age, the Balmer jump increases, first because of B stars, then due to A stars. The Balmer absorption peaks after a few hundred Myr when A stars dominate. Finally, F- and G-type stars with their characteristic Ca H and K lines take over.

Astronomers have developed techniques taking advantage of the various spectral indicators to measure cluster ages. The most important age tracers are discussed below. None of these age indicators require absolute flux measurements — they are all normalized quantities depending on the relative frequency of stars in one age group over younger and older stars.

Nebular emission lines Young stellar systems are embedded in gas. If O, B, and W-R stars are present, the gas will be ionized and excited. The observed electron temperature of the ISM results from an equilibrium between stellar heating and radiative, metallicity dependent cooling. Analysis of electron-temperature sensitive *line ratios* allows an independent determination of the stellar heating, and therefore of the stellar far-UV radiation field and its evolution with time. If the HI and HeI column densities are uniform and high enough for totally absorbing the ionizing photons, and in the absence of dilution by an underlying population, the *equivalent widths* of H and He lines are very useful age indicators (Stasińska & Leitherer 1996). The age range during which these techniques are sensitive coincides with the evolutionary timescale of O stars, i.e., $\sim$10 Myr. Evolutionary synthesis calculations coupled to photoionization models predict strong variations of the far-UV radiation field, in particular when hot stars with strong winds appear (e.g., Schaerer & Stasińska 1999). Different models agree with observations longward of the He^0 edge but discrepancies exist around and below the He^+ edge (Dopita *et al.* 2006). Possible reasons include the atmosphere and evolution models during the W-R phase (Smith *et al.* 2002).

UV stellar-wind lines The wavelength region between 1200 and 2000 Å is dominated by stellar-wind lines of, e.g., CIV λ1550 and SiIV λ1400, the strongest features of hot stars in a young population (e.g., Leitherer *et al.* 2001). In contrast, the optical and near-IR spectral regions show few, if any, absorption lines of hot stars when the cluster age is less than about 10 Myr. The reason is blending by nebular emission and the general weakness of hot-star features longward of 3000 Å. Hot-star winds are radiatively driven, with radiative momentum being transformed into kinetic momentum via absorption of extreme UV radiation by metal lines. Therefore the stellar-wind dominated epoch coincides with the nebular phase. Since the stellar far-UV radiation field softens with time for an evolving single stellar population, the wind strength decreases, and the lines change from being P Cygni wind profiles during the first few Myr to purely photospheric absorption lines after tens of Myr. This is illustrated in Fig. 9.8. If O stars with zero-age main-sequence masses above 30 $M_\odot$ are present, CIV shows a P Cygni profile. In the absence of such stars (after 10 Myr), the line becomes weaker and disappears due to the changing wind density and ionization conditions. Around 100 Myr, the population is B-star dominated, and singly and doubly ionized transitions of C and Si serve as age indicators (see below).

W-R emission bump The only unblended stellar features observable in the integrated optical spectrum of a young population are normally from W-R stars. D. Kunth first coined the term "W-R galaxy" from observations of W-R features in the starburst galaxy NGC 3125. W-R stars are extremely sensitive chronometers, but careful calibration of models is required (Schaerer 1999). The onset and termination of the W-R phase is subject to debate. In standard models, this phase lasts from 3 to 6 Myr but metallicity effects and mass transfer in close binaries can prolong or decrease the duration of the W-R phase. Given these uncertainties, one may wonder whether W-R stars could ever be better age tracers than other methods. However, W-R stars do become important, e.g., when nebular lines cannot be trusted as age indicators. Despite uncertainties in the evolutionary paths of W-R stars, it is established beyond doubt that they are related to very massive, young stars. Their detection in the central regions of AGN is direct proof of an O population, whose presence is difficult to establish directly in an AGN environment.

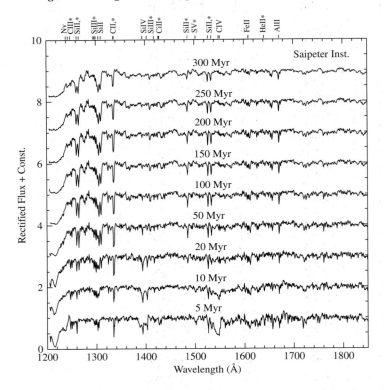

Fig. 9.8 Synthetic UV spectra for ages between 5 and 300 Myr. Instantaneous star formation. Excited lines (photospheric) are marked with asterisks. Reproduced from de Mello, Leitherer, & Heckman (2000) by permission of the AAS.

IR features from supergiants and giants   After about 5 Myr the most massive stars of a single population evolve toward cooler temperatures, forming RSGs. The RSG continuum will dominate the red and near-IR for the following tens of Myr. The photospheric lines of RSGs can be observed in the cluster spectrum as well. This marks the first occurrence of age-sensitive metal lines in the near-IR spectrum of a young cluster. Contributions from less luminous, but more numerous, giant stars must be carefully evaluated since both stellar components can be important, depending on age and star-formation history. The most prominent RSG features in the H and K bands are the first and second overtones of CO at 2.29 μm (2–0 band head) and 1.62 μm (6–3 band head) and the SiI absorption at 1.59 μm (Oliva *et al.* 1995). Dust-embedded populations can be detected in these lines with the obscuration reduced by a factor of 10 relative to the V band.

B-star UV lines   B stars, like O stars, have few strong spectral lines in the optical and near-IR so that the wavelength region of choice is the UV. Most of the discussion on O stars applies to B-star lines as well. B stars have weaker winds than O stars. The resonance line CII λ1335 is one of the strongest B-star wind lines, yet this and all other B-star lines are less conspicuous than lines from O stars in the UV spectrum of a population. Furthermore, lines from B-star winds are narrower due to lower wind velocities, and they come from lower ionization potentials. B-star lines, such as the CII resonance line, are often hard to distinguish from ISM lines, which tend to have similar ionization ranges, in particular if observed at low spectral resolution. In addition to wind lines, B stars have numerous relatively strong photospheric absorption lines in the UV region. These lines arise from excited levels and therefore cannot be produced in the ISM. A typical example is the SiIII multiplet around 1300 Å (see Fig. 9.8). The strong age dependence makes several of these lines useful clocks for populations of 100 Myr and older. Consider, for instance, the variation of the SiIII λ1300/SiIV λ1400 ratio between 5 and 100 Myr in Fig. 9.8.

Balmer absorption lines   The stellar hydrogen lines are among the strongest absorption features for a young population at most ages due to the ubiquitous presence of H. During the first 10 Myr in the evolution of a single population, most hydrogen absorption lines are strongly blended by nebular emission and not very useful for age determinations. After about 10 Myr, the ionizing O stars disappear, and a hydrogen absorption spectrum appears, increasing with age, and reaching maximum strength around several 10^8 yr when A stars dominate the spectrum (see Fig. 9.7). Balmer absorption lines have been used to age-date compact star clusters in cooling-flow galaxies, merging systems, and starburst galaxies (Whitmore 2003). Accurate ages are essential to tackle the question whether these clusters are progenitors of present-day globular clusters. Obtaining cluster spectra reaches the limits of major ground-based telescopes so that strong features, such as the Balmer lines, are the preferred tracers. In contrast, colors are often affected by uncertain extinction corrections and crowding. Balmer lines are costly in terms of observing time since the removal of nebular contamination requires high spectral resolution, but they have the advantage of being (almost) independent of metallicity and unaffected by reddening. The Balmer jump shows a similar behavior with age. It can be measured with intermediate-band filters, and thus may open the way for age-dating surveys of fainter clusters.

Optical and near-IR colors   Colors are the prime age indicators in intermediate and old stellar populations. Reddening and the optical color degeneracy of hot stars limit their use to ages of more than about 20 Myr. There is very little metallicity dependence in young systems due to the lack of strong metal lines in the optical and near-IR. Metallicity does enter via evolution models, in particular when RSGs affect the colors between 10 and 50 Myr. The main challenge in interpreting colors of young systems is to break the age-reddening degeneracy. The reddening vector is parallel to the evolutionary paths in most color–color diagrams, and the reddening correction uncertainty is often much larger than the desired precision of the age determination. Furthermore, broad-band colors are affected by strong nebular emission lines located within the filter passbands. Nevertheless, colors are the most widely used astronomical clock for clusters with ages of tens of Myr and higher.

The powering sources of giant HII *regions* are star clusters with sizes of a few to tens of pc. HII *galaxies* show nebular emission over kpc-sized areas (Telles, Melnick, & Terlevich 1997). A simple causality argument suggests that the stars are unlikely to have all formed simultaneously over such a large area. Star formation does not turn on instantaneously but propagates at finite speed. Therefore we expect to find stars having a range of ages in a kpc-sized star-formation region. The larger the region, the more the stars are mixed and in an apparent equilibrium between star birth and death. This equilibrium is reached after only about 10 Myr for O stars and 100 Myr for early-B stars.[5] In this case, the stellar-population properties are only mildly (or not at all) sensitive to age effects. This allows astronomers to determine star-formation rates from absolute spectral measures, such as continuum or line luminosities. In principle, one could attempt a corresponding mass measurement of a single ionizing star cluster from, e.g., the Hβ recombination flux. In practice, however, the Hβ flux of a cluster is so strongly dependent on the unknown age that it becomes impossible to decouple mass and age. Figure 9.9 illustrates this behavior. In an equilibrium state after a few Myr, N(LyC) is totally age independent and scales only with the star-formation rate. In marked contrast, N(LyC) of an individual cluster drops by more than a factor of 100 during the first 10 Myr.

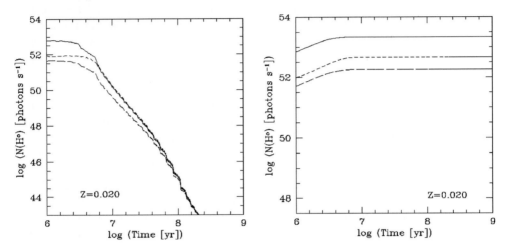

Fig. 9.9 N(LyC) vs. time for a single stellar population of mass $10^6\, M_\odot$ (left) and a population forming constantly at 1 $M_\odot$ yr^{-1} (right). Solid: Salpeter IMF; long-dashed: IMF with power-law exponent $\alpha = 3.3$; short-dashed: Salpeter IMF truncated at 30 $M_\odot$. Reproduced from Leitherer *et al.* (1999) by permission of the AAS.

In making these arguments, we implicitly assumed a universal IMF, a topic of much debate in the literature. We will review the evidence for or against this assumption in Section 10.2. Here we only mention that all the discussed star-formation indicators rely on the observation of massive stars and require an independent calibration of the low-mass stellar content. Our calibration is based on a Salpeter IMF for stellar masses between 1 and 100 $M_\odot$.

Integrated (i.e., taken through large apertures) spectra of galaxies along the Hubble sequence show distinct trends with respect to their stellar population properties (see Fig. 9.10). The typical population dominating the optical light has a younger age with later Hubble type. The nebular emission lines increase in strength with Hubble type as well. Although the spectra in Fig. 9.10 have a rather systematic behavior, they are clearly more complex than the cluster spectra of Fig. 9.7. For instance, the spectrum of the Sc galaxy NGC 6181 displays nebular

---

[5] Suppose the telescope aperture collects the light of 10 equally massive star clusters with ages of 1, 2, 3, ..., 10 Myr. The combined spectrum mimics a steady-state population of age 10 Myr in which the O stars are in equilibrium.

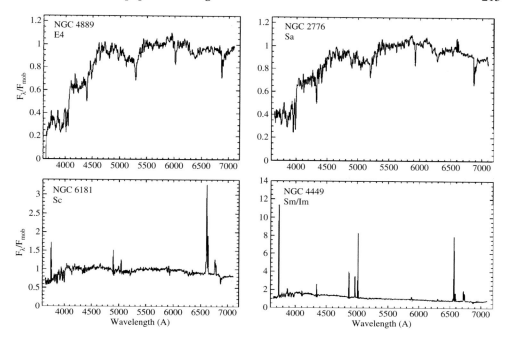

Fig. 9.10 Integrated spectra of an elliptical galaxy (upper left), an Sa galaxy (upper right), an Sc galaxy (lower left), and an irregular galaxy (lower right). The spectra show a progression from older to younger stellar populations. Reproduced from Kennicutt (1998a) by permission of the AAS.

emission lines, indicative of ionizing O stars, and a distinct Balmer series in absorption, characteristic of B and A stars. Comparison with Fig. 9.7 suggests that a single star cluster spectrum does not have these two components at the same time. Therefore the spectrum of NGC 6181 is produced by a mix of stars having different ages. We can derive the rate of star formation by assuming an equilibrium has been reached between star birth and death, an assumption which is more likely to be correct for massive stars with their short evolutionary timescales. If so, the emitted stellar radiation at a given wavelength is proportional to the star-formation rate $\psi(t)$. In the following we will assume that $\psi(t)$ does not vary over the time interval of interest and refer to the star formation rate as *SFR*.

Continuum luminosities   The most direct method to measure *SFR* is from the stellar continuum luminosity. Among all the diagnostics discussed below, it is the only one relying on *stellar* light itself. All other methods are indirect, utilizing radiation reprocessed by the ISM. The major drawback of a continuum luminosity is the required correction for dust reddening, which can be quite substantial. The monochromatic luminosity at 5500 Å, L_V, and *SFR* are related via

$$SFR\ [M_\odot\ \text{yr}^{-1}] = 6.92 \times 10^{-40} L_V\ [\text{erg s}^{-1}\ \text{Å}^{-1}]. \tag{9.2}$$

This and all other relations in this section were derived for a 100 Myr old population of solar metallicity with continuous star formation and a Salpeter IMF between 1 and 100 $M_\odot$.

L_V can easily be determined from an optical image or spectrum, but the validity of Eq. (9.2) is often compromised by a break-down of the assumption of star-formation equilibrium. Even in late-type galaxies, the light at 5500 Å can trace stars with ages of $\sim$1 Gyr. Significant variations of *SFR* can occur over that period. (An equivalent statement would be to say that the light is contaminated by an older, underlying population.) As the integrated light from a stellar population is weighted towards younger, more massive stars at shorter wavelengths, the solution is to observe galaxies at the shortest wavelengths possible. A widely used waveband is in the far-UV at 1500 Å. The relation between *SFR* and L_{FUV} is

$$SFR\ [M_\odot\ \mathrm{yr}^{-1}] = 4.06 \times 10^{-41} L_{FUV}\ [\mathrm{erg\ s}^{-1}\ \mathring{\mathrm{A}}^{-1}]. \tag{9.3}$$

The constants of proportionality in Eqs. (9.2) and (9.3) differ by a factor of 17. The factor is the luminosity increase between 5500 and 1500 Å of a spectrum with a power-law index of $\sim$−2.4. This spectral slope is typical of late-O/early-B stars, which are the main (but not the only!) contributors to the continuum. A simple extrapolation towards shorter wavelengths of the observed spectrum of NGC 6181 in Fig. 9.10 indicates a much smaller rise to the UV, confirming our concern of a significant contamination by an older population at 5500 Å.

Hydrogen recombination luminosities   The stellar ionizing flux is a measure of the most recent star-formation history. The steep luminosity drop in the Wien tail of the spectral energy distribution shortward of the Lyman edge in all but hot O stars excludes stars less massive than $\sim$10 $M_\odot$ from contributing to $N(\mathrm{LyC})$. The star-formation rate can be obtained from

$$SFR\ [M_\odot\ \mathrm{yr}^{-1}] = 4.56 \times 10^{-54} N(\mathrm{LyC})\ [\mathrm{s}^{-1}]. \tag{9.4}$$

$N(\mathrm{Lyc})$ itself is usually derived from hydrogen recombination lines or from the thermal radio flux. The strength of the hydrogen lines is determined primarily by the stellar ionizing radiation, with little dependence on the properties of the ISM. Therefore optical/near-IR emission lines such as Hα, Hβ, Paα, or Brγ are direct measures of the recent *SFR*. Caveats are the escape fraction of ionizing radiation and the absorption of stellar radiation by dust. For case B recombination, solar chemical composition, $T_e = 10\,000$ K, and no dust absorption, the conversion relations between $N(\mathrm{LyC})$ and the line luminosities are:

$$L(\mathrm{H}\alpha)\ [\mathrm{erg\ s}^{-1}] = 1.36 \times 10^{-12} N(\mathrm{LyC})\ [\mathrm{s}^{-1}], \tag{9.5}$$

$$L(\mathrm{H}\beta)\ [\mathrm{erg\ s}^{-1}] = 4.76 \times 10^{-13} N(\mathrm{LyC})\ [\mathrm{s}^{-1}], \tag{9.6}$$

$$L(\mathrm{Pa}\beta)\ [\mathrm{erg\ s}^{-1}] = 7.13 \times 10^{-14} N(\mathrm{LyC})\ [\mathrm{s}^{-1}], \tag{9.7}$$

$$L(\mathrm{Br}\gamma)\ [\mathrm{erg\ s}^{-1}] = 1.31 \times 10^{-14} N(\mathrm{LyC})\ [\mathrm{s}^{-1}]. \tag{9.8}$$

The ratio of the constants of proportionality in Eqs. (9.5) and (9.6) is the well-known ratio of the Hα/Hβ recombination coefficients for case B (recall Eq. (8.1)). As this ratio is almost exclusively determined by atomic physics, it can be used to derive the dust reddening of the ISM (recall Eq. (8.2)). $N(\mathrm{LyC})$ is related to the thermal radio luminosity as well. HII regions emit a characteristic bremsstrahlung spectrum with a spectral index of $\sim$0.1 and a moderate electron temperature dependence. The free–free radio luminosity L_{ff} is

$$L_{ff}\ [\mathrm{W\ Hz}^{-1}] = 1.6 \times 10^{-33} \nu^{-0.1} T_4^{0.45} N(\mathrm{LyC})\ [\mathrm{s}^{-1}], \tag{9.9}$$

with ν in GHz and T_4 in 10^4 K. Since dust attenuation is negligible at radio wavelengths, L_{ff} is an important diagnostic for evaluating the effects of dust on the derived star-formation rates. L_{ff} is difficult to measure in HII regions and galaxies because of a strong non-thermal component at cm wavelengths. The average ratio of the non-thermal to thermal (free–free) fluxes in spiral galaxies is

$$\frac{L_{\mathrm{nt}}}{L_{\mathrm{ff}}} \simeq 10 \left(\frac{\nu}{1\ \mathrm{GHz}}\right)^{0.1-\gamma}, \tag{9.10}$$

where L_{nt} is the non-thermal luminosity (see below) and γ is the spectral index of the non-thermal spectrum. Typical values are $\gamma \approx 0.8$. The non-thermal flux exceeds the thermal flux at all frequencies below ~ 30 GHz (1 cm) and is larger by a factor of 10 at 21 cm (Condon 1992).

Collisionally excited nebular lines   Hα and Hβ are redshifted out of the ground-based atmospheric transmission windows in galaxies at intermediate and high redshift. Therefore other strong nebular lines at shorter wavelengths are sometimes used to infer *SFR*. In most cases, the strongest line is [OII] $\lambda 3727$ (see Fig. 9.10). The strength of forbidden lines is not only determined by the stellar ionizing luminosity, but also (and even more so) by the excitation conditions and the chemical composition of the ISM. This makes forbidden lines only crude indicators of *SFR* and empirical calibrations are required to establish a relation between the line luminosity and *SFR*. Kennicutt (1998b) derived the mean relation for a sample of spiral and emission-line galaxies from a comparison with Hα derived star-formation rates. After scaling to our definition of the IMF, his relation becomes

$$SFR\ [M_{\odot}\ \mathrm{yr}^{-1}] = 5.6 \times 10^{-42} L([\mathrm{OII}])\ [\mathrm{erg\ s}^{-1}]. \tag{9.11}$$

Far-IR luminosities   The far-IR luminosity traces the ionizing and non-ionizing radiation from young OB stars, which are absorbed by dust and re-emitted at a peak wavelength of $\sim 60\,\mu$m. It is thus sensitive to the star formation history over tens of Myr. A far-IR selected sample of galaxies gives the least biased census of the global star formation since selection effects due to dust obscuration are avoided. If we assume the entire stellar radiation is converted, the far-IR simply becomes a measure of the total bolometric luminosity of the population because OB stars alone account for the entire radiative output. The relation between the far-IR luminosity and *SFR* is:

$$SFR\ [M_{\odot}\ \mathrm{yr}^{-1}] = 1.78 \times 10^{-44} L_{\mathrm{FIR}}\ [\mathrm{erg\ s}^{-1}]. \tag{9.12}$$

L_{FIR} can be computed, e.g., from the measured IRAS fluxes with the recipe given by Sanders & Mirabel (1996):

$$L_{\mathrm{FIR}}\ [\mathrm{erg\ s}^{-1}] = 2.1 \times 10^{39} D^2 (13.48 f_{12} + 5.16 f_{25} + 2.58 f_{60} + f_{100}) \tag{9.13}$$

where D is in Mpc, and f_{12}, f_{25}, f_{60}, and f_{100} are the IRAS 12, 25, 60, and 100 μm flux densities in Jy, respectively. L_{FIR} is a good approximation for the IR luminosity between 8 and 1000 μm, and therefore for the stellar radiation absorbed and re-emitted by dust. Equation (9.12) applies if the far-IR emission is dominated by warm dust heated by recently formed stars, as indicated by dust temperatures of 40–50 K. Lower dust temperatures can suggest an important contribution from an older disk population, complicating the interpretation of the far-IR luminosities.

Mid-IR luminosities   The contribution from mixed-aged, inhomogeneous populations to the bolometric far-IR emission from galaxies significantly limits the usefulness of Eq. (9.12) for deriving star-formation rates. The ISO and in particular the Spitzer satellites provide access to monochromatic luminosities in the mid-IR which turn out to be excellent tracers of the most recent star formation episode without contamination from older populations. The thermal dust emission at $24\,\mu m$ has been shown to tightly correlate with other independent star-formation indicators (Calzetti *et al.* 2007). Empirically, the most robust *SFR* measurement combines the observed Hα and $24\,\mu m$ luminosities as probes of the total number of ionizing photons. Calzetti *et al.* find:

$$SFR\,[M_\odot\,\mathrm{yr}^{-1}] = 3.31 \times 10^{-42}[L(\mathrm{H}\alpha) + (0.031 \pm 0.006)L(24\mu m)], \qquad (9.14)$$

where the luminosities are in erg s^{-1}, and $L(24\,\mu m)$ is expressed as νL_ν. The scaling factor in Eq. (9.14) is identical to that of Eq. (9.4) after conversion from $L(\mathrm{H}\alpha)$ to $N(\mathrm{LyC})$. The fundamental difference between the two relation is in the recombination line luminosities: $N(\mathrm{LyC})$ in Eq. (9.4) must be derived from the dereddened line fluxes, whereas $L(\mathrm{H}\alpha)$ in Eq. (9.14) is the observed, uncorrected value. Why would the scaling factor then be the same in both cases? The reason is the additional term $L(24\,\mu m)$ in Eq. (9.14), which is nothing else than the amount of reprocessed radiation that was removed from Hα and re-emitted in the mid-IR.

Supernova rates   Stars with initial masses above 8 $M_\odot$ end their lives as core-collapse supernovae, and the corresponding supernova rate ν_{SN} becomes a measure of *SFR*. Since the supernova progenitor mass is heavily weighted towards the least massive progenitor mass because of the IMF, ν_{SN} is almost completely determined by progenitors in the mass range 8 to 20 $M_\odot$. This is important because it is not clear whether all *very* massive stars explode as supernovae at all. Fortunately, this uncertainty does not affect the derived supernova rate due to the weighting by the IMF. The star-formation rate can be obtained from ν_{SN} from

$$SFR\,[M_\odot\,\mathrm{yr}^{-1}] = 49\,\nu_{\mathrm{SN}}\,[\mathrm{yr}^{-1}]. \qquad (9.15)$$

This relation accounts only for core-collapse supernovae (Type II and Type Ib/c), but not for Type Ia supernovae which arise from less massive binary stars. The most extreme starburst galaxies have star-formation rates of $10^2 - 10^3$ $M_\odot$ yr^{-1}. Even in these galaxies, the expected supernova rate is only a few per year, too low for direct counts. In addition, many supernovae will be missed because of dust obscuration. Therefore indirect tracers of ν_{SN} are needed. The most commonly used diagnostic is the non-thermal radio emission. Its dominant component is synchrotron radiation from electrons accelerated by supernova remnants. Consequently, the non-thermal radio flux depends on the supernova rate. Making quantitative predictions for the non-thermal luminosity (L_{nt}) from first principles is non-trivial as the synchrotron spectrum is strongly sensitive to the details of the acceleration process. Alternatively, an empirical approach can be chosen. Comparing the non-thermal radio luminosity in the Galaxy with the production rate of radio supernovae and/or the pulsar birth rate leads to the relation

$$L_{\mathrm{nt}}\,[\mathrm{W\,Hz}^{-1}] = 1.3 \times 10^{23}\nu^{-\gamma}\nu_{\mathrm{SN}}\,[\mathrm{yr}^{-1}] \qquad (9.16)$$

(Condon & Yin 1990). An independent method of constraining the supernova rate is the strength of [FeII] $\lambda 1.26$ emission. Iron as a refractory element is heavily depleted on dust in the diffuse ISM. Fast shocks propagating in the diffuse ISM, like those produced by supernova explosions, can destroy dust grains through sputtering processes or grain–grain collisions, and replenish the ISM with gas-phase iron. The gaseous iron is then collisionally excited in the cooling post-shock gas, and produces the observed infrared emission line. An empirical calibration of this process gives

$$L([\text{FeII}]) \ [\text{erg s}^{-1}] = 2.5 \times 10^{41} \nu_{\text{SN}} \ [\text{yr}^{-1}] \tag{9.17}$$

(Calzetti 1997). This relation is only intended to provide a very rough estimate of the supernova rate.

9.4   HII galaxies: stellar content and relation to starbursts

Extensive surveys of peculiar types of galaxies have been conducted since the middle of the twentieth century. The goal of these surveys was to identify unusual objects on the basis of their morphology, colors, or spectra. One of the pioneers among astronomers leading such surveys was Fritz Zwicky, who published a series of catalogs with the results of his efforts (e.g., Zwicky & Zwicky 1971). Interest in these surveys was revived by Sargent & Searle (1970), who demonstrated that some of Zwicky's galaxies showed spectra that were essentially identical to those of HII regions such as 30 Doradus or NGC 604. In their original publication, Sargent & Searle called these objects isolated extragalactic HII regions. Over time, the terminology evolved, and these galaxies are commonly referred to as HII galaxies.

The distinguishing properties of HII galaxies are their strong emission lines dominating over the continuum, their blue colors, compact sizes with characteristic diameters of hundreds of pc, and low chemical abundances. It became clear shortly after the discovery of these objects that their properties are most naturally explained by current vigorous star-formation activity. Specifically, the blue colors and strong emission lines suggest significant numbers of massive blue, ionizing stars.

One of the first HII galaxies studied in greater detail carries the name I Zw 18, which was first introduced in Chapter 5. Searle & Sargent (1972) immediately drew attention to the very low metal abundance of this object. They derived an oxygen abundance of $\log(\text{O/H}) + 12 = 7.2$, i.e., only a few percent of the solar value. Such a small oxygen abundance was not expected in galaxies with active star formation when the nature of I Zw 18 was uncovered some 35 years ago. As a result, I Zw 18 imposes strong constraints on models for the chemical evolution of galaxies, an issue we will return to in Section 11.3. Somewhat ironically, I Zw 18 is still the record holder among all star-forming galaxies, except for perhaps SBS 0335-052 (Izotov, Thuan, & Guzeva 2005). Despite numerous efforts, no star-forming galaxy with lower metal content has been found to date.

A recent image of I Zw 18 taken with HST is reproduced in Fig. 9.11. Dramatic morphological detail can be seen. The two major star-formation regions harbor thousands of massive stars which are responsible for the observed emission-line spectrum.

Owing to the fact that emission-line galaxies are easy to detect and therefore are useful as cosmological probes, several very deep surveys specifically targeting HII galaxies were conducted in recent years. Almost all such surveys have been carried out using the wide-field imaging capabilities of Schmidt telescopes and special detection techniques based on some tracer of galaxy activity. Examples of such tracers at optical wavelengths are a relative enhancement of the continuum radiation at short wavelengths and/or enhanced blue color and the existence of emission lines from hot gas near active regions. The technique of selecting

Fig. 9.11 HST ACS multiband image of I Zw 18. The field size is approximately 20″. Two major star-formation regions can be seen. The filamentary structures are shells powered by stellar winds and supernovae. Reproduced from Izotov & Thuan (2004) by permission of the AAS.

objects according to their excess UV emission on low-dispersion objective-prism spectra was introduced by Markarian (1967). More recently, the selection of galaxies according to the presence of emission lines in their low-dispersion, objective-prism spectra has become popular. Examples are the Universidad Complutense de Madrid survey (Alonso *et al.* 1999), or the Kitt Peak International Spectroscopic Survey (KISS; Salzer *et al.* 2000).

KISS is an objective-prism survey aiming to detect large numbers of extragalactic emission-line sources. The survey method is similar to previous surveys carried out with Schmidt telescopes and photographic plates. The main characteristic that distinguishes KISS from earlier work is the utilization of a CCD as the detector. With large-format CCDs, the areal coverage possible in combination with a Schmidt telescope allows for substantial improvements compared with previous photographic surveys. CCDs offer high quantum efficiency, low noise, good spectral response over the entire optical portion of the spectrum, and large dynamic range. CCDs also produce a high-quality survey whose selection function and completeness limits can be accurately quantified, therefore allowing a broad range of studies requiring statistically complete galaxy samples.

In Fig. 9.12 we have plotted the line ratios of [NII]λ6583/Hα and [OIII]λ5007/Hβ as found in a sample of KISS galaxies. The selected ratios are a powerful tool to distinguish HII

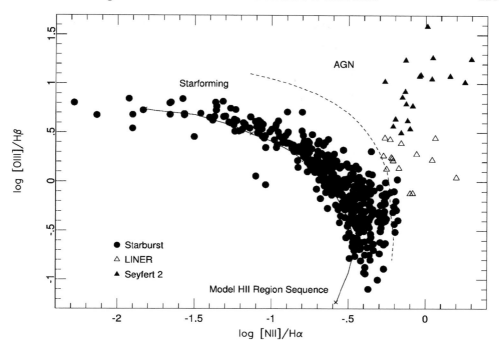

Fig. 9.12 Diagnostic diagram showing the relationship between [NII]λ6583/Hα and [OIII]λ5007/Hβ. Starburst galaxies are shown as filled circles, LINERs as open triangles, and Seyfert 2 galaxies as filled triangles. The dashed line roughly separates the star-forming galaxies from the active galactic nuclei. The solid line represents an HII model sequence at various metallicities, from 0.1 $Z_\odot$ at the upper left to 2 $Z_\odot$ at the lower right. Reproduced from Salzer *et al.* (2005a) by permission of the AAS.

galaxies from objects containing active nuclei. Star-forming galaxies follow a well-defined relation which is largely determined by the electron temperature of the ionized gas. The latter is, among others, a function of the metallicity of the gas. Lower metals leads to lower cooling of the gas so that low-ionization lines like [NII] become weaker. Conversely, the high-ionization line [OIII] increases in strength.

The KISS data allow a comprehensive characterization of the emission-line galaxy sample. Figure 9.13 summarizes color, reddening, and ionization properties. The data were cross-correlated with the 2MASS catalog, an IR-selected database. The optically selected sample tends to be bluer, less luminous and with a larger detected fraction of ionizing stars. (The latter follows from the behavior of the Hα equivalent width, which is a measure of the age.)

In many aspects, HII galaxies bridge the gap between giant HII regions at the low-luminosity end, and genuine starburst galaxies at the high end. One distinguishing property of the latter are their excessively high specific star-formation rates. We will discuss starburst galaxies in the next chapter. HII galaxies tend to be less dusty than the starburst class as a whole, and their properties are determined by young OB stars. This makes them useful training sets for applying the stellar population diagnostics for OB stars introduced in Section 9.3.

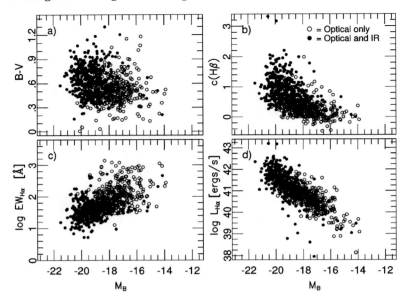

Fig. 9.13 Physical parameters for KISS star-forming galaxies plotted as a function of B-band absolute magnitude: (a) $(B - V)$ color; (b) the Balmer decrement reddening parameter $c(H\beta)$; (c) the logarithm of the $H\alpha$ equivalent width (in Å); and (d) the logarithm of the $H\alpha$ line luminosity (in erg s^{-1}). Filled symbols reflect objects detected by 2MASS in the J band, while open symbols indicate those that are not detected in the near-IR. Note the tendency for the lower luminosity galaxies to not be detected in the near-IR. Reproduced from Salzer *et al.* (2005b) by permission of the AAS.

The most spectacular starbursts are IR-bright or -luminous, suggesting the presence of a significant amount of dust which converts massive-star photons from ultraviolet to infrared wavelengths. Despite strong dust absorption, IR-bright starburst galaxies need not be UV-faint. Weedman (1991) estimated from a comparison of the far-IR and UV luminosity functions of a sample of Markarian galaxies that about 10% of the (non-ionizing) UV radiation escapes from starburst galaxies with large amounts of dust. This escape fraction may approach 100% in the absence of dust. A picket-fence model for the interstellar dust provides a natural explanation: the dust in the ISM is clumpy (Calzetti 2001) so that the escape probability for UV radiation is non-negligible even in dusty, IR-bright galaxies. HII galaxies, on the other hand, may reach escape fractions close to 100%.[6] This property makes them particularly suitable for UV studies, which can give us direct access to the *stellar* light. Consequently, observations of HII galaxies in the space-UV, mostly with the IUE, HST, HUT, FUSE, and Galaxy Evolution Explorer (GALEX) satellites, have dramatically increased our understanding of the starburst phenomenon in lower-mass, dust-free galaxies.

IUE was the first mission to collect a significant number of HII galaxy spectra. Over its lifetime, several hundred scientifically useful spectra of non-stellar objects were obtained, many of them from HII galaxies. An atlas with representative spectra was published by

---

[6] To avoid confusion, we emphasize that these numbers refer to the *non-ionizing* radiation at wavelengths above 912 Å. The fraction below 912 Å is much lower, as we will discuss at the end of this section.

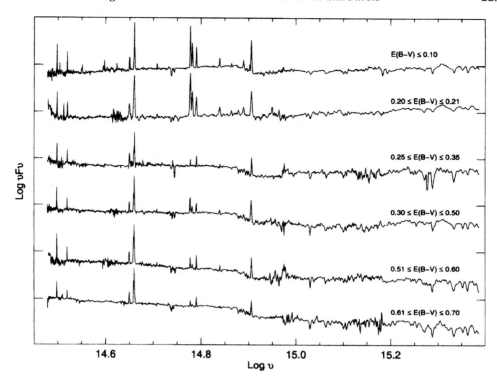

Fig. 9.14 Templates of starburst galaxies observed through similar optical and ultraviolet apertures. The six groups have different interstellar reddening. The wavelength range is approximately 10 000 to 1200 Å. Note the *spectral dichotomy*: an almost pure emission-line spectrum is seen longward of the Balmer jump, whereas an absorption-line spectrum is observed at shorter wavelengths. Reproduced from Kinney *et al.* (1996) by permission of the AAS.

Kinney *et al.* (1993). Kinney *et al.* (1996) utilized the same dataset to create template spectra of typical star-forming galaxies for a range of $E(B - V)$. The templates are shown in Fig. 9.14. Each spectrum is the sum of about five individual spectra of galaxies falling into the same $E(B - V)$ interval. The IUE data were combined with ground-based optical data obtained with an aperture of similar size. Galaxies with little reddening (uppermost spectrum in Fig. 9.14) have a rather blue continuum. The equivalent stellar spectral type would be about B0. In the absence of any reddening, a young stellar population of age 20 Myr or less has a predicted energy distribution $F_\lambda \propto \lambda^{-2.5}$ in the satellite-UV (Leitherer & Heckman 1995). The theoretically predicted spectral slope is quite insensitive to variations of the assumed underlying population. This is because the dominant contributors to the UV continuum light are stars in the mass range 10–30 $M_\odot$, which are observed near the Rayleigh–Jeans part of their spectrum. The fact that HII galaxies are often close to this slope is of course caused by a selection effect: most objects were selected from emission-line surveys, which by definition must have a dominant OB population. Once the population ages beyond 20 Myr, the spectrum becomes shallower.

Calzetti *et al.* (1994) made use of the lack of sensitivity of the UV continuum to population variations and ascribed deviations from the standard relation $F_\lambda \propto \lambda^{-2.5}$ to interstellar reddening. The redder continuum of the templates in the lower part of Fig. 9.14 is therefore predominantly due to larger interstellar reddening. This allowed Calzetti *et al.* to derive UV extinction laws in starburst galaxies which are "grayer" than the canonical LMC/SMC laws (recall Fig. 8.2). Their extinction laws should be appropriate, e.g., for application to star-forming galaxies at high redshift. In comparison with the SMC law, the "gray" law leads to lower star-formation rates.

The contrast between the optical and the UV *line* spectrum in Fig. 9.14 is striking. Superposed on a continuum due to B and A stars, a rich emission-line spectrum is observed in the optical. This is the classical HII region spectrum. In contrast, the UV shows essentially only stellar and interstellar absorption lines. It is only the UV where one can directly observe stellar lines from an OB-star population. The behavior of the spectra in Fig. 9.14 can easily be understood from the fact that massive stars have stellar winds and are surrounded by ionized gas. The winds give rise to the strongest UV absorption lines, and the ionized gas is the source of the emission-line spectrum.

HST is needed for a more detailed look at individual UV line profiles. An example is in Fig 9.15, where an HST spectrum of the blue compact dwarf galaxy He 2-10 is reproduced. From UV imaging, Conti & Vacca (1994) could resolve the star-formation activity into numerous small sub-units, whose characteristic sizes are 10 pc or less. The sizes and masses are typical for globular clusters. He 2-10 is forming extremely young globular clusters as the result of violent star formation. The ultraviolet spectrum is consistent with this suggestion. It shows pronounced P Cygni profiles of CIV λ1550 and SiIV λ1400 due to winds from massive stars.

The potential of stellar-wind lines, like CIV λ1550 and SiIV λ1400, for population studies was discussed in the previous section. An important consideration is the size of the entrance aperture of the spectrograph. IUE's capability to resolve starbursts spatially was limited by its $10'' \times 20''$ entrance aperture. At a distance of 10 Mpc, this corresponds to a linear size of $\sim$1 kpc, roughly the entire 30 Doradus region at that distance. On the other hand, R136, the central 30 Dor starburst cluster, would subtend an angle of only 0.05'' at 10 Mpc (see Table 9.2). HST's STIS is required to isolate regions that small. Since the star-formation process can be time-dependent, a loss of spatial information translates into a loss of information on the evolution of the starburst.

STIS on HST permits *line-profile* studies of HII galaxies on spatial scales matching individual star-forming regions. The main results are:

- stars as massive as $\sim$100 $M_\odot$ form in HII galaxies, and their IMF has a slope close to the Salpeter value;
- ages are between 3 and 6 Myr, consistent with emission-line results;
- the duration of an individual star-formation event (i.e., in a single cluster) is short in comparison with the stellar evolution timescale.

The *stellar* lines observed with HST in starburst galaxies are consistent with model predictions for standard massive-star populations. *Interstellar* lines, however, are stronger than expected if the same ISM conditions as in the vicinity of the Sun prevailed. For instance, Conti, Leitherer, & Vacca (1996) measured average equivalent widths of unblended interstellar lines in the HII galaxy NGC 1741 around 2 Å, which is fairly typical for other HII

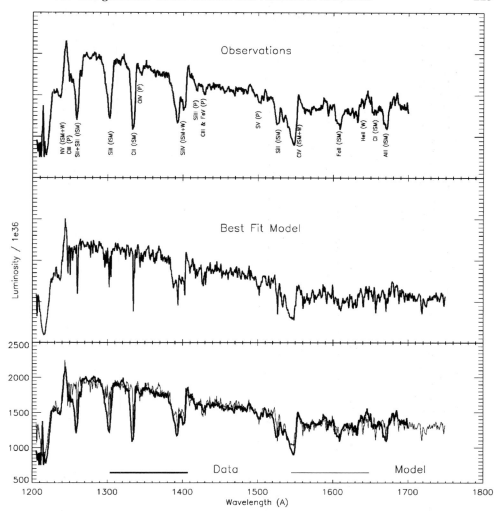

Fig. 9.15 UV spectrum of He 2-10, shown in the top panel with some important lines marked. "IS" indicates that the line comes predominantly from the interstellar medium, "W" indicates lines arising in stellar winds, and "P" indicates lines arising in stellar photospheres. The middle panel shows a best-fit instantaneous burst model (age of 5 Myr), and the bottom panel overplots the observations and best-fit model. Reproduced from Chandar *et al.* (2003) by permission of the AAS.

galaxies as well. At 1500 Å, this corresponds to a velocity dispersion $\sigma > 100$ km s^{-1} if the broadening were due to a single, virialized gravitational motion. Therefore, the implied gravitational masses would be in excess of 10^{11} $M_{\odot}$, too high for this galaxy. More likely, the observed equivalent width could be the result of many unresolved, unsaturated, narrow interstellar lines over a velocity range of a few hundred km s^{-1}. Shells and bubbles around individual starbursts, or even a large-scale outflow of the ISM in NGC 1741, could be responsible.

The situation in NGC 1741 (and most other HII galaxies as well) is somewhat inconclusive since any interstellar line in a young starburst could have a non-negligible stellar-wind contribution. A clearer picture emerges from galaxies whose star clusters are sufficiently old that a stellar population with strong winds is no longer present. The nearby ($D = 6$ Mpc) dwarf irregular NGC 1705 is an example of such a system (Meurer *et al.* 1992).

NGC 1705 hosts one of the brightest examples of a "super star cluster", a concentration of at least 10^5 to 10^6 $M_\odot$ of hot stars within less than a few pc (O'Connell, Gallagher, & Hunter 1994). We already introduced this cluster in Table 9.3, which compares R136 with other star-forming regions. A section of an HST GHRS spectrum of this cluster is shown in Fig. 9.16. The UV spectrum of this cluster is dominated mostly by interstellar and by some B-star lines. Superficially, especially when observed at low spectral resolution, the spectrum of the NGC 1705 cluster resembles a young, O-star dominated starburst cluster. However, a detailed study of the line spectrum suggests a pure interstellar origin with very little stellar contribution (Vázquez *et al.* 2004). The lines are *resolved* with the GHRS. Average full width at half-maximum (FWHM) values after deconvolution are $\sim$120 km s^{-1}. If interpreted in terms of a velocity dispersion, this corresponds to $\sigma \approx 50$ km s^{-1}. This is in conflict with independent measurements taken in the optical. Ho & Filippenko (1996) obtained a *stellar* velocity dispersion measurement of $\sigma \approx 10$ km s^{-1} using narrow red-supergiant lines in the 7000 Å region of the spectrum. The large discrepancy between the two values suggests that the interstellar lines indicate not virialized but rather random gas motion. This is consistent with their large equivalent widths, which suggest additional, non-gravitational forces as well

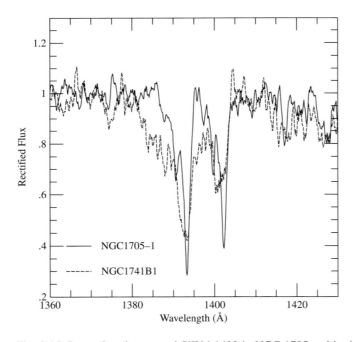

Fig. 9.16 Spectral region around SiIV λ1400 in NGC 1705 and in the brightest region of NGC 1741. The spectra were normalized and the wavelength scales are in the restframe of each galaxy. Note the blue wings in the NGC 1741 spectrum, indicating hot-star winds. Reproduced from Heckman & Leitherer (1997) by permission of the AAS.

(see above). The implication is that interstellar absorption lines in HII galaxies are not reliable probes of galaxy masses.

All interstellar lines in NGC 1705 have a blueshift of $\sim$80 km s^{-1} with respect to stellar (photospheric) lines. The ISM is driven out of NGC 1705 as a result of powering by stellar winds and supernovae during the starburst phase. Starburst-driven superwinds are predicted theoretically and have been observed via emission in hot (X-rays) and cold (optical recombination lines) gas (e.g., Heckman 2003). We will discuss superwinds in more detail in Section 10.4.

There is manifold evidence for turbulence, macroscopic motion, and large-scale outflows in HII galaxies. In some cases (e.g., Leitherer *et al.* 1996 for NGC 4214) evidence has been found for individual HII regions being density-bounded, and the cause has been ascribed to the porous state of the ISM. Are HII galaxies as a whole density-bounded? The answer to this question has far-reaching implications for the ionization of the early Universe (Madau & Shull 1996). Star-forming galaxies are the dominant contributor to the *non-ionizing* UV radiation field in the Universe. Are they a significant component of the *ionizing* background as well? Simple arguments might suggest otherwise. An HI column density of $\sim 1 \times 10^{18}$ cm^{-2} is sufficient to absorb essentially all the ionizing radiation. Since the measured extinctions in HII galaxies imply column densities that are three or four orders of magnitude higher than this, it might appear that essentially no ionizing radiation can escape. However, the porosity of the ISM seen in the non-ionizing continuum ($\lambda > 912$ Å) could very well extend below the Lyman edge and may dominate the shape of the emergent spectrum. The situation is sufficiently complex that the only way to determine the escape fraction f_{esc} of the ionizing radiation is via a direct measurement.

Attempts to measure f_{esc} fall into two categories: observations of local galaxies with a far-UV detector, or measurements using galaxies at cosmological redshift, which are accessible from the ground with 8-m class telescopes. Either technique has its advantages and disadvantages. The "local" approach faces the obvious challenge of extreme UV observations, whereas the "distant" measurement is affected by the radiative transfer in the intergalactic medium. In addition, a somewhat less direct method is to determine the Lyman continuum opacity from a comparison of the Hα and UV luminosity functions in the local Universe.

Almost all studies find more or less stringent upper limits on f_{esc} both in the low- and high-redshift Universe. The ISM in the observed galaxies is highly opaque, and very little stellar ionizing radiation escapes. Two notable exceptions stick out. In the local Universe, Bergvall *et al.* (2006) observed the blue compact dwarf galaxy Haro 11 with the FUSE satellite and detected significant emission below 912 Å. Their spectrum is reproduced in Fig. 9.17. The excess radiation in the Lyman continuum, if real and not an observational artifact, would translate into $f_{esc} = 4$–10%. What makes Haro 11 special among star-forming galaxies? We may speculate that Lyman radiation can only escape through few, small regions of low opacity, which then lead to a small solid angle of the radiation. This results in a small detection probability.

In the high-redshift Universe, Steidel, Pettini, & Adelberger (2001) detected significant Lyman-continuum flux in the composite spectrum of 29 Lyman-break galaxies with redshifts $z = 3.40 \pm 0.09$. If the inferred escaping Lyman-continuum radiation were typical of galaxies at $z \approx 3$, then these galaxies produce about five times more H-ionizing photons than quasars at this redshift, with the obvious cosmological implications. Shapley *et al.* (2006) obtained rest-frame UV spectroscopic observations of a different sample of 14 star-forming galaxies at

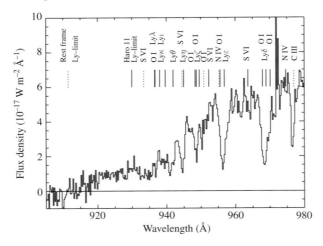

Fig. 9.17 FUSE spectrum of Haro 11 plotted vs. *observed* wavelength scale. The strongest spectral lines are identified and indicate the *intrinsic* wavelength scale. Note the detection of emission shortward of 912 Å at the intrinsic wavelength of Haro 11. From Bergvall *et al.* (2006).

redshift 3. They detected escaping ionizing radiation from two individual galaxies. However, the observed far-UV spectra of the whole sample exhibit significant variance, with no clear evidence of the factors controlling the detection or non-detection of escaping Lyman radiation. Understanding the source of this variance will be a major future objective of observational cosmology.

10

Starburst phenomena

The phrase "starburst" was used before in this volume, referring to a region with intense star formation. In this chapter we will define and discuss starbursts in more rigorous terms and relate them to the properties of the host galaxies. While luminous HII regions are often considered starbursts as well, starburst phenomena are more general and have profound cosmological implications. A major theme of this chapter, in contrast to the *local* focus of Chapter 9, will be the *global* interrelation between a starburst region and its environment. The starburst is nourished by the gas supply of the host, and it energizes the galaxy ISM with stellar winds and supernovae.

10.1 Definition of a starburst

One of the earliest uses of a term similar to "starburst" goes back to a seminal publication by Searle, Sargent, & Bagnuolo (1973). These authors added data points of many late-type galaxies to a theoretical two-color diagram which plots $(B-V)$ against $(U-B)$ color. Surprisingly, the galaxies had bluer colors than expected for a stellar population either evolving with steady, continuous star formation or with no star formation at all after the formation of the first stellar generation. This immediately suggested that some galaxies are observed in a special epoch: they are currently forming stars at a higher rate than previously, otherwise the bluer colors, indicative of an excess of hot stars, would not be understandable. Statistical arguments then favored brief "flashes" of star formation, a term which gradually evolved into a "burst" of star formation, or starburst. The latter terminology was coined by Weedman *et al.* (1981) and applied to NGC 7714. It has since become established in the literature.

It is not a coincidence that this discovery was made around 1970. While the colors of galaxies could be measured long before that, relating them to star-formation histories became feasible only after stellar evolution was understood and could be modeled numerically on large computers. Advances to link stellar evolution, star-formation histories, and spectrophotometric properties of galaxies are largely credited to Beatrice Tinsley's efforts (e.g., Tinsley 1968).

In general terms, starbursts are concentrated regions of ongoing or recent star formation. They range between 10^{12} and 10^7 in bolometric luminosity L_{Bol}, the high end overlapping with typical quasi-stellar object (QSO) luminosities, and the low end with values observed for very luminous HII regions. The broad range and the overlaps give rise to ambiguity and confusion, making starburst galaxies a mixed bag of morphological types. While there is no single agreed-upon starburst definition in the literature, three approaches are commonly used to characterize the starburst phenomenon.

Definition by observational selection technique   The most commonly used criterion to classify starbursts (and unfortunately the least rigorous) is to include all objects detected in flux- or volume-limited surveys which are sensitive to high-mass star formation. The major techniques are emission-line surveys, searches for objects with UV excess, and far-IR surveys with space missions, such as IRAS, ISO, Spitzer, or GALEX. Obviously these surveys are biased towards and against different subsets of the starburst class, with the most overlap between emission-line and UV surveys. Far-IR surveys favor luminous, dust-rich starbursts, which in turn tend to be under-represented in the other surveys. A famous example of an IR-selected starburst galaxy is M82. On the other hand, HII galaxies and blue compact dwarf galaxies are the typical members of emission-line and UV-excess catalogs.

Definition by the relation between starburst and galaxy host luminosity   We can define a starburst in terms of its *relative* luminosity with respect to that of its host galaxy. With this definition, the luminosity of a *starburst galaxy* is completely dominated by the starburst luminosity ($L_{\text{burst}} \gg L_{\text{galaxy}}$). On the other hand, if the starburst luminosity is substantial but smaller than or comparable to that of the host galaxy luminosity ($L_{\text{burst}} \lesssim L_{\text{galaxy}}$), a *starburst region* in a galaxy is observed. Finally, if $L_{\text{burst}} \ll L_{\text{galaxy}}$ for any individual star-forming region, the object is classified as a *star-forming galaxy*. This scheme was first proposed by Terlevich (1997). It has the advantage of being distance independent since it avoids basing the starburst definition on a relative luminosity.

Definition by the gas consumption timescale   The duration of a starburst must be short on a cosmological timescale. This can be seen from an illustrative example: IR-luminous starburst galaxies often have star-formation rates of order $100\,M_\odot\,\text{yr}^{-1}$. If these rates were sustained over 10^8 yr, about $10^{10}\,M_\odot$ of molecular gas would be consumed in the star-formation process. This exceeds the gas reservoir of all but the most luminous galaxies. However, the gas depletion argument is difficult to apply quantitatively for a variety of reasons. For instance, the unknown proportion of low-mass stars formed in the starburst makes estimates of the total (integrated over all masses) star-formation rate uncertain. A fundamental difficulty in determining the gas depletion time is the breakdown of a closed box model of a starburst, which ignores the various sources and sinks of material. Sources of gas replenishment are interstellar gas feeding the starburst and mass return by winds and supernovae. Conversely, supernova-driven outflows will remove material from the starburst site. The exact balance between these processes is poorly known. Therefore the definition of a starburst via the gas exhaustion timescale is often impractical.

In this volume we will adopt a starburst definition in terms of its luminosity relative to the galaxy host. Very often this approach can help unify the properties of starburst galaxies which initially appear strikingly different in their stellar content. The two objects NGC 7714 and M82 serve as illustrative examples. They have comparable bolometric luminosities of 2–$3 \times 10^{10}\,L_\odot$, as determined from their far-IR fluxes, and they both harbor central starbursts whose luminosities are comparable to the far-IR luminosity of the entire galaxy. Both are frequently considered "prototypical" starburst galaxies, NGC 7714 as a representative of a UV-excess galaxy, and M82 as an IR-bright galaxy. The two objects are complementary in their properties. Despite having similar chemical composition and luminosity, their apparent morphologies are strikingly different. The SBb galaxy NGC 7714 is seen face-on, thereby minimizing dust obscuration effects. This allows stellar population studies in all wavebands where stellar light is emitted. M82, in contrast, is observed almost edge-on, and high dust obscuration hides most of the *stellar* light in the UV and optical. This geometry, however, favors direct X-ray observations of starburst-driven interstellar *gas* along the minor axis. This outflow is very difficult to detect in the pole-on galaxy NGC 7714.

After taking into account the different aspect angles, the two galaxies become very similar in many aspects. What distinguishes them is the dust column along the line of sight. M82 is viewed almost edge-on, leading to a large dust column, and therefore large obscuration, whereas NGC 7714 is seen closer to pole-on, and the partially naked core is exposed. One can speculate whether NGC 7714 would resemble M82 if their inclinations were similar.

10.2 The starburst IMF

In a starburst, the total number of stars $N(\Delta M, T)$ in the mass interval M_1 to M_2, formed since star formation began at t_0 and until the present epoch T, can be written as

$$N(\Delta M, T) = \int_{M_1}^{M_2} \int_{t_0}^{T} \phi(M)\psi(t)\mathrm{d}M\mathrm{d}t, \tag{10.1}$$

where $\Delta M = M_2 - M_1$. This relation is valid only if the lifetime $\tau(M)$ of a star of mass M is smaller than T because otherwise the star would no longer be observable and it would be missed in our count. (Recall that $\phi(M)$ is the *initial* mass function which requires an extrapolation back to the ZAMS.) The left-hand term in Eq. (10.1) is the quantity accessible to observations, for instance by counting stellar types and assigning masses via evolutionary tracks. In order to derive the IMF, $\psi(t)$ must be determined as well, which can be a formidable task. In general, stars can form out of interstellar gas at a rate

$$\psi(t) = \psi(t_0)\mathrm{e}^{-t/\tau_{\mathrm{SF}}}, \tag{10.2}$$

where t_0 and τ_{SF} denote the onset and the duration of the star-formation (or starburst) event, respectively.

Two limiting cases are of particular interest. If the starburst duration is short compared to the age ($\tau_{\mathrm{SF}} \ll t$), we have a case where most stars form rapidly at some epoch, with little subsequent star formation. This is called an instantaneous starburst and corresponds to $\tau_{\mathrm{SF}} \to 0$ in Eq. (10.2). Instantaneous starbursts are often found in small clusters or in the very centers of galaxies where star formation is confined to small spatial sizes. Mathematically, $\psi(t)$ approaches a δ-function, and Eq. (10.1) can be simplified into a single integral over the IMF:

$$N(\Delta M) = C \int_{M_1}^{M_2} \phi(M)\mathrm{d}M. \tag{10.3}$$

The constant C follows from the IMF normalization and the total mass of the starburst. In other words, the star numbers are just proportional to the IMF.

Like Eq. (10.1), Eq. (10.3) applies only to starbursts whose age is less than the age of the most massive stars formed in the starburst, as these stars have the shortest lifetimes. If we observe an older starburst, we have no information about the properties of the stars whose lives have already ended as supernovae or black holes. In this case we can derive the *present-day* mass function (PDMF). The PDMF is identical to the IMF below a mass corresponding to the stellar mass whose $\tau(M)$ is smaller than the age of the starburst. Above this mass, the PDMF is a lower limit to the mass prediction by the IMF.

Another limiting case of the star-formation law assumes that the star-formation rate is constant with time ($\psi(t)$ = constant). Such a law is characteristic for large systems, like the disk of our Galaxy. The two cases of an instantaneous burst and a constant star formation will always bracket the real astrophysical case. Star formation proceeding at a constant rate can be theoretically approximated as a series of instantaneous events. Observationally, this may resemble a system composed of numerous individual starburst regions with an age range comparable to τ_{SF}. If $\psi(t)$ = constant, Eq. (10.1) can again be simplified. As before, we need to distinguish between observations at an epoch T younger or older than the evolutionary lifetimes τ of the least massive stars formed, respectively. If $T < \tau(M)$, Eq. (10.1) becomes

$$N(\Delta M) = \langle \psi(t) \rangle T \int_{M_1}^{M_2} \phi(M) \mathrm{d}M. \tag{10.4}$$

$\langle \psi(t) \rangle$ is the time-averaged star-formation rate. The physical interpretation of Eq. (10.4) is that all stars are still prior to their death and their number can be calculated with a relation that is the equivalent of Eq. (10.3) for an instantaneous burst. Alternatively, for $T \gg \tau(M)$ there is an equilibrium between star birth and death for all masses considered. We observe only stars formed after $t = T - \tau(M)$, and Eq. (10.1) becomes:

$$N(\Delta M) = \int_{M_1}^{M_2} \int_{T-\tau(M)}^{T} \phi(M) \psi(t) \mathrm{d}M \mathrm{d}t. \tag{10.5}$$

This equation can be re-written as

$$N(\Delta M) = \psi(T) \int_{M_1}^{M_2} \phi(M) \tau(M) \mathrm{d}M. \tag{10.6}$$

Here, the star numbers we observe are simply the product of the current star-formation rate and the IMF-weighted stellar lifetimes.

Before proceeding with the IMF in starburst galaxies, we will discuss representative results for the IMF in nearby, young star clusters, following on from results presented in Section 7.2. Clusters have relatively secure distances, a prerequisite for assigning masses from evolution models by placing the stars on the H-R diagram. Furthermore, cluster populations are close to being co-eval, i.e., they form a single stellar population without significant age spread during formation. This is the case $\tau_{SF} \to 0$ in Eq. (10.2). The duration of the star-formation event must be less than the evolutionary timescales of the most massive stars (~ 3 Myr) for this condition to be met.

The most comprehensive survey of high-mass star formation regions in the Galaxy was done by Phil Massey and his collaborators. One example is their study of the star clusters Trumpler 14 and 16. A near-IR image reproduced in Fig. 10.1 shows this spectacular massive-star nursery. The region, although minuscule in comparison with powerful extragalactic starbursts, is the birthplace of some of the most massive stars known in our Galaxy. The corresponding H-R diagram, constructed from spectroscopy and photometry of the most luminous stars, and the derived IMF are shown in Fig. 10.2. Despite being a rich young cluster by Galactic standards, the upper part of the H-R diagram is rather sparsely populated. A total of 82 stars more massive than $10\,M_\odot$ are found, and only a few are above $60\,M_\odot$.

Fig. 10.1 Near-IR image of the two galactic star clusters Trumpler 14 and 16. The clusters are in the upper right (Tr 14) and lower left (Tr 16) portions of the diagonal nebulosity in the image. This spectacular region contains an unusually high concentration of young massive stars. The region is the result of a burst of star formation about 3 Myr ago. Image obtained as part of the Two Micron All Sky Survey (2MASS), a joint project of the University of Massachusetts and the Infrared Processing and Analysis Center/California Institute of Technology, funded by NASA and NSF.

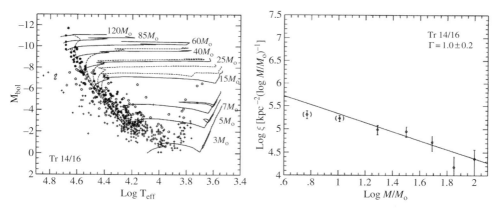

Fig. 10.2 Left: observational H-R diagram of the Trumpler 14/16 region. Superposed are theoretical evolutionary tracks. Right: derived IMF. Reproduced from Massey *et al.* (1995a) by permission of the AAS.

Massive stars are rare! The most massive stars have masses around $100 \, M_\odot$. Essentially all stars with good data are close to the ZAMS – consistent with the young age of the cluster. The derived massive-star IMF has a slope of $\alpha = 2.0 \pm 0.2$. The age is less than 3 Myr, suggesting that even the most massive stars have not yet evolved through to core-collapse supernovae. Therefore the PDMF is virtually identical to the IMF.

Other young clusters in the Milky Way give similar results: the IMF for masses above $\sim 10 M_\odot$ typically has a power-law slope with $2 < \alpha < 2.5$. A comprehensive summary can be found in Kroupa (2001). Very often this slope is referred to as the "Salpeter" slope, as this is the same slope derived in Salpeter's original study of the solar neighborhood IMF. The IMF can be reasonably well approximated by a power-law exponent, apparently to solar mass and below. Since a power law does not have a preferred mass, we call the power-law approximation a scale-free law. The IMF is scale-free over one to two orders of magnitude in mass at the massive end. Somewhere around spectral type late K to early M, corresponding to masses around $0.5 \, M_\odot$, the IMF flattens and eventually turns over. The extreme low-mass end of the IMF is quite uncertain, even in Galactic clusters, and the more so in starburst galaxies. Depending on the precise IMF behavior at low masses, one can deduce a preferred mass-scale from the condition $d\phi(M)/dM = 0$, i.e. the mass interval where the IMF is flat and falls off towards both lower and higher masses. The preferred mass in Galactic clusters is around $0.3 \, M_\odot$, suggesting that the star-formation process favors M dwarfs.

In Table 10.1 we illustrate the predominant output of light and mass by high- and low-mass stars, respectively. Consider a burst of star formation constantly forming stars observed at an age of 10 Myr. The stars have solar metallicity and follow an IMF with a slope $\alpha = 2.35$. From synthesis models we find that this population will emit $N(\mathrm{LyC}) = 2.2 \times 10^{53} \, \mathrm{s}^{-1}$ if the stars form at a rate of $1.0 \, M_\odot \, \mathrm{yr}^{-1}$ and if the total stellar mass is spread out between lower (M_low) and upper (M_up) cut-off masses of 1 and $100 \, M_\odot$, respectively. This reference model is listed in the first row of Table 10.1. Now we vary either M_low or M_up and calculate the star-formation rate required to reproduce the same $N(\mathrm{LyC})$ as our reference model.

Table 10.1 *Star-formation rates producing identical $N(\mathrm{LyC})$ for different IMF cut-off masses.* $Z = Z_\odot$, $\alpha = 2.35$, $t = 10 \, \mathrm{Myr}$.

M_low	M_up	SF rate
	$M_\odot$	$M_\odot \, \mathrm{yr}^{-1}$
1	100	1.00
1	60	1.59
1	30	4.67
10	100	0.310
5	100	0.462
3	100	0.601
0.3	100	1.66
0.1	100	2.55

Decreasing M_up from 100 to $60 \, M_\odot$ removes the hottest stars from the population. These stars have the highest photon output, and we need to increase the overall star-formation rate by a factor of 1.59 in order to match the original $N(\mathrm{LyC})$. Despite their high photon output, removing stars with masses above $60\,M_\odot$ is not all that significant because of their small number in the overall population. However, removing even more stars down to $M_{up} = 30\,M_\odot$ leads to a drastic drop in $N(LyC)$ (third row in Table 10.1): we have to increase the star-formation rate by a factor of 4.67 in order to make up for them.

We can simulate the effect of a "top-heavy" IMF by increasing M_{low} to 3, 5, and $10\,M_\odot$. This produces the opposite trend from before. The mass of newly formed stars is allocated to massive, luminous stars instead of less massive and less luminous stars which are negligible providers of $N(LyC)$. Therefore smaller star-formation rates are needed to reach $N(LyC) = 2.2 \times 10^{53}\,s^{-1}$. Conversely, a decrease of M_{low} to values below $1\,M_\odot$ requires larger star-formation rates. For instance lowering M_{low} from 1 to $0.1\,M_\odot$ corresponds to an increase of the star-formation rate by a factor of 2.55. In this case, we spread out the available mass to include a large number of non-ionizing stars. In order to meet our ionization requirement, we need a larger overall star-formation rate. In other words, including less and less massive stars in the IMF leaves $N(LyC)$ unchanged but adds to the total starburst mass.

With or without a turn-over at low masses, the mass contribution from low-mass stars is enormous when compared to that of high-mass stars. Even if the IMF were truncated at $1\,M_\odot$, most of the mass of a stellar population is locked in low-mass stars. For a Salpeter IMF, stars with masses between 1 and $10\,M_\odot$ contribute six times more *mass* to the total mass than all stars above $40\,M_\odot$. Conversely, most of the stellar *light* is provided by the most massive stars due to the steep stellar mass–luminosity relation. The relative mass proportions over different mass intervals can be calculated from

$$M(\Delta M) = C \int_{M_1}^{M_2} M\phi(M)\mathrm{d}M, \tag{10.7}$$

which is the equivalent to Eq. (10.3) but applies an additional IMF weighting over the mass. As we are interested in relative mass fractions only, we conveniently set the constant to 1 and integrate a power-law IMF between upper and lower limits M_2 and M_1, respectively:

$$M(\Delta M) = \frac{1}{2-\alpha} M^{2-\alpha} \Big|_{M_1}^{M_2}. \tag{10.8}$$

Using this relation, we find for a Salpeter IMF ($\alpha = 2.35$) with mass limits 1 and $100\,M_\odot$ that about half the stellar mass is below $4\,M_\odot$, and the other half is above.

Next we turn to the "Rosetta Stone" 30 Doradus, whose ionizing central cluster is reproduced in Fig. 9.3. The R136 cluster in the center of 30 Dor is the prototype of a small starburst. It is sufficiently massive that even the uppermost H-R diagram at $L_* > 10^5\,L_\odot$ is well populated. Such massive clusters are rare on a Galactic but frequent on a cosmological scale. Yet only R136 (together with a handful of other clusters in the Local Group of galaxies) is close enough for its massive-star content to be studied directly. Spectra of the bluest, most luminous stars indicate that many of these stars are of type O3, which are among the most luminous and most massive stars known. R136 alone hosts more O3 stars than the total inventory of this spectral class known elsewhere in the Galaxy and the Magellanic Clouds. Evolutionary models suggest that such stars must have ZAMS masses of at least $100\,M_\odot$. Does this excessively large number of very hot stars hint at a peculiar IMF? Massey & Hunter (1998) found the IMF to be completely normal, with a slope of 2.35. Their plot of the derived star numbers versus mass is in Fig. 10.3. The IMF was determined using the data reproduced in Fig. 9.5. R136 is rich in the most massive stars because it has formed a large number of stars of *all*

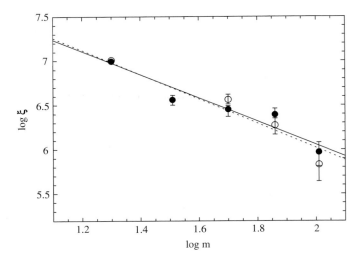

Fig. 10.3 IMF of R136. The quantity plotted is $\xi(M)$, the number of stars per logarithmic mass interval between 20 and 100 $M_\odot$. The derived slope is identical to a Salpeter slope with $\alpha = 2.35$. Reproduced from Massey & Hunter (1998) by permission of the AAS.

mass ranges. The similar IMF slope in R136 and in normal OB associations suggests that the distribution of masses is not affected by the stellar density. R136 is more than 100 times denser than a galactic OB association. To put the star density in perspective, we mention that R136 contains at least 10 000 massive stars, all concentrated in 1 pc. This is about the same scale as the distance between the Sun and our nearest neighbor α Centauri. Imagine 10 000 massive stars packed into the space between these two stars!

Even the closest genuine starburst galaxies are at distances making techniques such as those applied to R136 prohibitive. The "royal" way of obtaining an IMF is via star counts, preferably in connection with spectroscopy. Star clusters in galaxies outside the Local Group are no longer completely resolved into stars over a sufficient mass range, and the IMF must be derived from integrated light studies. In the following we will discuss case studies of the IMF in five representative galaxies whose main properties are summarized in Table 10.2. They are ordered by increasing distance from 2 to 50 Mpc and cover a broad range of luminosity and metallicity.

Table 10.2 *Parameters of the five galaxies with IMF determinations discussed in the text.*

Galaxy	d (Mpc)	$\log L$ ($L_\odot$)	Z ($Z_\odot$)
NGC 1569	2.2	9.2	1/4
M82	3.3	10.5	1/2
NGC 5253	4.1	8.9	1/6
I Zw 18	10	8.1	1/25
NGC 1741	50	10.3	1/4

The dwarf irregular NGC 1569 is one of the closest examples of a starburst galaxy. Most of its field population of stars with masses higher than a few $M_\odot$ was formed during a global star-formation episode beginning more than 100 Myr ago, and ending as recently as 10 Myr ago. The IMF in this mass range is a power law somewhat steeper than Salpeter, and the associated star-formation rate over the whole galaxy is about 1 $M_\odot$ yr^{-1} if the mass is distributed between 1 and 100 $M_\odot$. While this value may not sound too impressive (it is comparable to the total star-formation rate of the Milky Way), the normalized star-formation rate per unit surface reveals the extraordinary nature of this object. NGC 1569 forms stars at a rate of a few $M_\odot$ yr^{-1} kpc^{-2}, which is three orders of magnitude higher than in our Galaxy and a factor of 100 above the highest rates measured in dwarf irregular galaxies of the Local Group.

The low-mass end of the IMF in the NGC 1569 field is inaccessible to direct observations, even with the most powerful telescopes in existence. NGC 1569 hosts several "super star clusters", highly concentrated (half-light radius of a few pc), gravitationally bound stellar clusters which rival entire dwarf galaxies in luminosity ($M_B = -14$). The brightest super star cluster is one of the entries of Table 9.3 in the previous chapter. Faint red supergiant absorption-line features were detected and resolved in the integrated spectrum. The velocity dispersion is a measure of the gravitational well and allows a mass determination. Since the mass is dominated by low-mass stars, and the luminosity by bright, massive stars, the mass-to-light ratio is useful to constrain the proportion of low- to high-mass stars. This is a commonly used technique to infer low-luminosity objects whose observational signature is mainly gravitation. Bessel utilized it to infer a companion to Sirius in 1841, and the modern debate of dark matter in the Universe was preceded by the discovery of unseen matter in the rotation curves of galaxies. In the case of the NGC 1569 cluster, the mass and the mass-to-light ratio are consistent with an IMF having a Salpeter slope and with stars forming down to 1–2 $M_\odot$ (Origlia *et al.* 2001).

M82 is the prototype of a dusty starbursting dwarf galaxy. The terminology of a "top-heavy IMF" for this galaxy was coined in an influential paper by Rieke *et al.* (1980). "Top-heavy" indicates a higher proportion of red supergiants over giants and dwarfs than expected for a normal IMF. In terms of mass, a top-heavy IMF has an excess of stars in the mass range 10–20 $M_\odot$ over stars of 5 $M_\odot$ and less. Observational evidence for a top-heavy IMF in M82 comes from the relatively high K-band luminosity of the nucleus, together with its relatively low dynamical mass. This results in a low M/L ratio which can only be understood if red giants and dwarfs (which contribute mass) are deficient with respect to supergiants (which produce near-IR light). Low-mass cut-offs around 3–5 $M_\odot$ are required. Additional support for a top-heavy IMF has been found in some other galaxies as well, although the case is far from settled. Whether the M82 IMF is indeed top-heavy is critically dependent on the adopted dynamical mass, the extinction correction, and the star-formation history. If the starburst is not condensed in a central cluster but rather spread out over a few hundred pc as a result of sequential star formation, the observational constraints could in principle be satisfied with a standard Salpeter IMF extending to 0.1 $M_\odot$. However, such a model would use up 30% of the total dynamical mass. This is an uncomfortably large percentage since it is unlikely that the starburst accounts for most of the stellar mass, the constraint being the clear presence of an older pre-existing population.

Star clusters may provide indirect constraints on the low-mass IMF even if a dynamical mass determination is not available. The nearby blue amorphous dwarf galaxy NGC 5253 is an ideal laboratory to investigate starbursts down to pc sizes. At a distance of 3.3 Mpc, HST is capable of resolving structures as small as 1 pc. Although the outer regions of this galaxy are reminiscent of those in dwarf elliptical galaxies, star formation is active in the central kpc, as characterized by a high continuum surface brightness and strong Hα emission. The Hα luminosity within the central 2 by 2 kpc^2 suggests an average star formation rate of 0.1 $M_\odot$ yr^{-1}. High-resolution imaging resolves the starburst into multiple clusters with sizes, masses, and ages of 1–5 pc, 10^4–10^5 $M_\odot$, and tens of Myr, respectively. The two youngest clusters alone account for about half of the observed Hα emission. Star formation has been active for at least 100 Myr, with significant intensity variations at different locations in the central region. Spectra of individual clusters and of the diffuse field in NGC 5253 were obtained by Tremonti *et al.* (2001). Their UV spectra show the typical stellar-wind lines of CIV λ1550 and SiIV λ1400, indicative of hot stars in a young population. The spectra of the cluster and field population are quite different, as shown in Fig. 10.4. The cluster spectrum is characteristic of a single population with an age of a few Myr and containing massive stars

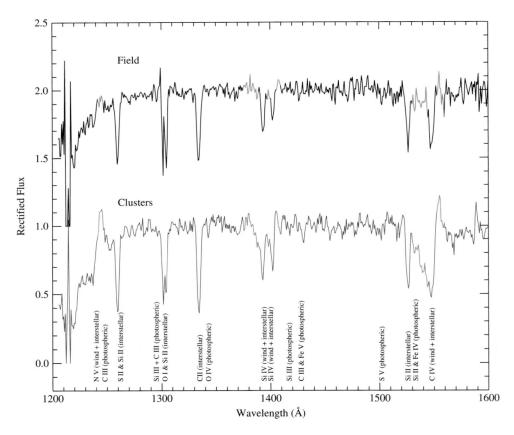

Fig. 10.4 Rectified spectrum of the field (top) vs. the combined spectrum of the clusters (bottom) in NGC 5253. The cluster spectrum shows the clear signatures of O stars in the broad stellar wind P Cygni features of NV λ1240, SiIV λ1400, and CIV λ1550. The corresponding features are greatly reduced in the B-star dominated field spectrum. Reproduced from Tremonti *et al.* (2001) by permission of the AAS.

up to ~ 100 $M_\odot$. In contrast, the field spectrum has weak wind lines, suggesting a deficit of very massive stars. The difference could be due to cluster evaporation in combination with the short lifetimes of massive stars. The short relaxation times due to the compactness of the clusters and the strong tidal fields in the center of NGC 5253 lead to their rapid evaporation. After cluster evaporation, the remaining less massive stars will be part of the surrounding field population. This picture would provide a natural explanation for an IMF with a higher proportion of very massive stars in clusters in comparison with the field.

Now we return to the extreme high-mass end of the IMF. W-R stars are a particularly sensitive tracer since they are considered the evolved descendants of the most massive O stars. As we explained in Chapter 5, W-R stars form out of more massive progenitors by rotation and stellar mass loss via radiatively driven winds. The mass-loss efficiency scales with the far-ultraviolet radiation field so that the most luminous, and therefore the most massive, stars have the strongest winds and are the most important evolutionary channel for W-R formation. W-R features have been detected in some fraction of HII galaxies. Their strength requires a massive star population following a Salpeter (or even slightly flatter) IMF. I Zw 18 is the most metal-poor star-forming galaxy known. The detection of W-R stars in this galaxy imposes rather stringent limits on the upper end of the IMF (Brown *et al.* 2002). Mass loss becomes less and less efficient at lower metallicity due to the weakening of the driving spectral lines. Therefore only the most massive main-sequence stars evolve into W-R stars at low metallicity. As a caveat, the assumption is made that W-R stars originate predominantly from single stars. If binaries are important, this argument would be weakened. Furthermore, W-R star evolution is not fully understood, even for single stars, and some bona fide W-R stars may in fact not be highly evolved, but extremely massive unevolved stars. Keeping in mind these caveats, quantitative modeling suggests that the observed W-R population in I Zw 18 evolved from progenitor stars with masses of at least 100 $M_\odot$. These masses are similar to those derived for individual hot stars in the R136 cluster, indicating a similar IMF at the high-mass end.

NGC 1741 is a luminous Magellanic-type interacting galaxy at a distance of 50 Mpc. The star formation is concentrated in a 500 pc large region where the star-formation intensity per unit surface is almost three orders of magnitude higher than in our Galaxy. Hot, massive stars with masses above 10 $M_\odot$ are currently forming at rate producing an ionizing photon output 100 times that of the 30 Doradus region. Yet, the stellar mass function in the NGC 1741 starburst and in 30 Doradus are very similar: both follow the classical Salpeter power law (Conti *et al.* 1996). The former was derived from the integrated light spectrum, and the latter from the stellar census described above in this section. The agreement between the results (unless they are fortuitous!) suggests confidence in these independent methods. In the optical, the morphology of NGC 1741 shows a disturbed disk composed of intermediate-age stars, with the starburst region off-set at the periphery. When observed in the UV at 2200 Å, the starburst completely dominates, and an irregular morphology is suggested. If NGC 1741 were observed at high redshift, its 2200 Å light would be shifted into the visual passband; NGC 1741 would look very similar to the irregular, clumpy galaxies currently observed in deep cosmological exposures, like the Hubble Deep Field (Williams *et al.* 1996).

Studies of starburst galaxies having a wide range of morphologies, chemical composition, and luminosities generally support the concept of a universal IMF which is rather insensitive to the conditions of the environment. The IMF slope is close to the Salpeter value in most cases, and stars with masses of at least 100 $M_\odot$ are universally detected. At the low-mass end, some evidence exists in a few objects for an IMF deficient of stars in the solar-mass range.

There is one class of starburst galaxies which may not follow this trend at the high-mass end. The most luminous starburst galaxies with luminosities in excess of 10^{11} $L_\odot$ are generally IR-selected because they are dust-rich, and therefore IR-bright and UV-faint, i.e. ultraluminous IR galaxies (ULIRGs). The high dust opacity makes direct detection of star light from hot stars a challenge, even at optical wavelengths. Stars are seen in the near-IR via absorption features from red supergiants. However, these features are quite insensitive to IMF variations. Massive stars can only be inferred via their ionizing radiation, e.g., measured from the Brγ recombination-line flux. The Brγ luminosities are proportional to the far-IR luminosities, at levels similar to those of star-forming regions. Since the far-IR luminosity is a good indicator of the overall star-formation rate (see Section 9.3), the proportionality suggests that star formation accounts for the bulk of the energy production. Alternatively, one could have suspected that additional energy sources, like an active galactic nucleus, may contribute and invalidate our assumption of a pure starburst. Agreement between the relative strengths of the Brγ, the K-band, and the far-IR luminosity is found only with models assuming a deficit of ionizing with respect to non-ionizing stars. In other words, the observed ionizing luminosity is relatively moderate and would be overpredicted by models if O stars were present in the same proportion as, e.g., in R136. While there is no doubt that the ionizing radiation field in these galaxies is softer than typically observed in less luminous and more metal-poor galaxies, the case for an anomalous IMF has not yet been convincingly made. It may very well be that stars with masses around 100 $M_\odot$ exist in normal proportions but that their environment is so opaque that they are completely hidden from view.

We close this section with some short, somewhat speculative thoughts on the IMF in the early Universe. This topic is increasingly attracting attention as the prospects for detecting the first generation of stars in the early Universe are improving thanks to the rapid technological progress in telescope instrumentation. What would be the properties of such Population III stars, as they are commonly called? This terminology is in analogy to the Population I and II stars, which are the later and the earlier stellar generation, respectively. Population III would be the very first, essentially metal-free population. We will discuss these objects in greater detail in Section 11.1. Here we will only focus on aspects related to their IMF.

Searches for Population III stars as the relics of the initial star formation in the Milky Way have been performed, all with a null result. The absence of extremely metal-poor stars in the halo of the Milky Way can in principle provide constraints on their IMF. Models for the formation of the bulge of our Galaxy predict how many such stars we should observe at a given metallicity. The fact that we observe no such stars could be interpreted as an IMF effect: the stars are not found because they never formed in the first place. Taking the models at face value would require a truncation of the IMF at masses as high as $\sim$10 $M_\odot$: the first stellar generation would be a pure O-star population.

There is theoretical support of an IMF biased towards very massive stars in a metal-free environment. While there is no evidence for a metallicity dependence of the IMF for Z down to a few percent of solar (corresponding to the metallicity of the most metal-poor galaxy known to host O stars), little is known below that metallicity threshold. The typical mass of newly formed stars is thought to be related to the Jeans mass (see Chapter 7), which is a function of the temperature of the ambient radiation field. This temperature, however, will almost certainly be higher in the early Universe for a variety of reasons. (i) The mean temperature of the cosmological background radiation scales like $T \propto T_0(z + 1)$, where $T_0 = 2.73$ K. If the first generation of stars formed at $z \approx 10$, the background radiation would be higher by an order of magnitude. (ii) The gaseous phase will be hotter at lower metallicity since the reduced opacity will remove the strongest emission lines, which are the dominant coolant of the ISM. (iii) The increased UV radiation of the newly formed, unblanketed metal-free stars will lead to larger heating of their environment, thereby again increasing the ambient temperature.

While these arguments are not yet supported by observations, they emphasize the fundamental importance of the concept of the IMF for the properties of not only starbursts but star formation in general.

10.3   The evolution of starbursts

In the previous section we discussed the mathematical formalism that is used to describe the rate of star formation and the distribution of stellar masses over the whole observed mass spectrum. We also provided empirical evidence for a fairly uniform mass spectrum among massive stars. In this section we will explore the micro-physics of a starburst. What is the triggering mechanism which sets off the sudden increase of star formation? What does a starburst look like at various spatial scales? How and why does it end? What is its ultimate fate?

A simple scaling law between the star formation rate and the gas density of the interstellar medium in the solar neighborhood was proposed as early as 1959 by M. Schmidt. He compared the scale-heights of gas and stars perpendicular to the Galactic disk and concluded that the star-formation rate must scale with the gas density with a power > 1 if the exchange of interstellar gas and newly formed stars should give the current star-formation rate. This reasoning was shown by Kennicutt (1998b) to be applicable to starburst galaxies as well. Combining the star formation rate surface densities, Σ_{SFR}, and the surface densities of the molecular gas, Σ_{Gas}, of normal disk galaxies and of starburst galaxies, a tight correlation between Σ_{SFR} and Σ_{Gas} over four decades is found:

$$\Sigma_{SFR} \propto \Sigma_{Gas}^{1.4 \pm 0.15}. \tag{10.9}$$

This equation is commonly called the "Schmidt Law". Figure 10.5 shows this relation for different classes of objects. Remarkably, the data are consistent with a single Schmidt Law extending over the whole surface density range from normal spiral galaxies to starburst regions. There are two important restrictions to the applicability of Fig. 10.5. First, the relation was derived by averaging over surfaces of order 1 kpc^2 for starbursts. This *global* relation cannot be extrapolated to scales smaller than hundreds of pc. Indeed, we expect larger and larger deviations from Eq. (10.9) for smaller and smaller star formation regions. Second, the relation will fail completely below a certain threshold surface density (Martin & Kennicutt 2001). This threshold may be related to the threshold for large-scale gravitational instability.

Figure 10.5 can be taken as empirical evidence for higher star formation rates in starbursts because the central gas densities are higher. Consequently, processes capable of increasing the central gas density can potentially trigger a starburst. For the gas to flow into the nucleus, angular momentum loss of the extra-nuclear gas must occur. Gravitational systems supported by rotation, like a disk galaxy, will suffer such angular momentum loss if deviations from axial symmetry occur. These deviations can be induced over large distances by gravitational torques. One of the theoretically best studied non-axisymmetric morphological feature in the nuclei of galaxies is bars. Many galaxies which had previously been thought to have

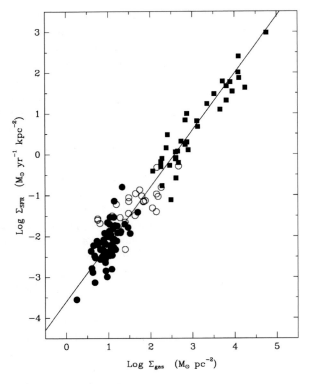

Fig. 10.5 Composite star formation law for normal disk (filled circles) and starburst (squares) galaxies. Open circles: centers of normal disk galaxies. The line is a least-squares fit with index $n = 1.4$. Reproduced from Kennicutt (1998b) by permission of the AAS.

symmetric nuclei have now been resolved into nuclear spirals driven by large-scale bars, triaxial bulges, and secondary bars. Progress was made possible largely due to the high angular resolution afforded by HST. These dynamical effects are thought to be responsible for generating the inner Lindblad resonance, a resonance between the velocity of the bar and the precession rate of stellar orbits. The Lindblad resonance typically occurs about 1 kpc from the galaxy nuclei and exhibits highly elevated star formation. Sometimes starbursts are not confined to the galaxy nucleus but occur wherever the gas density and gravitational force are favorable. On the other hand, circumnuclear rings can serve as reservoirs for cold gas and fuel central starburst activity.

The previous reasoning suggests that strong star formation and even starbursts may develop wherever bars are present. While the starburst phenomenon is wide-spread, it affects only a minority of galaxies. For instance, simple timescale arguments applied to the UV-selected starburst nuclei sample of Balzano (1983) imply that about 1% of all galaxies in the local Universe are currently undergoing a starburst phase. What then inhibits bar formation and subsequent starbursts? The stability of galaxy disks against bar formation may come from the gravitational effects of a massive central bulge or a dark halo. Both can stabilize the disk and suppress the inflow of gas. However, yet another mechanism may kick in to increase the gas flow: galaxy interaction.

Since the seminal work of Larson & Tinsley (1978) it has been known that star formation is enhanced in interacting galaxies. The colors of normal galaxies are consistent with a smooth, monotonically decreasing star formation rate over time. In contrast, galaxies classified as peculiar have colors consistent with short, intense bursts. In almost all cases, these galaxies show evidence of tidal interaction. Interacting systems without long tidal tails and therefore at early stages of dynamical evolution have colors indicative of the most recent bursts.

Barton, Geller, & Kenyon (2000) used the CfA2 survey to define a sample of galaxies in close pairs and searched for evidence for tidally triggered star formation. They found an anti-correlation between the Hα equivalent width (indicating the star-formation rate) and spatial and velocity separation (indicating age and strength of the interaction). This result is consistent with interaction leading to enhanced star formation over tens to hundreds of Myr.

The case for interaction as a starburst trigger is even more convincing at the highest luminosities. About 10% of galaxies with far-IR luminosities less than 10^{11} $L_\odot$ are in interacting/merging systems. Above 10^{12} $L_\odot$, this fraction becomes almost 100% (Sanders 1997). This number, however, is an upper limit to the occurrence of starbursts since an unknown fraction of ultraluminous galaxies is powered by an active galactic nucleus, rather than a starburst.

Numerical simulations do indeed support the suggestion of interaction induced starburst activity. Interaction leads to the loss of angular momentum of the disk gas by gravitational torque and dissipation. Angular momentum conservation then requires the gas to flow toward the center of the galaxy, with nuclear gas densities increasing by at least an order of magnitude. This mechanism is entirely analogous to the gravitational torque induced by galactic bars in isolated galaxies. Figure 10.6 is an example of an N-body simulation of the interaction and subsequent merger of two galaxies. This simulation is for a minor merger[1] change where the

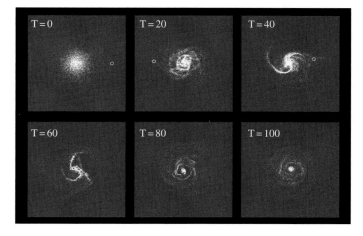

Fig. 10.6 Numerical simulation of a minor merger (Mihos, private communication). Shown is gas density. Each time unit T is about 10 Myr. The location of the gas-free low-mass companion is indicated by a circle.

---

[1] In a *minor* merger the primary galaxy is not highly disrupted during the encounter. In contrast, a *major* merger may fundamentally change and transform the morphology.

decay rate of the satellite's orbit is low due to its small mass. The nuclear gas density in the simulations increases significantly a few hundred Myr after the onset of the galaxy–galaxy interaction. Applying the Schmidt Law, the models predict nuclear star-formation rates of tens of $M_\odot$ yr^{-1}, in agreement with observations in starburst galaxies.

The N-body simulations are not self-consistent in the sense that the applicability of Eq. (10.9) is *assumed*. Mergers, interactions, and bars maximize only one of the necessary conditions for elevated star formation: high gas density. However, the physical mechanism governing star formation is not understood in detail. High *global* gas densities do not automatically induce *local* star formation. It is well known that star formation occurs on much smaller scales, and the connection between the larger and the smaller scales is not obvious. Therefore additional ingredients are required for a full description of the starburst trigger.

Many dense cloud cores may in fact be stable and never form stars. If encounters between the clouds are frequent, star formation may be induced by *collision*. Since the cloud–cloud collision process is proportional to the volume density of each cloud, and therefore to $\Sigma_{\rm Gas}^2$, this process could mimic a local Schmidt Law.

Gravitational instabilities can also be triggered by *stellar winds and supernovae* which develop only a few Myr after the formation of the first generation of stars. The resulting over-pressure can trigger the formation of new stars in adjacent regions, thereby inducing propagating star formation on a crossing timescale of the shock wave. For typical cloud sizes, the observed timescales are of order 10^7 yr.

What is known about the duration of the starburst? The duration of a star formation event, $\tau_{\rm SF}$, must not be confused with the age of the starburst, T. The former measures the time interval during which star formation occurs. The latter indicates the reference epoch of observation. For instance, a typical Milky Way globular cluster formed rapidly with $\tau_{\rm SF} \approx 0$ shortly after the formation of our Galaxy around $T = 10$–12 Myr ago. Conversely, the ionizing stars in the nuclear starburst galaxy NGC 7714 have an age of $T = 3$–5 Myr but the starburst has been ongoing over a time $\tau_{\rm SF} > 50$ Myr.

Star clusters are particularly well suited for a discussion of the starburst duration. As we will see further below, the definition of the duration is related to the definition of the spatial extent of the starburst. The observational advantage of clusters is the existence of well-studied local Galactic counterparts whose secure distances allow precise calibrations with theoretical isochrones. The astrophysically relevant issue is that a substantial fraction of the star formation in starbursts occurs in clusters. Meurer *et al.* (1995) estimate this fraction to be about 20%, but their value is likely to be a lower limit if some fraction of the clusters dissolves on a short timescale and the stars evolve into the field star population of starburst galaxies. Star clusters with sizes of a few pc and masses of order 10^5 $M_\odot$ have been found ubiquitously wherever starburst galaxies were studied at high enough spatial resolution. Therefore we can consider a star cluster as the smallest spatial scale which can still be resolved in starburst galaxies.

The star formation histories of numerous luminous star clusters in the Local Group of galaxies, including the Milky Way, can be studied with state-of-the-art color–magnitude analyses. An example is in Fig. 10.7. The galactic open cluster NGC 3603 is the most massive optically detected young star cluster in our Galaxy and possibly the closest counterpart to star clusters observed in starburst galaxies. Comparison with pre-main-sequence (PMS) and main-sequence isochrones suggests that all stars down to the detection limit at 0.1 $M_\odot$ are formed within less than $\tau_{\rm SF} \approx 1$ Myr. This timescale is shorter than the evolutionary timescale

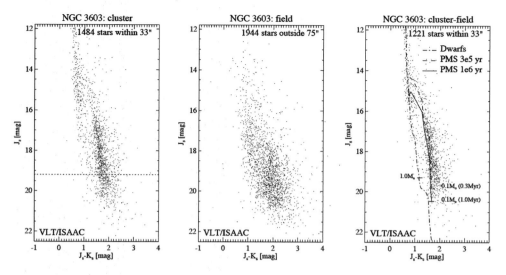

Fig. 10.7 Color–magnitude diagram (J-K, J) of NGC 3603. Left: all stars detected within the central 1 pc; center: surrounding field stars; right: cluster population with the field stars statistically subtracted. Theoretical isochrones of pre-main-sequence and main-sequence stars are plotted as well. From Brandl *et al.* (1999).

of the most massive stars observed in this cluster ($\sim$100 $M_\odot$, with a timescale of 3 Myr). Star formation at both the high- and low-mass end happens almost instantaneously. As discussed in the previous section, analogous results apply to less massive OB associations as well (e.g., Massey 2003). Extensive studies of Galactic and LMC/SMC star clusters and OB associations have established an almost instantaneous star formation mode. Typical age spreads are about 2 Myr or less. This is short in comparison with stellar evolutionary timescales, except for stars with the most extreme masses.

What is the relevance of these results to starburst galaxies whose *nuclear* star formation rates are at least an order of magnitude above those of the most luminous clusters in Local Group galaxies? The minimum expected duration of these starbursts is the crossing time for a disturbance to travel across the region occupied by the starburst. A rough estimate of the crossing time can be obtained from a plausibility argument. The size scales involved are of order 1 kpc. A typical velocity scale is the velocity dispersion, for which values around 100 km s^{-1} are measured. These two scales suggest crossing times of tens of Myr, more than an order of magnitude longer than τ_{SF} observed in nearby star clusters.

Could our estimate of the crossing time be an overestimate because the velocity dispersion is not a good proxy for the propagation speed of star formation? Fortunately, we can directly observe propagating star formation in nearby galaxies. 30 Doradus and its multiple stellar generations, each distinct and spanning an age range of tens of Myr, was mentioned before in Section 9.2. Similar evidence exists from the analysis of the integrated light in starburst galaxies. Puxley, Doyon, & Ward (1997) measured the Brγ recombination line flux and the CO 2-0 band strength around the center of the nuclear starburst in M83. These two age indicators are sensitive to ionizing and non-ionizing stars having ages of $\sim$5 and $\sim$20 Myr, respectively. Over the measured region, Brγ and the CO band exhibit the opposite behavior, and their strengths can be used to map the propagation of star formation. The data suggest an age gradient in the central region of M83 if the burst population dominates the emission. The position with the deepest CO absorption is significantly older than that having the largest Brγ equivalent width. The age range over which star formation has occurred is 10 to 20 Myr, reminiscent of the spatial morphology and age range observed in the 30 Doradus region.

HST imagery of the nuclear region of the starburst galaxy NGC 3049 is shown in Fig. 10.8. The galaxy is a typical metal-rich ($Z \approx Z_\odot$), low-luminosity ($L = 3 \times 10^9\, L_\odot$) HII galaxy hosting a nuclear starburst 10 to 100 times more luminous than 30 Doradus. The spatial resolution afforded by HST reveals complex sub-structure in the "nucleus". At least ten individual starburst clusters are detected, some of them with properties similar to those of 30 Doradus. Imaging of starburst galaxies with high spatial resolution generally does not suggest just one dominating super-massive cluster but rather the presence of many clusters having masses 10^4 to $10^6\, M_\odot$ with different ages. Most likely, the population of each individual cluster mirrors the behavior found in local clusters with star formation spreads of less than a few Myr. The starburst in NGC 3049 is fairly representative of many other low-luminosity starburst galaxies as well. We quoted an empirical fraction of 20% for star formation in clusters at the beginning of this section. These clusters are all very compact, having effective radii of a few pc and velocity dispersions of $\sim$10 km s^{-1}. Therefore their crossing times are of order 1 Myr, and it is reasonable to assume that they formed in a single short burst. In all respects, they have properties like nearby massive star clusters, such as NGC 3603 or R136. We do not know if *all* starburst galaxies behave like NGC 3049, whose luminosity is a factor of 1000 lower than those of the most luminous starburst galaxies. It is an open issue whether more luminous systems are just bigger, i.e., forming star clusters over a larger area.

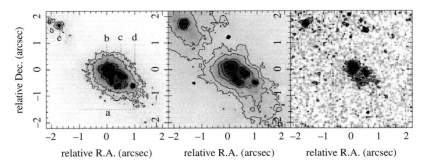

Fig. 10.8 HST/STIS images of the central $4'' \times 4''$ of the nuclear starburst NGC 3049 at UV wavelengths (left), optical wavelengths (center), and [OIII] emission line plus the underlying continuum (right). The nuclear region is dominated by a few luminous starburst clusters, each having less than 5 pc diameter. Numerous less luminous clusters are present as well. Reproduced from González Delgado *et al.* (2002) by permission of the AAS.

Observations of more distant starburst regions which can no longer be resolved into individual stars are consistent with these results *as long as the spatial resolution is high enough to isolate the starburst cells.* The strongest support comes from W-R galaxies, which contain W-R stars in large enough numbers to be detectable in their integrated spectra (Conti 1991). The observed strength of the W-R features in these galaxies can only be understood if we observe a single stellar population ($\tau_{SF} \approx 0$) just at the right time when W-R stars are present ($T \approx 5$ Myr). Of course, this is just a strong selection bias. W-R galaxies are detected by placing a narrow slit across the optically brightest region in a galaxy. Most likely, this will be an individual star cluster whose age is young. Since clusters fade rapidly over the first tens of Myr of their life (by about a factor of 5 between 3 and 20 Myr), the brightest cluster will either be the only one or it will dominate over any other older cluster within the spectrograph slit. This effect can lead to the impression that the entire starburst would be quasi-instantaneous. Figure 10.9 is an artist's view of the structure of a central starburst region illustrating this argument. The starburst has an age of at least 30 Myr, but some of the most luminous clusters have just formed recently. The distribution of the clusters is the result of propagating star formation. Note that the brightest young cluster is not necessarily the most massive cluster. Part of its brightness is due to its youth.

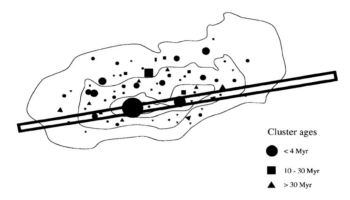

Cluster ages

● < 4 Myr

■ 10 - 30 Myr

▲ > 30 Myr

Fig. 10.9 Diagram of a typical slit orientation on a starburst. The contours represent the diffuse light distribution. The filled symbols are the clusters with symbol shape representing age, and size representing brightness. Reproduced from Meurer (2000) by the kind permission of the ASP Conference Series.

The lesson we have learned is that the duration and the size of a starburst are intimately connected. Star formation in starbursts occurs over kpc-sized regions, depending on the strength of the starburst. However, not all the star formation takes place simultaneously. Rather, it is concentrated in numerous massive clusters, each representing a single stellar population ($\tau_{SF} \approx 0$), with the population of clusters extending over an age range T of tens of Myr.

The subsequent evolution of the starburst clusters depends on their gravitational binding energy relative to the gravitational field of the host galaxy. The latter can lead to tidal disruption on a timescale ranging from only a few Myr to a Hubble age. Clusters are more likely to get disrupted if their mass is small and if they are near the galaxy center where the gravitational force is largest.

The disruption of clusters by gravitational forces can become the dominant effect governing the morphological evolution of a starbursts located in galaxy centers. Two well studied examples are the Arches and Quintuplet clusters in the center of the Milky Way (recall Section 7.4). These are newly formed (2 to 4 Myr), massive ($10^4 M_\odot$) highly obscured star clusters (Figer 2004). The short relaxation times due to the compactness of the clusters and the strong tidal fields near the Galactic Center lead to their rapid evaporation. After cluster evaporation, the remaining less massive stars will become part of the surrounding field population. However, this process alone cannot account for the *entire* high-mass star population in the Galactic Center as traced by their ionizing radiation. At least some very massive stars may form outside the most massive clusters.

The disruption of starburst clusters on a timescale much shorter than a Hubble time is relevant to the interpretation of young starburst clusters as precursors for old globular clusters. This evolutionary connection seems obvious because the masses and sizes of the brightest young clusters in merging galaxies are similar to those of the old globular clusters in the spheroids of galaxies. The major objection to this suggestion is the preferred mass scale seen in globular clusters. Their mass function has a turnover or peak at $\sim 10^5 \, M_\odot$ with a dispersion of only about 0.5 dex (Harris 1991). In contrast, the mass functions of the young starburst clusters have power-law form, with characteristic slopes of -2 between masses of 10^4 to $10^6 \, M_\odot$ (see Fig. 10.10). Two explanations have been proposed for the preferred mass scale of old globular clusters. The initial conditions could have favored the formation of objects with masses of $\sim 10^5 \, M_\odot$. Since the Jeans mass of the interstellar clouds will be different today from what it was at the epoch of globular cluster formation (see our discussion at the end of the previous section), newly formed clusters will differ in their mass spectrum as well. Alternatively, old globular clusters may have been born with a much wider spectrum of masses that was later modified by the selective destruction of low-mass clusters (Fall & Rees 1977). In this case, the power-law mass function observed for the young starburst clusters might evolve into a log-normal-like mass function. While the issue is far from being settled, dynamical modeling of young star clusters appears to favor the second hypothesis. If so, we are witnessing the progenitors of globular clusters like M13 or ω Centauri in the very centers of starburst galaxies.

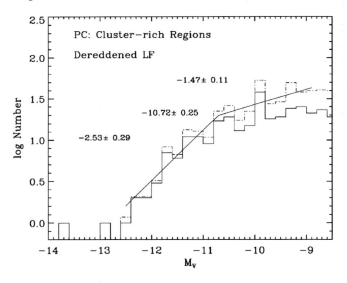

Fig. 10.10 Completeness-corrected cluster luminosity function of the Antennae galaxies. The bright end has a slope of –2.53, whereas the faint end is flatter with a slope of –1.47. Reproduced from Whitmore *et al.* (1999) by permission of the AAS.

To first order, the disruption of the starburst clusters does not terminate the starburst itself, as the stars continue to evolve in the field population. This of course leaves open the questions of if and how a dense cluster affects the ambient ISM, as opposed to a lower surface density continuum of stars. The ultimate upper limit to the duration of a starburst is the exhaustion of the gas reservoir which feeds the star formation process. The gas depletion is difficult to estimate quantitatively for a variety of reasons. First, the unknown proportion of low-mass stars formed in the starburst makes estimates of the total (integrated over all masses) star formation rate uncertain by at least a factor of two. The concept of a truncated (i.e., deficient in low-mass stars) IMF has sometimes been invoked just to extend the gas depletion timescale.

The fundamental difficulty in determining the gas depletion time is the break-down of a closed-box model of a starburst, which ignores the various sources and sinks of material. As discussed before, infall of gas to the starburst nucleus results from angular momentum loss due to gravitational torque and dissipation. Another source of gas replenishment is mass return by stellar winds and supernovae as the stars evolve. The energetic evolution of a starburst region, during and after the star formation, is determined by stellar evolution. For the first ~ 3 Myr, the starburst is "photon"-dominated, i.e., the dominant heating of the ISM is by energetic photons from hot main-sequence stars which emit about 1/3 of their luminosity shortward of the Lyman break. After evolving off the main sequence, the stars develop strong stellar winds capable of heating the surrounding ISM and creating giant wind-blown shells and bubbles. This phase is particularly pronounced when the W-R stage has its peak around 5 Myr. Subsequently, the first Type II supernovae appear, equaling and eventually surpassing the kinetic energy input by stellar winds. After 10 to 15 Myr, the kinetic energy release by winds and supernovae exceeds the heating by stellar ionizing photons, and the starburst becomes "matter"-dominated: supernovae shape the morphology and energetics of the ISM and may regulate subsequent star formation. In the absence of continuous star formation, this phase ends after about 50 Myr when intermediate-mass stars reach the main-sequence turn-off.

The impact of winds and supernovae on the mass budget of the gas reservoir can be significant: massive stars return about 50% of their mass almost instantaneously during the starburst. Therefore the gas depletion timescale will become almost independent of the rate of star formation, as more material is returned if more stars are formed. (This argument is no longer valid if low-mass stars form in large proportions since their mass return is small.) We should caution, however, that the stellar wind and supernova material is at coronal temperatures due to shock interaction with the ambient ISM. The cooling times are long in comparison with the age of the starburst so that this material may not immediately become part of the cold, star forming gas.

Galactic-scale outflows, or superwinds (Heckman 2003) constitute a significant sink for the gas reservoir. The combined effects of multiple stellar winds and supernovae are capable of initiating large-scale outflows of interstellar gas. Such outflows have been known from optical and X-ray imagery, and they have recently been analyzed by absorption line spectroscopy. The kinematics are compatible with a simple model of the adiabatic expansion of a superbubble driven by the supernovae in the starburst. Radiative losses are negligible so that the outflow may remain pressurized over a characteristic flow timescale of 10^8 to 10^9 yr, as estimated from the size and velocity. We will return to the physics of superwinds in more detail in the next section. For now we will focus on the effects of superwinds on the star formation process.

Estimates of the mass driven out of the region of star formation by large-scale outflows are rather uncertain but tend to give rates that are quite similar to the star formation rates themselves. Taken at face value, this suggests that the available gas reservoir will not only be depleted by the star-formation process but, more importantly, by removal of material due to galactic-scale outflows. If this is true in general, outflows are an important, or even the dominant, regulation mechanism for the duration and termination of the starburst. Starbursts may determine their own fate by their prodigious release of kinetic energy into the interstellar medium.

Few systematic studies of starburst regions exist which go beyond evolutionary phases dominated by massive OB stars. Photometric fading of the stellar population is significant and makes any survey of "post-starburst galaxies" prohibitive – except for globular clusters, the bona fide post-starbursts. (A 300 Myr old population is about 50 times fainter in V than a 3 Myr old population.) The so-called E + A galaxies (Dressler & Gunn 1983) have been suggested to be post-starbursts. Their spectra are dominated by strong Balmer lines in absorption and have negligible emission lines. This spectral morphology is indicative of a galaxy that has no significant current star formation but was forming stars in the past (<1.5 Gyr). The strong Balmer equivalent widths can only be understood by models seen in a quiescent phase soon after a starburst; for this reason the E + A spectra are often identified with post-starburst galaxies.

10.4   Starburst-driven superwinds

In Section 8.3 we provided the basic description for a wind-blown bubble around a single star. In reality, massive stars form in clusters. Very few stars are sufficiently isolated for a bubble to form around a single star. A typical wind-blown bubble is powered by a stellar population. An obvious modification to the previous line of arguments is the generalization to a stellar population following a specific IMF. The dynamics of such a so-called "superbubble" can be approximated by scaling up the wind-bubble theory for continuous energy input. Associations typically contain tens to hundreds of OB stars with powerful winds and the potential to explode as supernovae. The radius and the velocity of a superbubble powered by a time-dependent wind can be derived by combining the equations for energy and momentum conservation into a single, third-order dynamical equation:

$$\frac{d}{dt}\left[R(t)^3\frac{d^2}{dt^2}(R(t)^4)\right] = \frac{6R(t)^2}{\pi n_0}L_{\mathrm{mech}}(t),\tag{10.10}$$

where $L_{\mathrm{mech}}(t)$ denotes the time-dependent wind input from all stars (see Shull & Saken 1995). During the supernova phase (after about 10^7 yr) this energy input can be expressed as the product of the IMF, the mass–age relation, and the supernova energy. The mass–age relation for massive stars is $t(M_*) \propto M_*^{-\gamma}$ with $\gamma \approx 1.2$ (Schaerer *et al.* 1993). For a Salpeter IMF with $\alpha = 2.35$, we find that $L_{\mathrm{mech}}(t) \approx$ constant. This tells us that there are more and more supernova events for lower masses since there are more progenitors available (due to the IMF), but at the same time it takes longer for a star to turn into a supernova (due to the lifetime). The net effect is a nearly constant mechanical luminosity for the first 50 Myr during the evolution of a star cluster. For a constant wind luminosity, Eq. (10.10) has a self-similar solution:

$$R(t) = \left(\frac{125L_{\mathrm{mech}}t^3}{154\pi n_0}\right)^{1/5},\tag{10.11}$$

$$v(t) = 0.4R(t)t^{-1}.\tag{10.12}$$

These expressions are identical to those for a bubble around a single star, except for constants. Superbubbles are scaled-up versions of single bubbles. After the last supernova explodes, the structures dissipate, either because radiative cooling becomes important and the interior pressure is lost, or because the bubble bursts through the galactic disk and is disrupted. The latter phenomenon commonly occurs in starburst galaxies and is the topic of the following discussion.

Suppose a superbubble is powered by 10^5 massive stars, a value quite typical for a starburst nucleus. According to Eq. (10.11), such a superbubble will have a characteristic size of hundreds of pc. This is comparable to the vertical scale height in a disk galaxy, and the superbubble evolution will depend on the properties of the galaxy ISM, in particular the vertical density structure. If there is a density gradient, such as commonly found along the vertical axis of a disk galaxy, the superbubble will expand preferentially along the direction of the maximum pressure gradient. At some point the superbubble reaches the effective vertical scale height of the galaxy H_{eff}, defined as

$$H_{eff} = \frac{1}{n_0} \int_0^\infty n(z)dz. \tag{10.13}$$

n_0 is the ambient gas density. The characteristic *minimum* wind luminosity L_{min} determining the subsequent evolution is the luminosity for which the velocity at $H_{eff} = 1$ is equal to the isothermal sound speed c_0. This is the case for

$$L_{min} = 18 n_0 H_{eff}^2 c_0^3. \tag{10.14}$$

c_0 is the isothermal sound speed. All units in Eq. (10.14) are cgs. Typical values for galaxies with scale heights of a few hundred pc are $L_{min} \approx 10^{38}$ erg s^{-1} with this relation. Since a typical O star has a wind luminosity of order 10^{37} erg s^{-1}, the condition of Eq. (10.14) is met in a star cluster containing even a moderate number of massive stars.

The superbubble will continue to expand along the maximum pressure gradient. After the superbubble has reached several vertical scale heights, Rayleigh–Taylor instabilities develop, and the wall of the superbubble will dissipate. This allows the interior hot gas to blow out of the disk and leak into the surrounding galaxy halo. It is the onset of a phenomenon called galactic superwind (Heckman 2003). The evolution of the galactic superwind is determined by the importance of radiative losses of the hot phase. If the cooling time t_{cool} is large in comparison with the dynamical (or crossing) timescale of the starburst t_{dyn}, radiative losses are negligible, and the wind remains pressure-driven over the lifetime of the starburst. The ratio of t_{cool} over t_{dyn} in a starburst-driven wind embedded in a galactic disk–halo system with scale height H_{halo} (kpc) is given by

$$t_{cool}/t_{dyn} = 290 n_0^{-1.1} L_{mech}^{0.61} H_{halo}^{-1.7} (Z/Z_\odot)^{-1.6}, \tag{10.15}$$

where L_{mech} is the mechanical luminosity of winds and supernovae in 10^{43} erg s^{-1}. Starburst nuclei have typical parameters of $n_0 \approx 10^2$ cm^{-3}, L_{mech} of unity, $H_{halo} \approx 0.1$ kpc, and solar chemical composition. Therefore $t_{cool} \gg t_{dyn}$, and radiative cooling is negligible over the lifetime of the outflow (Heckman, Armus, & Miley 1990).

What is the observational evidence for this phenomenon? Lynds & Sandage (1963) were the first to draw attention to a large-scale outflow from the nucleus of the nearby starburst galaxy M82. At that time starbursts were yet unknown, and Lynds & Sandage speculated about an explosive event responsible for the outflow. NGC 253 is at the same distance as

M82 ($D \approx 3$ Mpc) and rivals M82 in the status of the proto-type starburst. Demoulin &
Burbidge (1970) found evidence for an outflow from the nucleus of NGC 253 on the basis
of long-slit spectroscopy. Again, the true nature of the starburst nucleus of NGC 253 was
unknown, but Demoulin & Burbidge suggested that the expanding gas was produced by a
"violent event" in the center of the galaxy. Ulrich (1978) performed an extensive kinematic
study of the nebulosity around NGC 253, confirming the existence of a massive galactic wind.
At that time the relevance of massive stars for the properties of the nucleus of NGC 253 had
been realized, and Ulrich proposed an "outburst of star formation" as a trigger of the flow.
Since then, starburst-driven superwinds have been recognized to be ubiquitously present.
X-ray searches with satellites like Einstein, ROSAT, ASCA, and Chandra have discovered
outflows in many starburst galaxies. Edge-on galaxies in particular have been searched for
because of their favorable geometry. They show evidence for hot gas in their halos, with
gas temperatures of a few times 10^6 to 10^7 K at distances of 10 kpc and more from the
galactic plane.

M82 has become one of the paradigms of a starburst-driven outflow seen in X-rays. An
X-ray image taken with the Chandra satellite is shown in Fig. 10.11. The bright spots in
the center are supernova remnants and X-ray binaries. These are some of the brightest such
objects known. The diffuse X-ray emission in the image extends over at least 10 kpc, and is
caused by $\sim 10^7$ K gas flowing out of M82. The spectral shapes of the point sources and the
extended gas are very different: the former have a hard spectrum, with most of the emission
coming from energies above 1 keV, whereas the emission from the outflow is softer and peaks
below 1 keV. The stellar disk of M82 is invisible in X-rays. For comparison, the X-ray data are
contrasted with an optical image of M82 reproduced at the same scale. This broadband image
delineates the galactic disk. The bipolar outflow is aligned with the minor axis, as defined
in the optical, and has an opening angle of approximately 40°. While the broadband optical

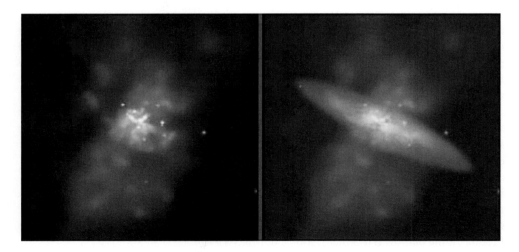

Fig. 10.11 Left: Chandra X-ray image of the starburst galaxy M82. The bright spots in the
center are supernova remnants and X-ray binaries. The diffuse X-ray emission extends over
several kpc, and is caused by $\sim 10^7$ K gas flowing out of M82. Right: optical image of the
galaxy superimposed on the X-ray image. Credit: NASA/CXC/SAO/PSU/CMU.

image does not reveal any extended nebular emission outside the galactic disk, deep narrow-band images in, e.g., Hα, show emission roughly coinciding with the X-ray emitting gas.

So far, we implicitly assumed that the extended soft X-ray emission is the principle phase filling the volume bounded by the optical emission-line gas. Ideally, the hot phase is homogeneous and enclosed by the cooler (10^4 K) gas. High-resolution X-ray images demonstrate that the actual situation is much more complex. The X-ray emitting gas is morphologically similar to the filamentary structures seen in the optical line emission. Furthermore, the X-ray gas shows evidence for limb-brightening and often lies just inside the Hα filaments. Therefore we may observe X-ray emission from regions where the interaction between the hot wind and the undisturbed ISM has resulted in substantial mixing of cold ISM gas into the hot wind. Alternatively, the filaments may be the relics of the ruptured bubble surface after Rayleigh–Jeans instabilities set in and enabled the outflow. If so, the optical emission may be the footprint of the forward shock into the galaxy halo, and the X-ray emission the reverse shock in the hot superwind.

Starburst-driven outflows have been detected in absorption lines as well. An example are the observations of NGC 1705 obtained with the FUSE satellite in Fig. 10.12. The absorption in NGC 1705 is blueshifted relative to the restframe velocity defined by the stars and the HI, suggesting the origin of the absorption in the outflow. Using interstellar absorption lines to probe the outflow offers several advantages. First, any ambiguity as to the direction of the flow is removed. Neither the X-ray nor the Hα emission data allow a direct probe of the sign of the velocity vector. In principle, some of the observations would be consistent with an *infall* of material as well. Infalling flows are indeed believed to occur in early-type galaxies. Such "cooling flows" as they are called, are predicted theoretically from the existence of the observed X-ray halos around elliptical galaxies and their inferred short lifetimes. In the absence of an outward force, these X-ray halos will cool and the material will flow *towards* the galaxy. It is mainly from measurements of absorption lines that cooling flows can be ruled out as the origin of the X-ray emission around starburst galaxies. Second, the strength of the absorption allows an estimate of the column density of the gas. The X-ray or optical surface brightness of the emitting gas is proportional to the emission measure. Therefore, absorption lines are a less biased probe of the gas densities than techniques relying on emission, which are strongly weighted in favor of the densest material, which may contain relatively little mass. Third, interstellar absorption lines can be used to study outflows in distant galaxies whose X-ray or optical emission may be undetectably faint.

Absorption lines are the preferred means of estimating outflow rates. The largest uncertainty in the derived rates comes from the fact that the strong absorption lines are usually saturated, so that their equivalent widths are determined by the velocity dispersion and covering factor, rather than by the column density. In those cases where the restframe-UV region can be probed, the total HI column can be estimated from the damping wings of Lyα, while ionic columns may be derived from the weaker, less saturated interstellar lines. The outflow rates inferred for the cool gas are similar to the star-formation rates in the starburst, typically of order 1 $M_\odot$ yr^{-1}. Outflows with the highest quality data allow crude estimates of the metallicities. Pettini *et al.* (2000) found a metallicity of about 1/4 solar in the lensed high-redshift galaxy MS1512-cB58. This determination provides stringent constraints on the chemical enrichment epoch in the early Universe: MS1512-cB58 is at a redshift of 2.7, corresponding to a cosmological look-back time of 90% of the age of the Universe. Already at that epoch, multiple stellar generations have enriched the ISM in MS1512-cB58 to a value close to solar.

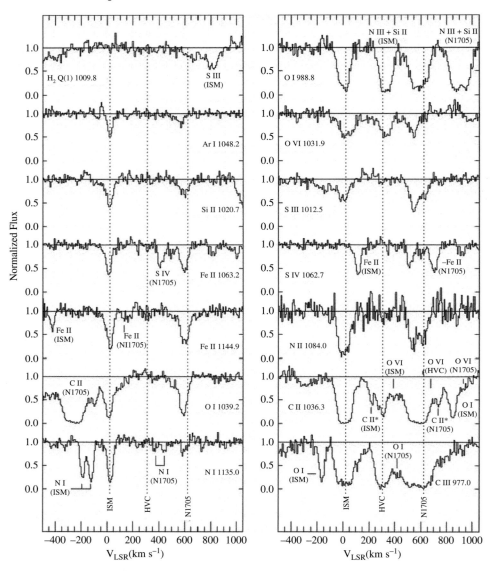

Fig. 10.12 Continuum normalized absorption line profiles of NGC 1705 versus velocity observed with FUSE. The vertical dashed lines indicate the velocities of the Milky Way ISM, of a high-velocity cloud (HVC) at 285 km s^{-1}, and the systemic velocity of NGC 1705. The absorption lines in NGC 1705 are blueshifted by 50 to 100 km s^{-1} relative to the systemic velocity. Reproduced from Heckman *et al.* (2001) by permission of the AAS.

Slightly subsolar chemical compositions of the cold gas have been found in the outflows of local starburst galaxies as well. (The composition of the hot phase is yet unknown. The X-ray emitting gas may very well have a higher metal abundance.) These results are consistent with the theoretical expectation that the cool component of the outflow is mostly ambient ISM accelerated by the wind rather than supernova material itself.

What is the ultimate fate of a starburst-driven superwind? Eventually, the outflow must stall due to radiative losses after a cooling timescale. The subsequent evolution depends on the starburst geometry, the mass of the galaxy, or the properties of the halo. It is important to realize that the available observational techniques probe wind material which is still located deep in the gravitational well of the galaxy's dark matter halo. Therefore the inferred outflow rates should not immediately be interpreted as the rates in which mass, metals, and energy are transported into the intergalactic medium. The material may fall back onto the galaxy disk or become unbound, with the second alternative being more likely in less massive galaxies.

The likelihood of the outflow's escape from the gravitational field of the starburst galaxy can be assessed by comparing the outflow velocities and the galactic escape velocities. The latter can be inferred from the rotation speed of the host galaxy. Typical outflow velocities are $\sim$500 km s^{-1}. The outflow velocities are independent of the rotation speed of the host galaxy for a velocity range corresponding to low-mass dwarf to massive disk galaxies. This is strong support for the suggestion that shallower galaxy potential wells are less likely to retain the chemically enriched starburst material. The selective loss of metal-enriched gas from shallower potential wells has profound cosmological consequences, such as the mass–metallicity relation in dwarf galaxies and the existence of radial metallicity gradients in elliptical galaxies and galaxy bulges. We will return to these issues in our discussion of the cosmological implications of starburst galaxies in Chapter 11.

10.5 The starburst–AGN connection

A close relation between the starburst and the AGN phenomenon can be inferred from different reasonings. A review of the arguments for and against such a relationship is useful in that it highlights the general physical concepts that apply to both starburst galaxies and active galactic nuclei. AGN are classified according to their observed properties (e.g., Peterson 1997).

- Radio luminosity: radio-loud AGN are strong radio emitters when compared to the optical, whereas radio-quiet objects have relatively little measurable radio flux.
- Bolometric luminosity: in low-luminosity objects we can observe the host galaxy around the AGN. Luminous AGN are so powerful that their nucleus outshines the host galaxy. These so-called quasars are quite rare; therefore they are on average more distant (there are none at non-cosmological redshift), and the surrounding galaxy can no longer be seen.
- Optical emission lines: *broad* spectral lines in the optical spectrum indicate high-velocity gas (moving at 10^3 to 10^4 km s^{-1}). These objects are known as "Type 1" AGN (including Seyfert1 galaxies). If only *narrow* lines are observed, we can see low-velocity gas between 10^2 and 10^3 km s^{-1}, either because the high-velocity gas is obscured by a torus of molecular gas or because of generally lower velocities. These are "Type 2" AGN, including Seyfert2 galaxies.

The so-called *Unified Model* for AGN (Fig. 10.13) is instrumental for our understanding of the physical mechanisms driving the observed processes and causing the observed morphological classes. The basic idea behind the Unified Model is that all observed AGN have the same internal structure in their nucleus. The apparent differences are caused by the viewing angle (i.e., the inclination of the accretion disk, the outer parts of the broad line region, and the molecular torus).

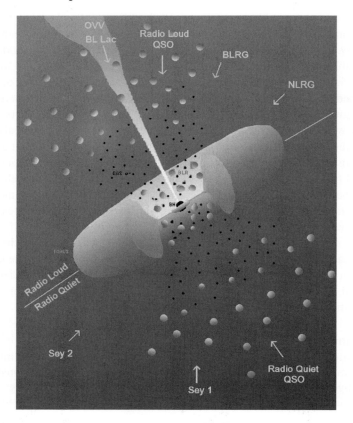

Fig. 10.13 Schematic of the Unified Model for AGN. Depending on the viewing angle, different morphological types are observed. Figure courtesy M. Gliozzi.

When observed edge-on, the central regions, including the black hole, accretion disk, and broad line region, are hidden from view. Only the molecular torus and the emission lines from the narrow-line region can be seen. Therefore the optical spectrum will be dominated by narrow emission lines; no broad lines are present. Light from the nucleus can be reflected towards the line of sight by hot gas above and below the torus. Observations of this reflected light in a few AGN have demonstrated that Type 1 nuclei are hidden in Type 2 objects. However, often the stellar light in Type 2 AGN is very significant, completely masking any reflected Type 1 component in the optical. The more the aspect angle favors pole-on observations, the higher the chances of seeing directly into the central regions. In this case, one can observe both broad and narrow lines in the optical spectrum, and the direct emission from the accretion disk.

The powering source of an AGN is gravitational energy. The tremendous energy output from AGN together with their small size places tight constraints on their powering source. Even with the biggest telescopes, an AGN is unresolved. Therefore it must be confined to a region much smaller than about 1 pc. This leaves the release of gravitational energy from material falling into a massive black hole as the most probable energy source. The evidence that black holes inhabit the centers of many, if not all, massive galaxies is increasing. Even our own Galaxy is supposed to harbor a relatively low-luminosity black hole. We can use Einstein's equivalence principle between mass and energy to estimate the energy production:

$$E = 0.1mc^2, \tag{10.16}$$

which tells us that material falling into the central black hole near the nucleus of a galaxy releases its gravitational potential energy at an efficiency of about 10%. This energy is transformed and redistributed among the electromagnetic spectrum and is responsible for the highly excited optical emission lines, the X-rays, the relativistically moving particles, etc., which produce the observed phenomena. The release of gravitational energy by a black hole with a mass of about 10^7 $M_\odot$ can be understood if one stellar mass is converted (or "accreted" by the black hole, as this process is often called) every ten years. The prototypical Seyfert2 galaxy NGC 1068 does indeed have an inferred accretion rate of 0.1 $M_\odot$ yr^{-1} (Veilleux 2001). This is close to the star formation rate in a typical starburst galaxy like M82. Therefore it is natural to suggest that infall of gas to the galaxy center could feed both a starburst and an AGN, and the arguments in our discussion of the starburst triggering mechanism apply to AGN as well. In order to drive gas from the outer disk to the nucleus of a spiral galaxy, from scales of tens of kpc to tens of pc, the angular momentum of the gas must be reduced by about six orders of magnitude. A wide variety of phenomena involving torques and dissipation can drive gas from the outer disk into the inner kpc region of a galaxy with an AGN. As in starburst galaxies, non-axisymmetric features such as bars, as well as tidal interactions and mergers, can be relevant. The details of the transport of the gas to the innermost (pc scale) region of the galaxy are less well understood. There is growing evidence for bars again being crucial. A smaller nuclear bar nested within the large primary bar (a "bar within a bar") can be an efficient mechanism for driving gas closer to the galactic center to fuel both a central starburst and an AGN.

The connection between the stellar properties of a galaxy and those of its central black hole is particularly striking in the correlation between the black hole mass and the mass of the galaxy bulge (Ferrarese & Merritt 2000; Gebhardt *et al.* 2000). Black holes appear in galaxies over a wide range of luminosity. The only requirement is on the ellipsoidal shape of the host: AGN are found either in elliptical galaxies or in the bulge of a disk galaxy, like our Milky Way. Galaxies without bulges lack black holes. The black hole mass is roughly proportional to the mass of the bulge as determined from its luminosity. On average, the black hole mass is about 0.1% the mass of the galactic bulge. Even the Milky Way follows this trend; its modest central black hole mirrors its relatively small bulge. Surprisingly, the mass of a black hole is also related to the average velocity of stars within the bulge, even outside the region where the black hole's gravity dominates. The implication is that the mass of the central black hole is determined by the process that determines the size of the bulge. The latter is thought to be set initially during bulge formation: once star-formation has occurred, a preferred scale is created. Subsequently, interactions or mergers with other galaxies may trigger renewed star formation and affect the growth of the bulge. If so, the AGN and starburst phenomenon would have a strong causal connection.

What do the overall spectral energy distributions of starbursts and AGN tell us? We are used to describing astronomical phenomena in terms of their output of optical light. This, however can be very misleading. Although hot, massive stars are brightest in the UV and emit a significant fraction of their energy at optical wavelengths, their light in starburst galaxies is

often obscured by dust. Simple energy conservation then dictates that the obscured light must
be re-emitted in a different waveband. The dust grains scattering and absorbing the stellar
light have typical sizes of about 0.1 μm, i.e., comparable to the wavelength of the photons
that interact with the dust. The dust is heated to temperatures of tens of K and radiates more
or less like a blackbody. Since a blackbody with this temperature has its peak emission in the
far-IR around 60 μm, the net effect for a dust-obscured stellar population is a redistribution
of the stellar UV/optical flux to the far-IR. This can be seen in Fig. 10.14 where we have
plotted the observed average energy distributions between 6 cm and 1 keV for a sample of
starburst galaxies. The sample is subdivided into high- ($E(B-V) > 0.4$) and low-reddening
($E(B-V) < 0.4$) starbursts. Both species have similar overall spectral characteristics, except
in the far-IR and the UV, which are anti-correlated: objects with lower reddening have higher
UV and lower far-IR flux, whereas the opposite holds for objects with higher reddening.
The interpretation is obvious from our previous reasoning: starbursts with larger reddening
have higher dust content and are more effective in absorbing UV photons and re-emitting the
processed radiation in the far-IR. The flux difference between the two data sets at ∼1500 Å
is about a factor of 5 to 10. This factor is related to the escape fraction of UV radiation. On
average, about 80% of the stellar UV photons are hidden from direct view and can only be
recovered as reprocessed light in the far-IR. The direct escape fraction is only about 20%.

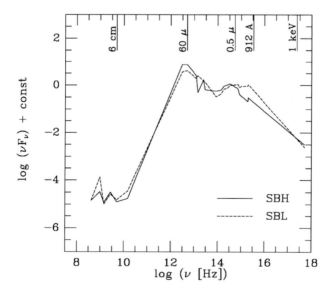

Fig. 10.14 Average spectral energy distributions from the radio to X-rays for high- (SBH)
and low-reddening (SBL) starburst galaxies. The spectra are normalized to $\log(\nu F_\nu) = 0$ at
5500 Å. From Leitherer (2000).

The starburst SED peaks *always* around 60 μm, irrespective of the amount of reddening.
Even small quantities of dust are sufficient to produce significant radiation reprocessing.
This property has important consequences for the prospects of detecting starburst galaxies
at high redshift. This restframe wavelength, if observed at $z > 3$, is accessible to telescopes
operating at sub-mm wavelengths. Since far-IR fluxes provide an unbiased, reddening-free

starburst count and since k-corrections[2] work in favor of detecting fainter objects, sub-mm observations of distant starburst galaxies can provide important constraints on the cosmic star-formation history, as we will explain in Chapter 11.

The photons seen in starburst spectra originate either in stars or in the ISM. Star light dominates from the near-IR to the far-UV (2.2 μm to 912 Å). All other parts of the energy distribution in Fig. 10.14 are due to interstellar dust (60 μm) and gas (6 cm and X-rays). The entire spectrum is of course powered by *stellar* energy input, and the radio and X-ray radiation have their origin in the non-thermal stellar luminosity related to winds and supernovae. Figure 10.14 can be used for a rough estimate of the ratio between the radio and X-ray luminosity over the bolometric luminosity, assuming the peak at 60 μm is indicative of L_{Bol}. The average starburst galaxy has $L_X/L_{Bol} \approx 10^{-3}$ and $L_{radio}/L_{Bol} \approx 10^{-5}$. We already discussed the origin of the X-ray luminosity: galactic-scale superwinds at temperatures of 10^7 K and reaching distances of tens of kpc from the galactic disk.

Since galactic superwinds are powered by stellar winds and supernovae, we can use Fig. 10.14 for a rough estimate of the efficiency of this process. A population of massive stars provides about 1% of its energy release non-thermally, via winds and supernovae. This is the value predicted from stellar evolution and population synthesis. If we take the observed X-ray luminosity in starbursts as a measure of the energy release in galactic superwinds, the fraction of $L_X/L_{Bol} \approx 10^{-3}$ together with the prediction of 1% then suggests that the efficiency of this process is about 10%. In a similar fashion we can utilize the value of the 6 cm radio flux in Fig 10.14 to constrain its physical origin. As opposed to the X-ray emission, the bulk of the radio flux is spatially coincident with the optical/near-IR light. The source of the radio emission is synchrotron radiation from relativistic electrons accelerated in magnetic fields.[3] Although details are uncertain, the emission is (at least partially) related to compact supernova remnants. These are the relics of the supernovae interacting with the surrounding ISM. Calculating the amount of synchrotron radiation is very model dependent. Therefore we rather take the observed properties of individual supernova remnants in our Galaxy and in the LMC as a guideline. These objects have $L_{radio}/L_X \approx 10^{-2}$, surprisingly similar to the value observed in the starburst galaxy population as a whole. This supports the hypothesis that the galactic radio emission has its origin in a population of thousands to millions of compact supernova remnants.

The global spectral energy distributions of Type 2 AGN are similar to those of starbursts in the radio to near-IR waveband, fainter in the visual to ultraviolet, but stronger in the X-rays. Less extreme Type 1 AGN follow the same trend. These AGN all have in common a relatively strong obscuration of the central black hole and a correspondingly important contribution from the narrow-line region and from stellar light. As in starburst galaxies, the mid/far-IR emission is the most important contributor to the bolometric luminosity. The increased level

---

[2] The k-correction is the corrective term that needs to be applied to the observed flux in a certain waveband due to the effect of redshift. A flat-spectrum source would have the same observed flux, irrespective of redshift. In contrast, a source with a spectrum rising towards longer wavelength will be relatively brighter when observed at high redshift because the observable window moves towards longer wavelengths. The corresponding correction factor is called the k-correction.

[3] Radiation at cm wavelengths can be non-thermal (synchrotron) and thermal (bremsstrahlung) in starburst galaxies. The former is more important at longer wavelengths, and the latter at shorter. In most starburst galaxies, the emission at 6 cm is purely non-thermal.

of X-ray flux can be understood as due to a hard component arising in the vicinity of the central black hole.

There are strong *theoretical* arguments in favor of accretion as the powering source of AGN. On the other hand, circumnuclear starbursts can have bolometric luminosities that rival even powerful QSOs. The close connection between starbursts and AGN has often been emphasized (e.g., Terlevich 1994) but *observational* proof, via spectroscopy or direct imagery, has remained elusive until very recently. It was largely the spatial resolution of the Hubble Space Telescope and its UV sensitivity which could unambiguously establish the co-existence of powerful starbursts and AGN within a few pc of the galaxy centers.

An example is shown in Fig. 10.15, which contrasts HST WFPC2 and FOC 2200 Å imaging of the central region of the UV-bright Seyfert2 galaxy NGC 7130. The 2200 Å passband contains no strong nebular emission lines and traces the stellar continuum of the recently formed OB stars. The nucleus, which was previously thought to be point-like, displays a complex morphology, suggestive of a circumnuclear starburst ring. Spectroscopy confirms that the UV and optical light in the central 500 pc of NGC 7130 (and many other galaxies) is entirely dominated by star light and not by emission associated with the AGN.

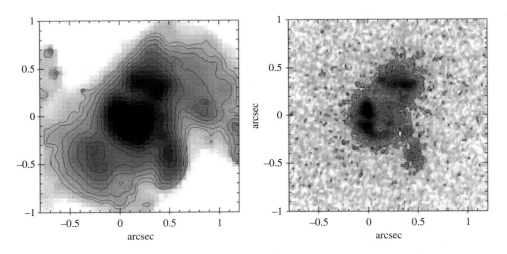

Fig. 10.15 Central $2'' \times 2''$ (620×620 pc^2) of NGC 7130 in optical (left) and UV (right) light. Reproduced from González Delgado *et al.* (1998) by permission of the AAS.

A related study was performed for a sample of UV-bright low-luminosity AGN, and the results are reproduced in Fig. 10.16. Here, HST's UV sensitive spectrograph was used to record the spectrum within tens of pc of the AGN. Massive, hot stars have few strong spectral features in the optical and near-IR so that their presence is easily hidden by a strong non-stellar continuum and by emission lines. This is the reason for the difficulty of performing such a test in the more easily accessible optical region. The situation changes in the UV, where a population of hot stars, if present, has unique broad SiIV $\lambda1400$ and CIV $\lambda1550$ features. The AGN spectra in Fig. 10.16 are compared with the spectrum of the starburst galaxy NGC 1741. We discussed this starburst galaxy in detail in Section 10.2. Its underlying stellar population is dominated by OB stars, with a standard IMF and an age of a few Myr. Therefore this spectrum is an ideal template spectrum representing pure stellar light without any additional

AGN contribution. The AGN spectra in Fig. 10.16 show clear absorption lines of massive stars, indicating a stellar origin for the UV continuum. The similarity between NGC 4569 and NGC 1741 is particularly striking. The compact central UV continuum source that is observed in these galaxies is a nuclear star cluster rather than a low-luminosity AGN. This also implies that any reflected light from the nucleus by hot gas above and below the torus must be negligible. Otherwise the stellar absorption lines would be diluted by the reflected featureless continuum, and their depth would be much decreased.

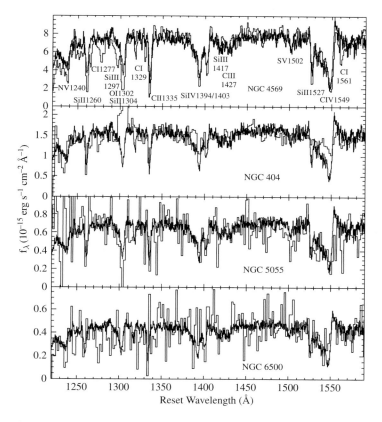

Fig. 10.16 HST spectra of the four low-luminosity AGN (heavy lines). Overlaid in each case is the spectrum of the starburst galaxy NGC 1741. Reproduced from Maoz *et al.* (1998) by permission of the AAS.

HST UV spectroscopy directly shows that hot stars provide most of the UV light in about 50% of the brightest Type 2 AGN. The population of hot stars in these AGN is typically obscured and reddened by dust, with attenuation factors of 5–10 at 1500 Å. The implied UV luminosities of the starbursts range from 10^8 to $10^9 L_\odot$ in low-luminosity AGN and 10^{10} to $10^{11} L_\odot$ in Seyfert galaxies. Massive stars are clearly energetically significant in these AGN. However, very few AGN could be studied with this technique, and these are biased in favor of cases with high UV surface brightness.

Clearly these results must not be generalized to all AGN; these galaxies were deliberately chosen to maximize the detection probability of hot stars and have a weak AGN by definition. On the other hand, low-luminosity AGN are found in almost 50% of disk galaxies. Therefore, in terms of ubiquity, low-luminosity AGN are the most common species among active galaxies. An important implication of this result is the co-existence of a central black hole and one or more young star clusters within a few pc of the very center of the host galaxy. The different levels of AGN activity, produced by the different ionizing continuum, could be understandable as the result of the age and mass of the star clusters on the one hand, and of the accretion rate and the mass of the black hole on the other. A low-mass black hole with a low accretion rate would emit at a bolometric luminosity of $\sim 10^6 \, L_\odot$. A central young star cluster with an age of a few Myr would make a substantial, or even dominant, contribution to the energy output. Once the young stars evolve and reach an age of 10 Myr, their light output decreases, and their contribution to the ionizing luminosity becomes negligible. From then on, the central region of the host galaxy would be dominated by the radiation from the accreting black hole.

Analysis of stellar features is preferable over nebular diagnostics, which can often be degenerate. Nevertheless, since the nebular emission lines are by far the strongest features in an AGN spectrum, numerous efforts have been made to utilize them as a tracer of the starburst population. The fundamental challenge is to disentangle the extreme-UV radiation field of stars from the power-law spectrum emitted by the gas accreting onto a black hole. Often, both types of ionizing spectra can produce very similar optical and IR emission lines.

A case in point is attempts to interpret the emission-line spectra of low-luminosity AGN in terms of an ionizing stellar radiation field. Shields (1992) demonstrated that their general characteristics can be understood in terms of photoionization by relatively hot, yet normal main-sequence O stars. His calculation results show improved agreement with observed spectra if the irradiated plasma is relatively dense and has solar abundances with a dust component. Such dense circumstellar media may be a natural consequence of high interstellar pressures, which may explain the preference of AGN phenomena for the nuclei of large and early-type disk galaxies.

One of the most influential studies on this subject was published by Terlevich & Melnick (1985). They proposed a model of an AGN without a black hole, an idea which evolved into the concept of a pure starburst energy source for AGN. This model had been the subject of much controversy since it was first proposed and is now believed to be not applicable in its most extreme version. Nevertheless, the main hypothesis of a central starburst in many AGN has found support by observations, contrary to early belief, and many of the original ideas in the model have been incorporated in more recent "hybrid" starburst plus black hole scenarios.

The fundamental assumptions of the starburst model for AGN are that hot stars can power the ionization in Seyfert2 galaxies and that their evolutionary end products are associated with compact supernova remnants in a nuclear starburst. The model is generally referred to as the "warmer" model because it relies on a young, massive metal-rich cluster containing "warmers", extreme W-R stars reaching temperatures of about 150 000 K. These stars are required to account for the high ionization seen in the nebular emission lines. Terlevich & Melnick's calculations demonstrated that a cluster containing warmers could have a sufficiently hard ionizing spectrum to photoionize the surrounding gas and mimic a Seyfert2 spectrum. Although some very hot W-R stars have been found, modern stellar atmosphere models suggest that very few W-R stars have such high *surface* temperatures. The strong stellar winds and high continuum column densities make W-R stars opaque to their ionizing radiation at atmospheric depths where the radiation temperature would be in excess of 10^5 K. The warmer discussion is a perfect example of the relevance of different temperature definitions when applied to phenomena outside the field of stellar atmospheres. This is not to say that W-R stars are unrelated to AGN. The existence of luminous AGN showing W-R features suggests a connection between the feeding mechanism of the black hole and the star-formation process. However, the AGN and the W-R stars are not causally related.

An alternative approach was taken by Taniguchi, Shioya, & Murayama (2000) who proposed an identification of some low-luminosity AGN as *post-starburst galaxies*. In the post-starburst model, the ionization sources are planetary nebula nuclei with temperatures of $\sim 100\,000$ K that appear in the evolution of intermediate-mass stars with mass between 3 and $6\,M_\odot$. This phase lasts until the death of the least massive stars formed in the starburst, which is about 5×10^8 yr for a stellar IMF truncated at $3\,M_\odot$. The top-heavy IMF and the absence of less massive stars produces the required spectrum resembling that of a low-luminosity AGN.

Photoionization models of optical emission lines are plagued by this degeneracy: a carefully fine-tuned starburst spectrum can often mimic the same nebular emission line spectrum as does a power law spectrum from an AGN. An important result of the ISO mission (Genzel & Cesarsky 2000) was a set of mid-IR diagnostics which can break the degeneracy. ISO, the Infrared Space Observatory, was an Earth-orbiting observatory observing at near- to far-IR wavelengths. Figure 10.17 shows representative ISO SWS spectra of 30 Doradus

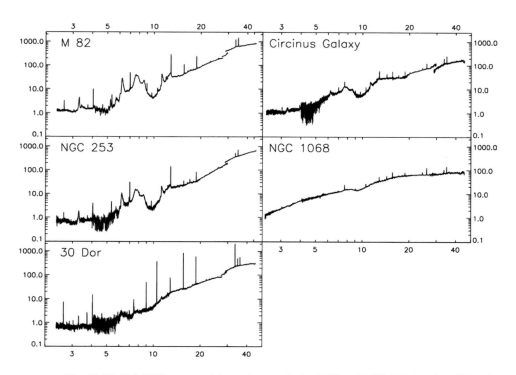

Fig. 10.17 ISO SWS spectra of the starburst galaxies M82 and NGC 253, the giant HII region 30 Doradus, and the Seyfert2 galaxies Circinus and NGC 1068. From Sturm *et al.* (2000).

(representing a giant HII region), M82 and NGC 253 (pure starbursts), and NGC 1068 and the Circinus galaxy (two Seyfert2 galaxies with circumnuclear starbursts). The differences between the spectra are striking. The starbursts and 30 Doradus have strong low-excitation fine structure lines, few and/or weak high-excitation lines, little dust emission below 10 μm, and very strong PAH bands. PAHs (polycyclic aromatic hydrocarbons) are large carbon molecules formed in dense, neutral interstellar clouds. In contrast, the two Seyfert2 galaxies have fainter emission lines, but with much higher excitation (e.g., the much stronger [OIV] 25.9 μm line). The dust continuum is much stronger in the AGN-type spectra. Note that NGC 1068 and the Circinus galaxy both have circumnuclear starbursts. Therefore they show starburst features as well.

Diagnostic diagrams are powerful tools for an empirical characterization of the excitation state of a source. Genzel *et al.* (1998) introduced the [OIV] 25.9 μm to [NeII] 12.8 μm line flux ratio versus the strength of the PAH bands to discriminate between starbursts and AGN. Figure 10.18 demonstrates that this ISO diagnostic diagram clearly separates known star forming galaxies from AGN. The diagram is insensitive to dust extinction if the emission lines and the bands are affected by a similar amount of extinction. [OIV], although weak, is often detected in starburst galaxies. Its origin is not yet completely understood. Fast, ionizing shocks powered by supernovae and stellar winds have been considered. In addition, hot W-R stars can meet the ionization budget in some galaxies as well.

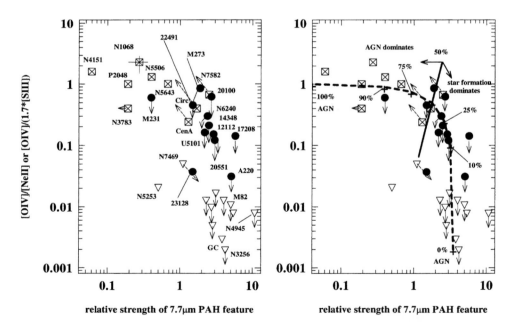

Fig. 10.18 Diagnostic diagram showing the [OIV]/[NeII] line ratio vs. the strength of the 7.7 μm PAH feature. Starburst galaxies: open triangles; ULIRGs: filled circles; AGN: crossed rectangles. Left: basic data with individual sources marked. Right: simple linear mixing curve, made by combining various fractions of total luminosity in an AGN and a starburst. Reproduced from Genzel *et al.* (1998) by permission of the AAS.

The diagnostic diagram of Fig. 10.18 is most powerful when applied to dusty, IR-luminous galaxies, where the nature of the underlying source is not known a priori. Ultraluminous IR galaxies are of particular interest, as the starburst versus AGN debate is still ongoing in these objects. When studied with this method, starbursts dominate up to luminosities of about $2 \times 10^{12} \, L_\odot$, above which AGN become energetically more and more important. The most luminous starbursts are detected in galaxies with luminosities close to $\sim 10^{13} \, L_\odot$. This limit corresponds to the maximum star formation rate of $\sim 10^3 \, M_\odot \, \mathrm{yr}^{-1}$ of a gas-rich spheroid undergoing a monolithic collapse on a dynamical timescale.

There is now ubiquitous evidence for starbursts in AGN over several decades in luminosity. While the precise evolutionary scheme is still unknown, it seems likely that the gas supply feeding the central AGN is capable of sustaining the circumnuclear starburst as well.

11

Cosmological implications

Less than a few hundred thousand years after the Big Bang, the temperature was high enough that cosmic gas consisted of protons, free electrons and light nuclei. Once the Universe cooled to about 3000 K, the electrons and protons were moving sufficiently slowly that they combined to form hydrogen atoms. With scattering of photons much reduced, they were able to move in straight lines indefinitely, and may be seen redshifted into the microwave part of the spectrum as the 2.7K CMB. So began the era of recombination, or so-called "dark ages" when the IGM became mostly neutral. Within the current cold Dark Matter model for the hierarchical formation of structure, mini-halos of mass $\sim 10^6$ $M_\odot$ (Couchman & Rees 1986) provided the gravitational seeds for the first stars at $z \approx 20$–30, ending the "dark ages" through re-ionization of the IGM. A comprehensive review of the astrophysical role of dark matter is provided by Jungman, Kamionkowski, & Griest (1996).

Galaxies formed as baryonic gas cooled in the centers of dark matter structures, from which galaxy mass built up via mergers of halos and proto-galaxies (White & Rees 1978; Davis *et al.* 1985). Since most present-day galaxies are relatively old, it follows that they formed at $z \geq 2$. The timescale over which galaxies assembled remains unclear, particularly the bulges and disks which are the main components of present-day galaxies. Old, super metal-deficient stars – whose iron abundances are 100 000 times less than our Sun – which have been discovered in the halo of our Galaxy, in principle provides tight constraints upon the Milky Way star formation at high z. Finally, high redshift gamma-ray bursts (GRBs) – intimately connected to the deaths of some massive stars (Woosley & Bloom 2006) – may also be used as probes of the circumstellar and interstellar environment within their host star-forming galaxy.

11.1   Population III stars

Once individual sources began emitting ionizing radiation, isolated HII (and HeIII) regions formed, and eventually overlapped, as illustrated in Fig. 11.1, after which the IGM became transparent to Lyα radiation. Residual absorption continued as discrete, redshifted Lyα absorbers known as the Lyα forest, shortward of the HI and HeII Lyα lines at $1215(1+z)$ Å and $303(1+z)$ Å, respectively. As the H and He$^+$ opacity increases, these regions grow into black absorption troughs, known as the Gunn–Peterson effect, following the method outlined by Gunn & Peterson (1965).

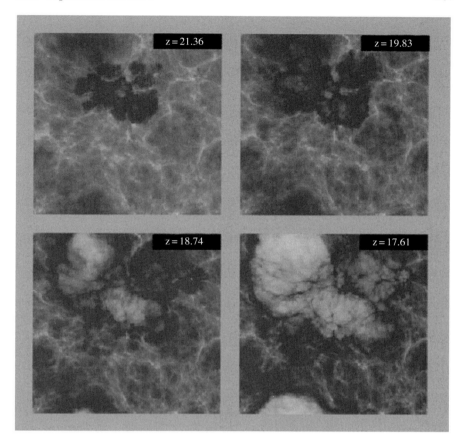

Fig. 11.1 Simulations between $z = 21.4$ and 17.6 showing the effect of the HII regions produced by the first stars (bright zones) upon the neutral gas (filamentary structure). Each box measures 0.25×0.25 Mpc^2 in co-moving coordinates. From Sokasian *et al.* (2004).

The epoch of re-ionization may be deduced by analysis of weak transmission in the quasar continuum troughs shortward of Lyα, for which the Sloan Digital Sky Survey (SDSS) discovered suitable high-redshift quasars, as illustrated in Fig. 11.2. Analysis of these data reveals $z \sim 6.1$ for HI (Becker *et al.* 2001; Fan *et al.* 2004). Historically, quasars were expected to be the prime source of ionizing photons in the early Universe. However, the number of quasars declines rapidly above a redshifts of $z \sim 5$, such that early generations of stars are believed to be responsible for the re-ionization of the Universe above $z \sim 6$ (Bromm & Larson 2004). The situation is less clear for helium, for which far-UV FUSE observations indicate re-ionization occurs at $z \sim 2.8$ for HeII (Kriss *et al.* 2001). Recent observations of high-redshift GRBs with Swift, to be discussed later in this chapter, indicate that they too may be important probes of the IGM during re-ionization.

Cosmic Background Explorer (COBE) first discovered the small-scale anisotropy of the CMB, at a level of one part in 100 000, revealing how matter and energy were distributed when the Universe was still very young, leading to the large-scale structure of the present Universe.

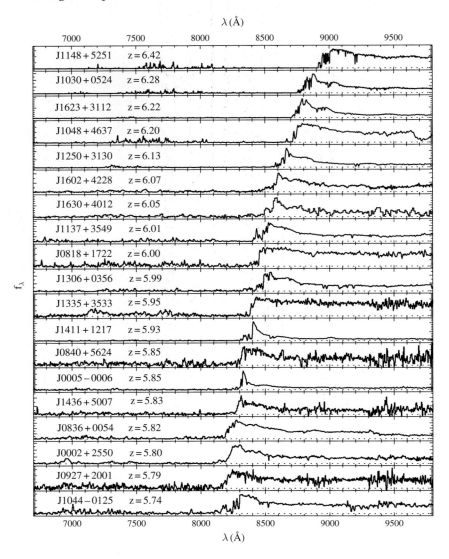

Fig. 11.2 Moderate resolution spectra of nineteen quasars at $5.74 < z < 6.42$. The sample was drawn from the Sloan Digital Sky Survey. From Fan, Carilli, & Keating (2006).

Its success led to other missions such as the Wilkinson Microwave Anisotropy Probe (WMAP) which have refined the CMB anisotropies, allowing a variety of cosmological parameters to be derived to high precision, of which the most familiar are the Hubble constant, $H_0 = 72$ km s^{-1} Mpc^{-1}, matter density $\Omega_m = 0.29$, and baryon density, $\Omega_b = 0.05$. In addition, WMAP measured the polarization of the CMB, revealing an optical depth to electron (Thompson) scattering of $\tau_e = 0.17^{+0.08}_{-0.07}$ according to Spergel *et al.* (2003). Simulations indicate that an IGM ionized at $z \approx 6$ according to the Gunn–Peterson observations would contribute only $\tau_e = 0.04$, in which the excess must have arisen from scattering of CMB photons off free electrons at higher redshifts. This provoking extensive theoretical study, suggesting that the re-ionization of the Universe began at relatively high redshift of $z \approx 17$ according to Ciardi, Ferrara, & White (2003), is shown in Fig. 11.3.

Analysis of the WMAP observations after 3 years revealed modest revisions to the cosmological parameters, $H_0 = 73$ km s^{-1} Mpc^{-1}, $\Omega_m = 0.24$, and $\Omega_b = 0.04$, plus a *reduced* optical depth of $\tau_e = 0.09 \pm 0.03$ (Spergel *et al.* 2007). From Fig. 11.3, this revision suggests the peak of re-ionization at $z \approx 11$, reducing the interval over which re-ionization occurred. Further progress is anticipated using new low-frequency radio telescopes such as the Low Frequency Array (LOFAR) to search for emission or absorption from the well-known hyperfine 21 cm transition of atomic hydrogen, redshifted to > 1.5 m at $z > 6.5$.

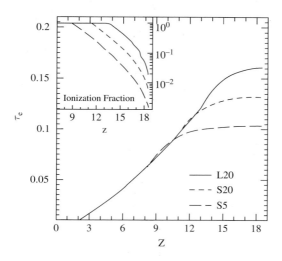

Fig. 11.3 Simulations showing the redshift evolution of the Thompson scattering optical depth τ_e for various initial mass functions and photon escape fractions (L = Larson, S = Salpeter, 5 or 20 = 5% or 20% escape fraction). In the inset the redshift evolution of the volume averaged ionization fraction is shown for the three cases. From Ciardi *et al.* (2003).

Observationally, there are two established populations of stars within the Milky Way (Baade 1944). Population I stars are relatively young, formed from gas with a similar composition to that of the Sun, i.e. of order 1–2% metals by mass. Population II stars are relatively old, formed from gas deficient in metals by a factor of several with respect to the solar case, as a result of reduced chemical enrichment at the time of their formation. In extreme hyper-metal poor cases, old halo stars have been found to be deficient in iron peak elements by several orders of magnitude, e.g. HE0107-5240 and HE1327–2326 with [Fe/H] = −5.2 and [Fe/H] = −5.4, respectively (Christlieb *et al.* 2002; Frebel *et al.* 2005). Abundance patterns in these stars are shown in Fig. 11.4, revealing that although they are extremely metal-poor in iron, they are only moderately metal-poor in C, N, and possibly O. Although details of Population II and Population I stars differ, they are fundamentally similar in nature.

In addition to these two stellar populations there is a third class, known as Population III stars, which represent the very first generation of stars, formed out of pristine material in the

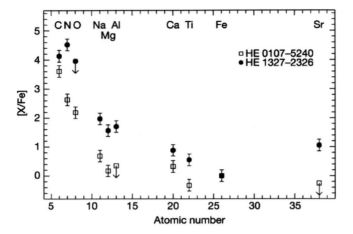

Fig. 11.4 Abundance patterns of two hyper-metal-poor halo stars HE0107–5240, with [Fe/H] = −5.2 and HE1327–2326 with [Fe/H] = −5.4. The figure illustrates an extreme over-abundance of CNO elements with respect to iron, providing important limits upon the early chemical evolution of the Milky Way. From Frebel *et al.* (2005).

very early Universe, and quite unlike anything seen in the present-day Milky Way. Numerical simulations of the collapse of primordial gas clouds have been carried out (Bromm, Coppi, & Larson 1999; Abel, Bryan, & Norman 2000). The virial temperatures of dark matter mini-halos are below the $\sim 10^4$ K threshold for efficient cooling by atomic hydrogen. Regardless of the initial conditions, the primordial gas achieves temperatures (~ 200 K) and densities ($\sim 10^4$ cm^{-3}) for which the fragmentation scale is dictated via cooling dominated by H$_2$. In the absence of dust grains to permit their formation, molecules have to form in the gas phase via

$$H(e^-, \gamma)H^-(H, e^-)H_2.$$

Gravitational collapse occurs when the mass exceeds the Jeans mass, or more strictly the Bonnor–Ebert mass, M_{BE} (Bonnor 1956; Ebert 1955) in the context of collapsing gas cores, i.e.

$$M_{BE} \sim 700 M_\odot \left(\frac{T}{200 \text{ K}}\right)^{3/2} \left(\frac{n}{10^4 \text{ cm}^{-3}}\right)^{-1/2} \tag{11.1}$$

(Clarke & Bromm 2003). A pre-stellar clump of mass $M \geq M_{BE}$ is the progenitor of a single star or multiple system. As a result, primordial stars are preferentially massive – perhaps as large as several hundred solar masses, for which 700 $M_\odot$ is an approximate upper limit. The characteristic mass of primordial stars, however, remains rather controversial.

In addition, the absence of metal opacities implies that Population III stars are extremely compact. The radii of Population III stars are much smaller than present-day massive stars, such that they possess high effective temperatures which greatly enhances their ionizing photon production, particularly in the HeII Lyman continuum, as illustrated in Fig. 11.5.

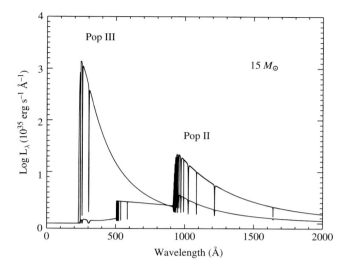

Fig. 11.5 Comparison of theoretical spectra calculated for Population II ($Z = 0.001$, $T_{eff} = 36\,000$) and Population III ($Z = 0$, $T_{eff} = 63\,000$) stars of 15 $M_\odot$. Reproduced from Tumlinson & Shull (2000) by permission of the AAS.

From an evolutionary perspective, Population III stars would not be expected to lose their H-rich envelope via a conventional line-driven wind, due to the absence of metals. As a consequence, Population III stars are expected to remain massive at the end of core H-burning, which a characteristic lifetime of 2–4 Myr above $40M_\odot$ (Schaerer 2002). This occurs via the relatively inefficient p-p chain rather than the CNO cycle, due to the absence of metals. Consequently, very massive Population III stars will be intense sources of hard ionizing radiation, but the absence of stellar winds would imply that chemical yields depend solely upon the nature of their demise.

According to Heger *et al.* (2003), Population III stars would end their life as conventional core-collapse SN up to 25 (neutron star) or $40M_\odot$ (black hole), above which they either undergo direct collapse to a black hole, or form pair-production supernovae for stars with initial masses in the range 140–260 $M_\odot$, which would leave behind no compact remnant (Section 5.6). Population III supernovae were responsible for the initial enrichment of the IGM (Gnedin & Ostriker 1997) such that chemical abundances in old metal-poor stars in principle permit the characteristic mass of Population III stars to be established.

The apparent absence of true Population III stars in the Milky Way halo (Bond 1981) suggests there may be a critical metallicity, $Z_{crit} \leq 10^{-4}\ Z_\odot$, below which low-mass star formation is inhibited/excluded depending on whether cooling by fine-structure lines of carbon (CII) or oxygen (OI) occurs (Hollenbach & McKee 1989; Bromm *et al.* 2001), for which the characteristic temperature may reach 10 K, rather than 200 K in the presence of solely H_2.

Regarding the potential for the direct detection of Population III stars at $z > 7$, strong nebular Lyα and HeII 1640 emission would be expected, albeit diluted by a strong nebular continuous emission (Schaerer 2002). Dust formation is a cause of concern for the feasibility of direct observation of integrated Population III stellar signatures. Detection of UV signatures from such stellar populations is one of the principal aims of the James Webb Space Telescope (JWST), the successor to the Hubble Space Telescope. The anticipated sensitivity of the JWST in the near-IR is ideally suited to the detection of UV emission from sources at a redshift of $z \approx 10$.

Considering the ionizing budget necessary to re-ionize the IGM – subject to an uncertain escape fraction of ionizing photons from young galaxies – suggests a peak of the stellar IMF below Z_{crit} in the regime of core-collapse SN, according to Tumlinson (2006). Chemical abundance patterns in old halo stars and the IGM also suggest a peak in the mass function for the first stars closer to 10 $M_\odot$ than 100 $M_\odot$.

How is this reconciled with the numerical simulations for the collapse of primordial gas? The presence of deuterium in the primordial gas may permit the cooling to temperatures below that possible solely via H_2 line cooling. It is possible that primordial gas shocked either the first SN explosions of Population III stars, or from merging dark matter halos in the very early Universe, could form so-called Population II.5 stars (Mackey, Bromm, & Hernquist 2003). Cooling via HD down to temperatures of $\sim$100 K would provide a characteristic mass of $\sim$10 $M_\odot$. If such stars were to dominate over classical $\sim$100 $M_\odot$ Population III stars, chemical evolution models may be reconciled with the observed abundance patterns of the oldest stars and the IGM (Johnson & Bromm 2006).

11.2   Lyman-break galaxies

The distant Universe at redshifts beyond $z > 1$ has finally become accessible to observations. As opposed to quasars, only very few galaxies at $z > 1$ were known prior to the last decade of the twentieth century. At the beginning of the new millennium, thousands had been discovered and studied in detail, many of them at spectroscopic resolution. Their numbers are likely to increase by orders of magnitude as a result of several ongoing and planned redshift surveys in the next several years. Understanding this rapidly growing sample of high-redshift galaxies has become a major goal of contemporary astronomy. Much of the analysis focuses on the properties of the newly formed massive stars. The results are required to answer basic questions, such as when did the stars in the Universe form, or how are the sites of star formation related to the evolving perturbations in the distribution of matter?

The finite velocity of light allows astronomers to use galaxies at progressively higher redshifts (and therefore distances) as probes of the astrophysical environment in the remote past. It is now possible to routinely observe star-forming galaxies up to cosmological look-back times of more than 90% of the age of the Universe for redshifts up to $\sim$7. The correspondence between redshift and look-back time depends on the cosmological model of the Universe. In Table 11.1 we list this relation for a flat Universe with cosmological parameters $H_0 = 70$ km s^{-1} Mpc^{-1}, $\Omega_m = 0.3$.[1] Recall that Ω_m is the contribution provided by baryons to the total density of the Universe.

Why is there such an interest in studying galaxies when the Universe was at its infancy? Apart from the quest for ever fainter and more distant galaxies per se, we believe that such observations hold the key for relating the properties of present-day galaxies to the formation of the first stellar generations in the early Universe. The galaxies in the local Universe display a wide variety of shapes and sizes. Roughly 50% of the stars in the present-day Universe are

---

[1] From Ned Wright's Cosmology Calculator http://www.astro.ucla.edu/wright/CosmoCalc.html

Table 11.1 *Look-back time as a function of redshift.*
$H_0 = 70 \ km \ s^{-1} \ Mpc^{-1}$; $\Omega_m = 0.3$; $\Omega_\Lambda = 0.7$.

z	T (Gyr)	T/T_∞
0	0	0
0.5	5.04	0.37
1	7.71	0.57
2	10.24	0.76
3	11.35	0.84
4	11.95	0.89
5	12.31	0.91
10	13.00	0.97
∞	13.46	1.00

found in "old" systems, like elliptical galaxies and the bulges of spirals. The conventional view is that these galaxies formed a long time ago, probably when the Universe had not yet reached 50% of its present age. Table 11.1 suggests that this corresponds to redshifts of at least about 1. Most old systems formed relatively rapidly, whereas the assembly of the disks of spiral galaxies, including the Milky Way, occurred over a longer period of time. One would expect the formation of ellipticals and bulges to be associated with a period of intense star formation, similar to today's starburst galaxies.

Identifying such star-forming galaxies at cosmological distances had been a major quest in the second half of the twentieth century. Star-forming galaxies in our vicinity have very characteristic optical spectra. Between 3500 and 7500 Å, the spectra are dominated by strong nebular emission lines produced in the ionized gas associated with newly formed massive stars. In comparison, the underlying stellar continuum radiation is rather faint. At redshift $z > 2$, however, the portion of the spectrum accessible to an optical telescope is the rest frame UV below 3000 Å. Early searches for bona fide star-forming galaxies targeted what was expected to be the strongest nebular emission line: Lyα. Despite massive efforts, these searches resulted in failure (Thompson, Djorgovski, & Trauger 1995). In retrospect, it is easy to understand why the detection rate of Lyα emitting galaxies is so low. Although copious amounts of Lyα photons may be produced in the nebulae ionized by hot stars, their escape probabilities can be rather low. Similar to the situation in a stellar atmosphere, Lyα can be trapped in the nebula by multiple scattering events by neutral hydrogen. The longer the path length, the higher is the likelihood of absorption by dust grains. Only if the Lyα photons can escape from this trapping process, for instance if large-scale velocity gradients are present in the ISM (note again the similarity to an expanding stellar atmosphere), is a strong Lyα emitter observed. The net result of these processes is a generally much weaker Lyα for a given star-formation rate. Modern Lyα surveys (e.g., Rhoads & Malhotra 2001) do indeed detect star-forming galaxies, although at smaller numbers than originally thought.

One method which has succeeded in finding large numbers of star-forming galaxies at high redshift was developed by Steidel & Hamilton (1993). It makes use of a fundamental property of the Lyman series limit at 912 Å: under most conditions, a strong discontinuity is produced in the far-UV spectrum of a star-forming galaxy. Three mechanisms are responsible for this "Lyman-break".

- Hot OB stars have a pronounced intrinsic Lyman-break of about a factor of 3 when measured immediately shortward and longward of the Lyman series limit. The break is weaker for hotter stars and for stars with very extended atmospheres, such as W-R stars. However, these stars are generally not numerous enough to dominate the spectrum of a population. Therefore a hot-star population spectrum will always exhibit some sort of Lyman break.
- The galaxy ISM is highly opaque in the Lyman continuum. Therefore almost all of the stellar Lyman photons are absorbed, generating a Lyman-break of ∞. While some HII regions are density-bounded and therefore transparent to ionizing radiation, there is manifold evidence for a highly opaque ISM. Recall that no O star in the solar neighborhood has ever been detected at wavelengths below 912 Å because of the ISM opacity.
- Finally, the intervening IGM is a significant opacity source. This last effect has been quantified by Madau (1995) from the statistics of QSO absorption line spectra. Somewhat unexpectedly, the radiative transfer in the IGM turns out to be the overriding factor determining the colors of high-redshift galaxies. When studying a Lyman-break galaxy (LBG) below 912 Å, we actually measure the absorption of Lyman photons along the line of sight, and not necessarily in the galaxy itself. The IGM opacity can become so strong that at $z > 3$–4 the dominant break is not caused by the Lyman edge but by intergalactic Lyα along the sightline. This phenomenon is called the Lyα forest.

Steidel & Hamilton (1993) introduced a filter system with wavelengths shortward and longward of 912 Å *in the restframe UV*. The filter system is called $U_n GRI$ and has effective wavelengths of 3650, 4750, and 6930 Å. In deep images of the sky taken through these filters, galaxies around and above a redshift of about 2.5 stand out from their lower-redshift counterparts by their color excess. Combined with the expanding Universe, these filters become tunable. Galaxies at the right redshift will be very bright in one filter and very faint in another. Galaxies at too low or too high redshift will be fainter and/or brighter in all filters because their Lyman-breaks are shifted away from the wavelength which produces optimum color contrast. A schematic color–color diagram in the $U_n GRI$ system identifying candidate LBGs is shown in Fig. 11.6. The strongest candidates are in the shaded area enclosed by the solid polygon in the upper left. The colors of the galaxies are consistent with standard stellar population models of continuous star formation, altered by the galaxy ISM and the statistical opacity of the IGM, and reddened by intrinsic dust.

Very deep HST imaging such as that provided by the Hubble Deep Field (Williams *et al.* 1996) has also proved to be very effective in detecting LBGs. The HST data reach much deeper but are confined to very small areas on the sky; therefore, the ground-based and space-based surveys have been largely complementary.

The Lyman-break technique successfully identified a large population of star-forming galaxy *candidates* at high redshift. Follow-up spectroscopy with 10-m class telescopes is required for unambiguously confirming the nature of the photometrically selected objects. Almost all of the robust candidates are indeed high-redshift galaxies, the remainder being mainly faint galactic stars. However, even with the light-gathering power of large telescopes, a

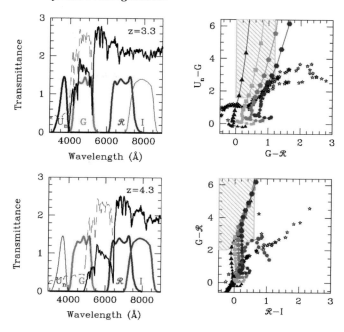

Fig. 11.6 Redshifted synthetic model spectra which include the effects of cosmic opacity are plotted together with the $U_n GRI$ filters' transmittance (left panels). The two upper panels show the U-band dropouts at $z \approx 3$, defined with the $U_n GR$ filters; the lower ones the G-band dropouts at $z \approx 4$, defined through the GRI set. The candidates are selected from their position in the color–color space, as shown in the right panels. The curves represent galaxies placed at progressively higher redshifts, starting at $z = 0.5$ with step $\Delta z = 0.1$. The four tracks correspond to model spectra with different amounts of dust obscuration. The shaded areas indicate the color selection criteria adopted for ground-based surveys. From Giavalisco (2002).

typical star-forming galaxy at $z \approx 3$ is often too faint to yield a high signal-to-noise spectrum. For reference, an $L^\star$ galaxy[2] has an R magnitude of about 24.5 at a redshift of 3. Thus, more detailed studies of the physical properties of these early episodes of star formation require an additional observational aid, namely, the light magnification produced by gravitational lensing. A few examples have been found serendipitously already and targeted searches in the fields of foreground clusters of galaxies may detect more candidates.

One of the best-studied case of a lensed star-forming galaxy is MS 1512-cB58. This object was discovered in the CNOC cluster redshift survey (Yee *et al.* 1996). Its exceptional nature was immediately apparent from its brightness of $R \approx 20.5$ at a redshift $z = 2.7$. HST WFPC2 V- and R-band data of the foreground cluster MS 1512+36 (Fig. 11.7) together with a galaxy lens model demonstrated that MS 1512-cB58 is not extraordinarily luminous intrinsically but lensed into a gravitational arc by the cluster. The arc has an axis ratio of 1:7, is marginally resolved in width and about 3 arcsec long. The magnification factor was found to be about 40, or 3.4 magnitudes. Thus, MS 1512-cB58 is a typical $L^\star$ LBG fortuitously

---

[2] $L^\star$ is the turn-over point in the "Schechter" (1976) luminosity function $N(L) = (N^\star/L^\star)(L/L^\star)^\alpha \exp(-L/L^\star)$. In this function, $N^\star$ is the normalization, and α is the slope at the faint end. $L^\star$ represents a characteristic galaxy with properties similar to those of the Milky Way.

Fig. 11.7 HST WFPC2 image of the core of the galaxy cluster MS 1512+36. The cluster is dominated optically by its cD galaxy in the center. MS 1512-cB58 is the elongated galaxy denoted by the arrow 5″ from the cD galaxy. A face-on spiral can be seen on the cD galaxy's opposite side. From Seitz *et al.* (1998).

made accessible to detailed spectroscopic studies by the presence of the foreground cluster MS 1512+36 at $z = 0.37$.

Figure 11.8 shows a restframe UV spectrum of MS 1512-cB58 obtained with the 10-m Keck telescope on Mauna Kea. For a redshift of $z = 2.7$, we sample the spectral region from below Lyα to just beyond [CIII] λ1909. Somewhat ironically, this is one of the best UV spectra of a starburst galaxy obtained at *any* redshift, including local examples studied with HST, such as NGC 1741 whose proto-typical spectrum was discussed in Section 9.4. The spectrum is extremely rich in absorption lines, mostly originating from stars and interstellar gas in MS 1512-cB58, but also from intervening gas from the IGM and the Milky Way along the line of sight.

The strong P Cyg profiles of NV λ1240, SiIV λ1400, and CIV λ1550 immediately suggest recent star-formation activity. Evidently, the UV spectral properties of this high-redshift galaxy are very similar to those of local starbursts. The profiles are best understood in terms of continuous star formation with a Salpeter IMF extending well beyond 50 $M_\odot$. Both a flatter IMF and an IMF lacking the most massive stars can be excluded. The UV continuum is redder than that expected for OB stars, most likely as a result of dust extinction. The difference between the slope of the spectrum measured and inferred from the stellar lines suggests a reddening of $E(B - V)$=0.1−0.3. The amount of reddening is similar to results found in local starbursts.

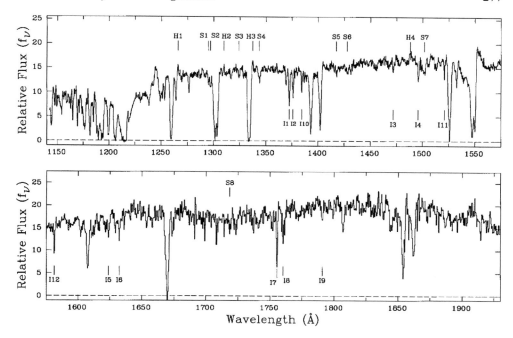

Fig. 11.8 Keck spectrum of MS 1512-cB58 reduced to the systemic redshift of the galaxy, $z = 2.7268$. Marks above the spectrum identify weak stellar lines and weak emission lines that are attributed to HII gas. Marks below the spectrum show the positions of intervening absorption lines in the IGM. Reproduced from Pettini *et al.* (2000) by permission of the AAS.

Both stars and gas show evidence of a relatively high degree of metal enrichment from the absorption line strengths. The strength of the interstellar absorption lines suggests a metallicity of about 1/3 solar. This surprisingly high metallicity is the result of multiple stellar generations prior to the current star-formation episode. Since almost all interstellar lines in star-forming galaxies are saturated, their usefulness as metallicity indicators may appear questionable: the observed equivalent widths W are on the flat part of the curve-of-growth,

$$W \propto b \left(\ln \frac{N_{\mathrm{ion}}}{b} \right)^{\frac{1}{2}}, \tag{11.2}$$

where W is quite insensitive to column density (N_{ion}) variations. W becomes primarily a measure of velocity, as expressed via the Doppler line-broadening parameter b. The fact that there is a correlation between the line strength and metallicity can therefore be only caused by a correlation between metallicity and velocity. The physical basis is the correlation between non-thermal energy input from the starburst and Z: more metal-rich starbursts are more vigorous and lead to more "stirring" of the ISM. This picture is consistent with the relative velocities of interstellar absorption lines, stellar photospheric lines, and HII region emission lines, as well as the highly asymmetric Lyα profile. The mechanical energy deposited by the starburst has produced a shell of swept-up interstellar matter that is expanding with a velocity of 200 km s^{-1}. The mass outflow rate of 60 $M_\odot$ yr^{-1} is comparable to the star formation rate derived from the UV luminosity. The galactic wind of MS 1512-cB58 is quite similar to the outflows seen in low-redshift starburst galaxies. It could be the mechanism that regulates star formation, distributes the metals over large volumes, and allows the escape of ionizing photons into the IGM.

It turns out that MS 1512-cB58 is not unusual in its spectral properties. By constructing a database of almost 1000 LBG spectra, Shapley *et al.* (2003) could create an extremely high S/N template representative of the LBG population as a whole. This composite spectrum is reproduced in Fig. 11.9. Features attributable to hot stars, HII regions, dust, and outflowing neutral and ionized gas can be seen. LBGs with stronger Lyα emission have bluer UV continua, weaker low-ionization interstellar absorption lines, and lower star formation rates. Our current knowledge of normal, unobscured galaxies at high redshift relies heavily on observations in the restframe-UV region, where we see the integrated light of OB stars and where the strongest spectral features are interstellar *absorption* lines. Paradoxically, the situation is quite different at low redshift, where the UV spectra of star-forming galaxies have become accessible only with the advent of satellites like IUE, HST, HUT, or FUSE. For most of the twentieth century, star-formation regions were studied first and foremost at optical wavelengths through the rich emission-line spectrum produced by their HII regions.

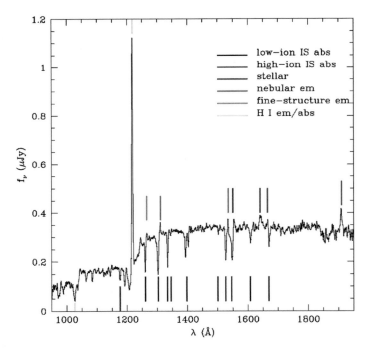

Fig. 11.9 Composite rest frame UV spectrum constructed from 811 individual LBG spectra. Dominated by the emission from massive O and B stars, the overall shape of the UV continuum is modified shortward of Lyα by a decrement due to intergalactic HI absorption. Several different sets of UV features are marked: stellar photospheric and wind, interstellar low- and high-ionization absorption, nebular emission from HII regions, SiII* fine-structure emission whose origin is ambiguous, and emission and absorption due to interstellar HI. Reproduced from Shapley *et al.* (2003) by permission of the AAS.

There is a strong desire to obtain restframe *optical* spectra for galaxies at $z > 2.5$ as well. The luminosity of the Balmer lines gives a measure of the star-formation rate which can be immediately compared with the values deduced in local surveys. Furthermore, since the optical emission lines are less affected by dust reddening than the UV continuum, the relative luminosity of a galaxy in these two tracers of star formation is a reddening indicator. When integrated over an entire galaxy, the line widths of nebular emission lines indicate the velocity dispersion of the HII regions within the overall gravitational potential. This allows an estimate of the kinematic mass, as opposed to the situation in the UV. At UV wavelengths, the interstellar absorption lines are sensitive to gas accelerated to high velocities by winds and supernovae, while the stellar-wind lines from O stars already have intrinsic line widths of hundreds of km s^{-1}. Finally, nebular emission lines are the foundation of chemical abundance diagnostics. Abundance measurements are much more difficult in the UV, where the equivalent widths are primarily sensitive to the turbulent velocity of the gas and not to the column density of the absorbing ions.

At $z > 2.5$, the traditional HII region lines are redshifted into the near-IR, where observations are much more challenging due to the restricted atmospheric transmission. Nevertheless, with a large sample of galaxies such as that produced by the Lyman-break technique, it is possible to isolate redshifts that place the transitions of interest in suitable atmospheric windows.

The star-formation rates of LBGs derived from the luminosity of Hβ agree with the values from the UV continuum luminosity at 1500 Å without applying any dust-reddening corrections. There is no indication for higher rates from Hα than from the UV, as expected from the shape of the reddening law that rises from the optical to the UV. Evidently, any such differential extinction must be small.

The widths of the nebular lines suggest one-dimensional velocity dispersions of $\sim 10^2$ km s^{-1}. If the widths reflect the relative motions of HII regions within the gravitational potential of the galaxies, the implied masses are of order 10^{10} $M_\odot$. Since the measurements refers to a radius where the rotation curves are likely to be still rising, these masses are lower limits to the total galaxy masses. More important, however, is the physical origin of the measured velocity dispersions. Similar to the situation in local, individual giant HII regions, the observed line broadening may be caused not only by gravity but by macroscopic turbulence induced by multiple generations of supernovae.

The central problem with the available data on LBGs is that the total energy produced by star formation is emitted across more than six decades of frequency, from the far-UV to the radio, while the restframe UV and optical probe only a narrow range of frequencies. Therefore the bolometric energy output is never directly observed. UV/optical surveys detect the portion of massive-star emission that is not absorbed by dust. Conversely, sub-mm surveys detect the fraction of the portion that is. Correlating the results from surveys at such widely spaced wavelengths is frustratingly uncertain. Therefore it is of major interest to explore additional properties of these galaxies, in particular at longer wavelengths, to be used in concert with the existing UV/optical measurements.

Such complementary data are necessary for understanding the range of objects selected with the Lyman-break technique, and the relationship of these objects to present-day galaxies. Different cosmological models provide quite divergent descriptions of the nature and fate of LBGs. LBGs could be bright in the restframe UV because they are experiencing a merger-induced starburst. The intense starburst occurs on relatively short timescales of order 100 Myr and produce much less stellar mass than seen in a typical L^* galaxy. These low-mass bursting objects would then be the precursors of local low-mass spheroids, or they merge with similar objects to form more massive systems (Somerville, Primack, & Faber 2001). Alternatively, LBGs reside in massive dark matter halos and form stars relatively quiescently over timescales of Gyr. Eventually they could evolve into the ellipticals and spiral galaxies at the bright end of the local luminosity function (Steidel *et al.* 1996).

11.3   Massive stars and cosmic abundances

Until recently, our knowledge of abundances at cosmological redshifts ($z > 2$) was limited to bright quasars and the material along their sightlines. Over the past 5–10 years, the number and nature of accessible objects has evolved dramatically, and "normal" star-forming galaxies at high redshift, such as LBGs (see the previous section), have become targets of abundance studies (Pettini 2004). LBGs are sufficiently bright for obtaining spectroscopic data ($25 > M_R > 21$) — a necessary prerequisite for abundance analyses. They are relatively dust-poor with typical visual attenuations of $A_V \simeq 1$ mag and are fairly luminous, but not ultraluminous ($L \simeq 10^{11} L_\odot$). Since LBGs are actively star-forming at rates of order $10^2 M_\odot$ yr^{-1}, they have correspondingly short recycling times. Therefore the chemical compositions of the newly formed stars and the interstellar gas are identical.

Most LBGs are in the redshift range $2 < z < 3.5$. The corresponding lookback times in a standard Universe are 10.3 Gyr $< t <$ 11.8 Gyr, or about 20% of the age of the Universe after its formation (see Table 11.1). In this section we are going to explore the determination of abundances in these galaxies, their nucleosynthetic production in massive stars, their relation to the ambient IGM, and how they fit into the overall cosmic chemical history.

Abundance determinations typically fall into two categories, either relying on indicators in the restframe optical, or on those in the restframe UV. The restframe optical wavelength region has traditionally been used to determine galaxy abundances from nebular emission lines. At a redshift of $z = 3$, the restframe optical is observed in the near-IR H and K bands. Spectroscopic observations of LBGs in the near-IR have become technically feasible (e.g., Pettini *et al.* 2001) but abundance analyses are still challenging. Only the strongest lines such as Hα, Hβ, [NII] λ6584, and [OIII] λ5007 are detectable at sufficient S/N. Even when good-quality spectra are available, the atmospheric windows usually restrict the wavelengths to a narrow range, which precludes commonly used techniques such as the classical R23 strong-line method (McGaugh 1991).

The need for alternative variants of the classical strong-line method led Pettini & Pagel (2004) to re-address the usefulness of the N2 and O3N2 ratios. The former is defined as the ratio [NII] λ6584 over Hα and was recently discussed by Denicoló, Terlevich, & Terlevich (2002); the latter includes the oxygen line for the ratio ([OIII] λ5007/Hβ)/([NII] λ6584/Hα) and was originally introduced by Alloin, Collin-Souffrin, & Joly (1979). After calibrating the two abundance indicators with a local HII region sample, Pettini & Pagel found that O3N2 and N2 predict O/H to within 0.25 dex and 0.4 dex respectively, at the 2 σ confidence level.

The observed frame optical wavelength region corresponds to the restframe UV of LBGs. The UV contains few nebular emission lines in star-forming galaxies (Leitherer 1998) and has rarely been used for chemical composition studies in *local* galaxies of this type. Figure 11.10 compares the UV spectrum of the LBG MS 1512-cB58 with theoretical spectra (Leitherer *et al.* 2001). Three groups of lines can be distinguished: (i) Interstellar absorption lines, most

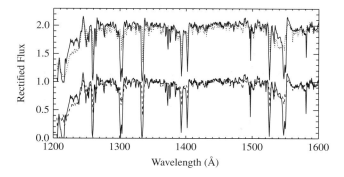

Fig. 11.10 Comparison between the observed spectrum of MS 1512–cB58 (solid) and two synthetic models with $1/4\ Z_\odot$ (lower; dashed) and $Z_\odot$ (upper; dotted). The models have continuous star formation, age 100 Myr, and Salpeter IMF between 1 and $100\ M_\odot$. The stellar spectral lines are weaker in the metal-poor model. Reproduced from Leitherer *et al.* (2001) by permission of the AAS.

of which are strong and heavily saturated. Only in very few cases can unsaturated absorption lines in LBGs be used for an abundance analysis. (ii) Broad stellar-wind lines with emission and blueshifted absorption. These lines are the telltales of massive OB stars whose stellar winds are metallicity dependent. (iii) Weak photospheric absorption lines which can only be seen in high-quality spectra. Abundance studies from stellar lines in restframe UV spectra must rely either on suitable template stars or on extensive non-LTE radiation-hydrodynamic models which are only beginning to become available (Rix *et al.* 2004).

An initial, rough estimate of the heavy-element abundances can be obtained from the equivalent widths of the strong UV absorption lines. Heckman *et al.* (1998) pointed out the close correlation of the SiIV λ1400 and CIV λ1550 equivalent widths with O/H in a sample of *local* star-forming galaxies. This correlation seems surprising, as these stellar-wind lines are deeply saturated. The reason for the metallicity dependence is the behavior of stellar winds in different chemical environments. At lower abundance, the winds are weaker and have lower velocity, and the lines become weaker and narrower. As a result, the equivalent widths are smaller at lower O/H. A similar, somewhat weaker correlation exists between O/H and the equivalent widths of the strongest *interstellar* lines. This is even more unexpected because the equivalent widths of saturated lines have essentially no dependence on the column density $N_{\rm ion}$ (see Eq. (11.2)). Therefore the correlation must be caused by the velocity dependent b factor. More metal-rich galaxies are thought to host more powerful starbursts with correspondingly larger mechanical energy release by stellar winds and supernovae. The energy input leads to increased macroscopic turbulence and higher gas velocities at higher O/H (Heckman *et al.* 1998). If the same applies to star-forming galaxies in the *high-redshift* Universe, their measured equivalent widths again indicate an oxygen abundance of about 1/3 the solar value (Crowther *et al.* 2006c).

Pettini *et al.* (2001) determined oxygen abundances in five LBGs from emission lines in restframe optical spectra. The redshift range of the sample dictated the use of the R23 method. The galaxies turned out to be rather metal-rich, with O/H somewhat below the solar value. This is roughly in agreement with restframe UV results, and an order of magnitude above the metallicities found in damped Lyα absorbers (DLA; see below) which are found at the same redshift. Because of the double-valued nature of the R23 method, the possibility exists but is deemed less likely that the sample has oxygen abundances of only 1/10 of the solar value.

The lensed LBG MS 1512-cB58 and its bright restframe UV spectrum can be studied at sufficiently high S/N and resolution to detect and resolve faint, unsaturated interstellar absorption lines. Pettini *et al.* (2002) measured numerous transitions from H to Zn covering several ionization stages. Abundances of several key elements could be derived. The α-elements O, Mg, Si, P, and S all have abundances of about 40% solar, indicating that the interstellar medium is highly enriched in the chemical elements produced by Type II supernovae. In contrast, N and the Fe-peak elements Mn, Fe, and Ni are all less abundant than expected by factors of several. In standard chemical evolution models, most of the nitrogen is produced by intermediate-mass stars, whereas Type Ia supernovae contribute most of the Fe-peak elements. Since the evolutionary timescales of intermediate- and low-mass stars are significantly longer than those of massive stars producing the α-elements, the release of N and the Fe-group elements into the interstellar medium is delayed by $\sim 10^9$ yr. MS 1512-cB58 may be an example of a star-forming galaxy in its early stage of chemical enrichment, consistent with its cosmological age of only about 15% of the age of the Universe. Mehlert *et al.* (2002) provided similar arguments to explain variation of the CIV $\lambda 1550$ line relative to SiIV $\lambda 1400$ in a small sample of LBGs. CIV appears to decrease in strength relative to SiIV from lower to higher redshift, which may reflect the time delay of the carbon release by intermediate-mass stars.

Detailed studies of weak interstellar lines such as that done for MS 1512-cB58 remain a technical challenge, even for high-throughput spectrographs at the largest telescopes. Furthermore, the results for Fe-peak elements carry some uncertainty because of the a priori unknown depletion corrections. Abundance analyses using stellar lines are not affected by depletion uncertainties. However, the existence of non-standard element ratios precludes the use of locally observed template spectra for spectral synthesis. Therefore theoretical template spectra must be calculated from model atmospheres. An example is in Fig. 11.11 where several faint stellar blends around 1425 Å and 1978 Å are used as a metallicity indicator (Rix *et al.* 2004). The 1425 Å feature is a blend of SiIII, CIII, and FeV, and the 1978 Å absorption is mainly FeIII. The synthesized spectra for five metallicities are compared with the observed restframe UV spectra of MS 1512-cB58 and Q 1307-BM1163. The model having 40% solar metal abundance provides the best fit to the data, in agreement with the results from other methods.

A variety of independent techniques lead to consistent results for the chemical composition of LBGs. While each method by itself is subject to non-negligible uncertainties, the overall agreement of the results gives confidence in the derived abundances. LBGs at $z \simeq 3$ have heavy-element abundances of about 1/3 the solar value.

Star-forming galaxies at $z \simeq 3$, at an epoch when the Universe's age was only 15% the present value, display a high level of chemical enrichment. What does their chemical composition tell us about their relation to other galaxies at lower redshift and to other structures found at $z = 3$?

Galaxies at somewhat lower redshift have only recently become accessible for detailed study due to the combined challenges of instrumentation and the galactic spectral properties. Shapley *et al.* (2004) obtained K-band spectroscopy of seven UV-selected star-forming

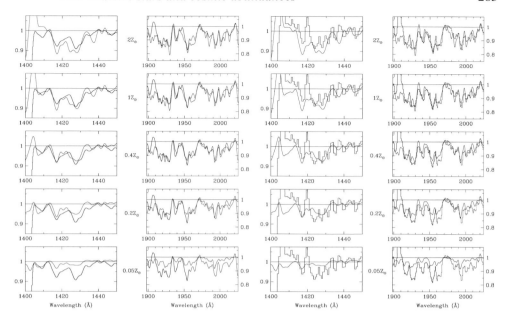

Fig. 11.11 Left pair of panels: comparison of the observed spectrum of MS 1512-cB58 (thick) with fully synthetic spectra (thin) for five different metallicities, from twice solar to 1/20 solar. First panel: region around 1425 Å; second: region of the FeIII blend near 1978 Å. Each pair of panels is labeled with the metallicity of the synthetic spectrum shown. Right pair of panels: same as left pair, but for Q 1307–BM1163. Reproduced from Rix *et al.* (2004) by permission of the AAS.

galaxies at redshifts between 2 and 2.5. The N2 method calibrated by Pettini & Pagel (2004) was used as an abundance diagnostic. When compared with the original higher-z LBGs, the $z \simeq 2$ sample is more metal-rich. This can be seen in Fig. 11.12, where O/H of the $z = 2$ galaxies is compared with that of LBGs at $z \simeq 3$ and of local star-forming galaxies over a range of blue luminosities. The latter were analyzed with the R23 method. The $z = 2$ sample has almost solar chemical composition but is still less metal-rich than local late-type galaxies *with comparable luminosities*. As a caveat, the comparison rests on the assumption that the N2 and R23 calibrations have no significant offset. The difference between the average redshift of the LBG sample and of the $z = 2$ galaxies translates into a mean age difference of about 1 Gyr. Both the chemical properties and the masses of the $z = 2$ galaxies and LBGs are consistent with standard passive evolution models.

Kewley & Kobulnicky (2005) followed the metallicity evolution of star-forming galaxies with comparable luminosities from $z = 0$ to 3.5. O/H was determined from restframe optical emission lines using the strong-line method in four homogeneous galaxy samples. The samples were taken from the CfA2 survey, from the Great Observatories Origins Deep Survey (GOODS) field, from Shapley *et al.* (2004), and from the LBG sample, covering $z \approx 0, 0.7$, 2.1–2.5, and 2.5–3.5, respectively. The average oxygen abundance in the local Universe, as defined by the CfA2 sample, is about solar. O/H decreases with redshift to approximately 1/3 solar at $z = 3$.

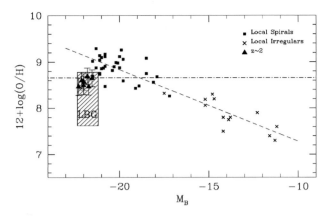

Fig. 11.12 Metallicity–luminosity relationship. Data for *local* spiral and irregular galaxies are from Garnett (2002). The $z = 2$ objects are overluminous for their (O/H) abundances, derived using the N2 calibration of Pettini & Pagel (2004) but lie closer to the relationship for the local galaxies than $z = 3$ LBGs. Reproduced from Shapley *et al.* (2004) by permission of the AAS.

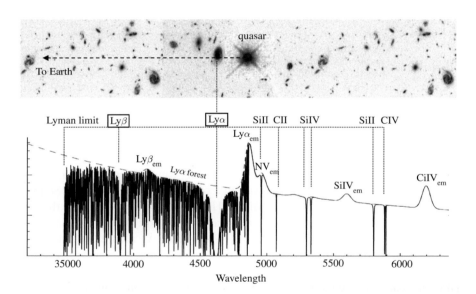

Fig. 11.13 The technique of QSO absorption-line spectroscopy. The light from a background QSO intercepts galaxies and the IGM which happen to lie along the line of sight (and are therefore at lower absorption redshifts z_{abs} than the QSO emission redshift z_{em}). Gas in these structures leaves an imprint in the spectrum of the QSO in the form of narrow absorption lines (J. Webb, private communication).

It is instructive to compare the heavy-element abundances of LBGs with those of other structures found at similar redshifts. The commonly employed technique of QSO absorption-line spectroscopy is explained in Fig. 11.13. QSOs are used to probe DLAs and the Lyman-forest at the same redshift (Pettini 2004). DLA systems have metallicities of about $1/15\, Z_\odot$

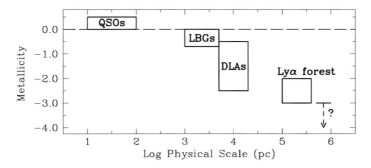

Fig. 11.14 Summary of our current knowledge of abundances at high redshift. Metallicity is plotted logarithmically on the vertical axis relative to the solar value; the latter is shown as the broken horizontal line at 0.0 and corresponds to approximately 2% of the baryons in elements heavier than helium. The horizontal axis shows the typical linear dimensions of the structures to which the abundance measurements refer, from the central regions of AGN on scales of 10–100 pc to the IGM traced by the Lyα forest on Mpc scales. Reproduced from Pettini (2006) with the kind permission of the ASP Conference Series.

and are thought to be the cross sections of the outer regions and halos of (proto)-galaxies seen along the sightlines of quasars. The abundances determined for these structures are compared with those of LBGs in Fig. 11.14. Although the properties of LBGs and DLAs do not immediately support a close relation between the two classes of objects, at least some link seems likely. If so, the observed outflows in LBGs may provide the metal enrichment of the halos. The Lyman forest is predicted by Cold Dark Matter models to result from structure formation in the presence of an ionizing background. The Lyman forest had long been thought to be truly primordial, but metal enrichment of $1/100–1/1000 \, Z_\odot$ has recently been detected (Aguirre *et al.* 2004). This relatively high metal abundance early in the evolution of the Universe could have been produced by a first generation of Population III stars. Such stars can account for the amount of metals, and at the same time could have provided copious ionizing photons, as metal and photon production are closely correlated. Alternatively, star-forming galaxies at high redshift could be the production sites of the metals seen in the intergalactic medium *if* superwinds are capable of removing the newly formed metals from galactic disks.

Galactic superwinds are, in principle, able to transport dust and chemically enriched material from the star-formation sites into the galactic halo. The ultimate fate of the superwind depends on the starburst geometry, the mass of the galaxy, and on the properties of the halo. The material may fall back onto the galaxy disk or become unbound, with the second alternative being more likely in less massive galaxies.

After a characteristic time the outflow will reach pressure equilibrium with the ambient IGM at some distance R_e from the galaxy. Hydrodynamic models predict $R_e \approx 50$ kpc. This is small in comparison with the separation of objects with mass $\sim 10^{10} \, M_\odot$ formed in the early Universe according to Cold Dark Matter models, which is of order 1 Mpc. Therefore, in the absence of an additional dispersal process, the metals transported out of the galaxies by superwinds essentially stay close to their birthplace from a cosmological point of view. It is not unreasonable to assume that additional dispersal mechanisms exist. Large-scale diffusion processes in the IGM, peculiar motions of galaxies, and galaxy–galaxy mergers and interactions all tend to disperse the ejected metals into larger volumes.

Whatever dispersal mechanism operates, an approximate picture for the star formation, metal production, and pollution of the IGM at high redshift is beginning to emerge.

- Late-type/irregular galaxies with peculiar morphologies at redshift around 3 are forming stars at high rates, but otherwise have properties similar to those in the local Universe.
- Stellar winds and supernova explosions are an efficient mechanism to transport heavy elements from the galaxy centers to the halos, and possibly into the IGM.
- Depending on the existence of a dispersal mechanism, the metals may be distributed over large volumes and pollute the IGM early in the evolution of the Universe.
- Alternatively, most of the metals found in the IGM may have been produced by the first stellar generation that existed before the population of stars we are currently observing in LBGs.

The state of our knowledge of star formation from low to high redshift allows a rather fundamental consistency test. One can calculate the total mass of stars that have ever formed during the history of the Universe. Figure 11.15 shows an updated version of a plot first introduced by Madau *et al.* (1996). The figure gives the cosmic star-formation history by following the redshift evolution of the luminosity density of star-forming galaxies. Integrating the curve along the lookback time gives the total mass in stars in the Universe. This number turns out to be in good agreement with what we observe in galaxies of all morphological types in the present Universe. Next we can calculate the total mass in metals produced by all stars formed. Stellar evolution models can provide the necessary conversion factor. Comparing the derived value with the actually observed number of metals in the IGM leads to a startling discrepancy: the metals currently observed in the IGM (and the ISM inside galaxies) can account for no more than 10–15% of what we expect to have been produced by all stars that have ever formed. Where are these missing metals? It is likely that these metals are hidden in gas phases which have hitherto remained unexplored. The gas may have escaped detection because of high temperatures, yet it may make a major contribution to the overall metallicity. Astronomers often refer to this elusive gas phase as the cosmic web. Its exploration is a major quest of observational cosmology.

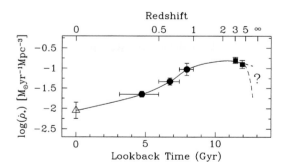

Fig. 11.15 The co-moving star-formation rate density $\dot{\rho}_*$ vs. lookback time compiled from wide-angle, ground-based surveys. Reproduced from Pettini (2004) by permission of Cambridge University Press.

11.4 Gamma ray bursts

Gamma ray bursts were discovered using US Department of Defense Vela satellites during the "Cold War", designed to monitor Soviet nuclear testing. Instead, regular bursts of gamma ray radiation came from outer space, for which a Galactic origin was favored initially. The major leap forward came in 1991 when the Burst and Transient Source Experiment (BATSE) aboard the Compton Gamma Ray Observatory (CGRO) was launched. BATSE established that their distribution was isotropic – indicating that GRBs were either located in the galactic halo or of extragalactic origin. In addition, GRBs formed two fairly distinct groups, long ($\sim$20 s) with a relatively soft energy distribution and short (~ 0.2 s) with a harder energy distribution (Kouveliotou *et al.* 1993). Statistics from the fourth BATSE catalog are presented in Fig. 11.16. A duration of 2 s is commonly used to distinguish between short and long bursts, but the figure illustrates that the populations overlap in burst duration.

The next breakthrough for the long, soft variety of GRBs was the detection of an optical counterpart for GRB 970228 (van Paradijs *et al.* 1997) using the Italian–Dutch BeppoSAX satellite, confirming their extragalactic origin, see Fig. 11.17. Initially, the (isotropic) energies involved, $E_{\gamma,\text{iso}}$, were believed to be several orders of magnitude greater than supernovae.

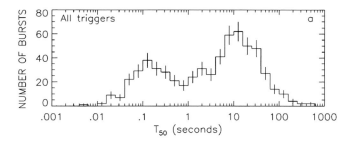

Fig. 11.16 Burst duration distribution from the fourth BATSE catalog, indicating distinct peaks at $\sim$20 s and $\sim$0.2 s. Reproduced from Paciesas *et al.* (1999) by permission of the AAS.

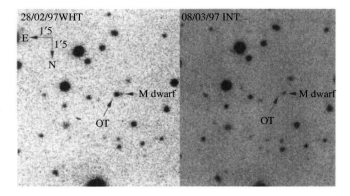

Fig. 11.17 90$\times$90 arcsec V-band images of the first detection of a GRB optical transient (denoted "OT") from 28 Feb. 1997 (left, WHT) and 8 March 1997 (right, INT) confirming their extragalactic nature. From van Paradijs *et al.* (1997).

However, subsequent observations implied that the radiation is highly collimated, with an estimated beam opening angle of ~5 degrees (Frail *et al.* 2001), such that the energies involved equate to $E_\gamma = 10^{51}$ erg, as illustrated in Fig. 11.18, i.e. comparable to the energy released in supernovae. The opening angle of the burst is obtained through achromatic steepening of the afterglow light curve, which is attributed to synchrotron emission from ultra-relativistic ejecta.

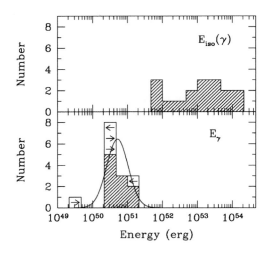

Fig. 11.18 Distribution of the apparent isotropic GRB energies for cases with known redshifts (top panel) versus geometry-corrected energy for those GRBs whose afterglows exhibit the signature of a non-isotropic outflow (bottom panel). Reproduced from Frail *et al.* (2001) by permission of the AAS.

The extragalactic origin of long-soft GRBs greatly restricted the possible physical causes – the most likely causes involve a neutron star rotating close to centrifugal break up, or an accretion disk surrounding a rotating black hole, produced most likely in a "collapsar" model. The short duration of GRBs indicates a relatively compact source, such that rotating Wolf–Rayet stars were the preferred progenitors of the latter scenario, in preference to, e.g., red supergiants (Woosley 1993).

Observationally, the next landmark was provided by GRB 980425 which was observed to be spatially coincident with a Type Ic supernova SN 1998bw in a spiral arm of ESO 184-G82 at $z = 0.008$. SN 1998bw was extremely bright (10^{52} erg) and possessed very broad lines of ~27 000 km s^{-1} – such extreme supernovae are often termed "hypernovae". The optical spectrum of SN 1998bw is compared with normal Type Ib and Type Ib SN in Fig. 11.19. This provided support for the "collapsar" model of MacFadyen & Woosley (1999) involving a rotating massive star, although the GRB itself was unusually dim. No spherically symmetric model is capable of accelerating sufficient mass to such high speeds, and a spherical explosion is inconsistent with the beaming observed in GRBs, such that this SN/GRB must involve the non-spherically symmetric death of a massive star.

Hammer *et al.* (2006) detected W-R features in the host galaxy spectra of three nearby GRBs including GRB 980425. The spectroscopic confirmation of W-R stars in GRB hosts provides strong support for a connection between long GRBs and massive stars. Figure 11.20 shows

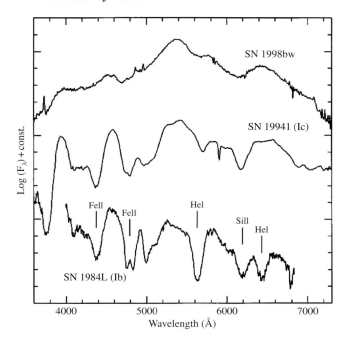

Fig. 11.19 Representative spectra of SN 1998bw (GRB 980425) at visual maximum together with Type Ic SN 1994I and Type Ib SN 1984L. From Galama *et al.* (1998).

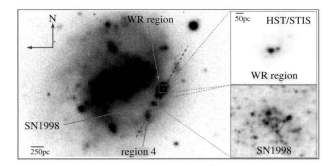

Fig. 11.20 Ground-based image of the host galaxy of SN 1998bw/GRB 980425, including insets showing HST/STIS images of the W-R region (top) and SN/GRB (bottom). The location of the SN is indicated by a circle. From Hammer *et al.* (2006).

that the W-R cluster in ESO-184-G82, the host galaxy of GRB 980425, is located several hundred pc from the SN/GRB, leading Hammer *et al.* to suggest it may be a runaway star.

Other nearby GRB supernovae have since been observed, notably the High Energy Transient Explorer (HETE-II) detection of GRB 030329 for which an optical spectrum of the afterglow revealed a supernova spectrum (SN 2003dh) at $z = 0.1685$ which is remarkably similar to SN 1998bw. In this case, the GRB was of cosmological brightness, supporting the "collapsar" model. Other examples include the X-ray Flash (XRF) 060218, with a much softer gamma-ray signature, which was also observed to be associated with the Type Ic supernova SN2006aj (Pian *et al.* 2006; Mazzali *et al.* 2006). As the optical afterglow from the GRB fades, a late photometric brightening in the light curve from more distant GRBs after 20–30 days has also been interpreted as the signature of SNe at visual maximum, emerging from the decaying optical afterglow (e.g., Della Valle *et al.* 2003).

Morphologically, the host galaxies of cosmological GRBs are fainter and more irregular than the hosts of high-z core-collapse SNe observed as part of the GOODS survey. This can be understood in terms of the mass–metallicity relation for galaxies. Most star formation, and therefore most core-collapse SNe, occur in massive (metal-rich) spiral galaxies, while GRBs are biased towards low-metallicity hosts, which are likely to be low mass irregulars. Comparisons are presented in Figs. 11.21 and 11.22. In addition, GRBs are far more concentrated in the brightest regions of their host galaxies than core-collapse SNe. The most common core-collapse SNe originate from 8–20 $M_{\odot}$ stars (recall Section 5.6), which may be formed in modest star clusters throughout a spiral galaxy, which would appear relatively faint in high-z galaxies. In contrast, GRBs appear to be descendents of the most massive stars, which solely form in the most massive, dense star clusters, and explode at such a young age that the star cluster has not yet faded away.

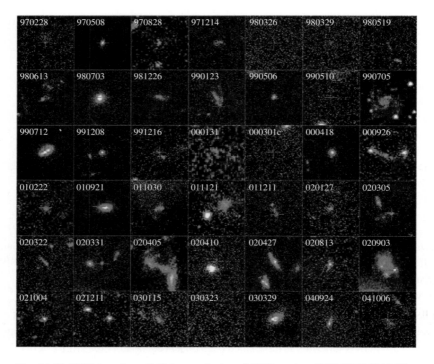

Fig. 11.21 HST imaging of 42 host galaxies of GRBs using STIS/WFPC2/ACS, centered upon the GRB (3.75×3.75 arcsec^2). Owing to the redshifts of the hosts, these images generally correspond to rest frame blue or UV light, i.e. sensitive to light from massive stars. From Fruchter *et al.* (2006).

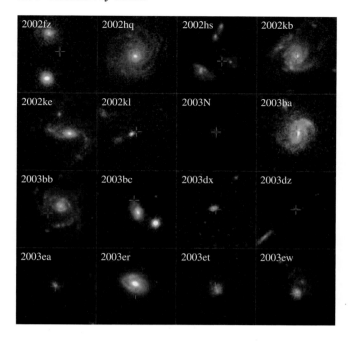

Fig. 11.22 HST/ACS imaging of host galaxies of 16 GOODS core-collapse supernovae (7.5×7.5 arcsec^2). Owing to the redshifts of the hosts, these images generally correspond to rest frame blue or UV light, i.e. sensitive to light from massive stars. From Fruchter *et al.* (2006).

Theoretically, rotation is ignored in standard core-collapse supernovae models. Instead, models for long-soft GRBs rely on very rapid rotation to produce the "collapsar", in which the rotational axes provides a preferred direction for the jet. Since only a fraction of one percent of SNe involve a GRB, even after correcting for their narrow beams, the physics leading to a GRB is not necessarily the same as for normal supernovae. A further puzzle was presented by GRB 060614, an apparently nearby long burst, but without an associated SN (Gehrels *et al.* 2006). Either this represents a rare category of massive star collapse in which the black hole forms via fallback – rather than direct collapse – as predicted by Fryer, Young, & Hungerford (2006), expelling little ^{56}Ni (which powers the SN light curve), or it is a short burst with an extended emission tail.

Mészáros (2002) provides a theoretical consideration of prompt emission and afterglows from GRBs. Zhang, Woosley, & MacFadyen (2003) provide simulations for the break out of relativistic jets from a collapsing 15 $M_\odot$ massive star, whose surface is rotating at 10% of critical velocity, and is assumed to have shed its hydrogen envelope. No more than the inner 2 $M_\odot$ iron core is assumed to be able to collapse directly to a black hole. Rotation and infall modify the core structure relative to a non-rotating case within the immediate environment of the black hole, with the density in the equatorial plane ten times greater than that in the pole at a radius of 2000 km. Between one and several solar masses of material then accrete onto the black hole during the early phase of the GRB.

Such accretion is an incredibly efficient energy source. If perhaps one tenth of a percent of the rest mass energy goes into jet outflows, one obtains a few 10^{51} erg, in reasonable agreement with observations. Since the duration of a long-soft GRB is ∼10 s, the jet power is of order 10^{50} erg s^{-1}. The passage of the jet through the star leads to additional collimation. Once it breaks out of the star it encounters the remnant W-R stellar wind. Snapshots in a similar simulation are shown in Fig. 11.23.

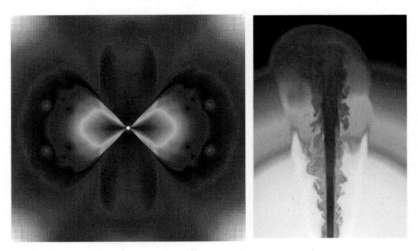

Fig. 11.23 Simulation during the collapse of a rapidly rotating $14 M_\odot$ helium core produces a black hole and accretion disk. The left-hand panel shows the density structure (spanning 10^7 to 10^9 g cm^{-3}) 20 sec after the black hole has formed and commenced accretion at a rate of 0.1 $M_\odot$ yr^{-1} for the last 15 seconds (current mass 4.4 $M_\odot$). For this collapsar model, jets exit the star, as indicated on the right panel, 8 seconds after the jet originated in the center. From Woosley & Janka (2005).

As we have already seen in Chapter 5, cores of massive single stars at solar metallicity spin down during either the red supergiant or W-R phase, exploding as a conventional core-collapse SN, leaving behind them pulsars rotating at a typical rate of 10–20 milliseconds. A rapidly rotating core may occur during core-collapse at sufficiently low metallicity in the case of homogeneous evolution. Alternatively, rapid rotation may occur via spin-up during tidal interactions or possibly a merger within a common envelope phase (Chapter 6). The binary scenario could plausibly reproduce GRBs from a statistical perspective, whilst the single star case would suggest that GRBs are favored in low metallicity environments.

Of course, the ejecta strongly interacts with the circumstellar material, probing the immediate vicinity of the GRB itself (Chevalier & Li 2000). This provides information on the progenitor, for which one expects $\rho \propto r^{-2}$ for W-R winds. The apparent metallicity-dependent W-R winds argue that one would potentially expect rather different environments for the afterglows of long-duration GRBs that were dependent upon the metallicity of the host galaxy. Indeed, densities of the immediate environment of many GRBs suggest values rather lower than typical solar metallicity W-R winds. Half of long GRBs apparently occur in uniform density environments, favoring a post-common envelope binary merger model.

Observationally, the improved soft energy sensitivity of HETE-II and Swift with respect to BATSE permits the detection of GRBs at large redshift. Indeed, after the first 18 months of operations, the mean redshift of long-soft GRBs detected by Swift is $z \approx 2.5$, with equivalent (isotropic) energies of 10^{53} erg (recall Fig. 11.18). Conversely, only a minority of Swift GRBs are short bursts. Nevertheless, the rapid automated slewing of Swift (1 deg s^{-1}), based upon US Department of Defense technology, permits the detection of X-ray (e.g. GRB 050509B: Gehrels *et al.* 2005) and optical afterglows (e.g. GRB 050709: Hjorth *et al.* 2005) to be detected. The host galaxy redshifts of the first few detected short bursts of $z \leq 0.5$ suggest equivalent (isotropic) energies of 10^{49} erg. Short GRBs are associated with both late-type star-forming galaxies (e.g. GRB 050709) and early-type galaxies (e.g. GRB 050509B), for which no evidence of associated supernovae are seen at late epochs. This reinforces their origin via the merger of either two extremely compact objects, i.e. two neutron stars, or a neutron star and a black hole, for which coalescence timescales are expected to be long ($\sim$Gyr). Larger samples of short burst GRB hosts suggest somewhat higher $0.4 \leq z \leq 1.1$ redshifts, from which higher isotropic energies of $E_{\gamma,\mathrm{iso}} \sim 10^{51}$ erg s^{-1} have been inferred (e.g. Berger *et al.* 2007). The high sensitivity of Swift with respect to, e.g., BATSE has enabled extended emission tails to be seen in some "short" bursts, following the initial prompt emission, with durations of up to tens or hundreds of seconds. Do such short bursts represent the accretion of a fragmented neutron star onto a more massive black hole? It is possible that some nearby (Virgo cluster) short GRBs detected by BATSE are hyperflares from magnetars, analogous to the Dec. 2004 event from SGR 1806-20.

Routine observations of high-z GRB afterglows also permit their use as probes of their circumstellar environment and the ionization and metal-enrichment histories of the IGM during the epoch of re-ionization. Indeed, the current record holder, GRB 050904 at $z = 6.3$, was observed optically at high spectral resolution three days after the burst, revealing a characteristic black absorption trough from the Gunn–Peterson effect. Such intense beacons permit studies of re-ionization, unhindered by the faintness of normal proto-galaxies or quasars at high-redshift. Metallicities of the host galaxies of long GRBs, as deduced from medium to high spectroscopy obtained immediately after the burst, are metal-poor. This favors the single star case, but it is not firmly established whether the observed GRB statistics are consistent with this scenario.

Type Ia supernovae have been successfully used in a cosmological role to $z \sim 1$ by, e.g., Perlmutter *et al.* (1999) as a record of the cosmic expansion history. If GRBs can be established as standard candles, they will extend and complement Type Ia supernovae to the highest redshifts. However, their large dispersion in isotropic energies, $E_{\gamma,\mathrm{iso}}$, appears to rule out a cosmological standard candle role for GRBs (Fig. 11.18). The beam corrected burst energetics E_γ correlates well with the peak energy of the prompt emission, E_{pk} (Amati *et al.* 2002), providing the afterglow is followed to derive the cone opening angle. Other correlations have been identified by Ghirlanda, Ghisellini, & Lazzati (2004) and Firmani *et al.* (2006), although a physical basis for these relations is lacking. Nevertheless, it is hoped that they could provide a means of constraining the cosmological parameters solely from observations of GRBs.

References

Abbott, D. C. 1982, *ApJ*, **263**, 723

Abbott, D. C. & Lucy, L. B. 1985, *ApJ*, **288**, 679

Abel, T., Bryan, G. L., & Norman, M. L. 2000, *ApJ*, **540**, 39

Aguirre, A., Schaye, J., Kim, T. S., *et al.* 2004, *ApJ*, **602**, 38

Allen, D. A., Hyland, A. R., & Hillier, D. J. 1990, *MNRAS*, **244**, 706

Allen, D. A., Swings, J. P., & Harvey, P. M. 1972, *A&A*, **20**, 333

Alloin, D., Collin-Souffrin, S., & Joly, M. 1979, *A&AS*, **37**, 361

Alonso, O., García-Dabó, C. E., Zamorano, J., *et al.* 1999, *ApJS*, **122**, 415

Alves, J. & Homeier, N. 2003, *ApJ*, **589**, L45

Amati, L., Frontera, F., Tavani, M., *et al.* 2002, *A&A*, **390**, 81

Arnould, M. & Goriely, S. 2003, *Phys. Rep.*, **384**, 84

Arnould, M. & Takahashi, K. 1999, *Rep. Prog. Phys.*, **62**, 395

Asplund, M., Grevesse, N., Sauval, A. J., *et al.* 2005, *A&A*, **435**, 339

Azzopardi, M. 1987, *A&AS*, **69**, 421

Azzopardi, M. & Vigneau, J. 1982, *A&AS*, **50**, 291

Baade, W. 1944, *ApJ*, **100**, 137

Bakker, C. J. & van de Hulst, H. C. 1945, *Nederlands Tijdschrift voor Natuurkunde*, **11**, 201

Baldwin, J. A., Ferland, G. J., Martin, P. G., *et al.* 1991, *ApJ*, **374**, 580

Balzano, V. A. 1983, *ApJ*, 268, 602

Barlow, M. J. & Cohen, M. 1977, *ApJ*, **213**, 737

Barlow, M. J., Roche, P. F., & Aitken, D. A. 1988, *MNRAS*, **232**, 821

Barton, E. J., Geller, M. J., & Kenyon, S. J. 2000, *ApJ*, **530**, 660

Bastian, N., Saglia, R. P., Goudfrooij, P., *et al.* 2006, *A&A*, **448**, 881

Baumgardt, H. & Makino, J. 2003, *MNRAS*, **340**, 227

Beals, C. S. 1929, *MNRAS*, **90**, 202

Beals, C. S. & Plaskett, H. H. 1935, *Trans. IAU*, **5**, 134

Becker, R. H., Fan, X., White, R. L., *et al.* 2001, *AJ*, **122**, 2850

Becker, S. R. & Butler, K. 1990, *A&A*, **235**, 326

Benjamin, R. A., Churchwell, E., Babler, B. L., *et al.* 2003, *PASP*, **115**, 953

Berger, E., Chary, R., Cowie, L. L., *et al.* 2007, *ApJ*, **664**, 1000

Bergvall, N., Zackrisson, E., Andersson, B. G., *et al.* 2006, *A&A*, **448**, 513

Bertelli, G., Bressan, A., Chiosi, C., *et al.* 1994, *A&AS*, **106**, 275

Beuther, H., Schilke, P., Sridharan, *et al.* 2002, *A&A*, **383**, 892

Bibby, J. L., Crowther, P. A., Furness, J. P., & Clark, J. S. 2008, *MNRAS*, **386**, L23

Bica, E., Alloin, D., & Schmitt, H. R. 1994, *A&A*, 283, 805

Bieging, J., Abbott, D. C., & Churchwell, E. B. 1989, *ApJ*, **340**, 518

Biehle, G. T., 1994, *ApJ*, **420**, 364

Bionta, R. M., Blewitt, G., Bratton, C. B., *et al.* 1987, *Phys. Rev. Lett.*, **58**, 1494

Bjorkman, J. E. & Cassinelli, J. P. 1993, *ApJ*, **409**, 429

Blaauw, A. 1961, *Bull. Astron. Inst. Neth.*, **15**, 265

Blöcker, T., Balega, Y., Hofmann, K. H., *et al.* 1999, *A&A*, **348**, 805

Blum, R. D., Damineli, A., & Conti, P. S. 1999, *AJ*, **117**, 1392

Blum, R. D., Damineli, A., & Conti, P. S. 2001, *AJ*, **121**, 3149

Blum, R. D., Barbosa, C. L., Damineli A., *et al.* 2004, *ApJ*, **617**, 1167

Bohannan, B. 1997, in *Luminous Blue Variables: Massive Stars in Transition*, eds. A. Nota & H. J. G. L. M. Lamers (San Francisco: Astronomical Society of the Pacific), 3

Bohannan, B. & Conti, P. S. 1976, *ApJ*, **204**, 797

Bohlin, R. C., Savage, B. D., & Drake, J. F. 1978, *ApJ*, **224**, 132

Böhm-Vitense, E. 1981, *ARA&A*, **19**, 295

Bond, H. 1981, *ApJ*, **248**, 707

Bond, J. R., Arnett, W. D., & Carr, B. J. 1984, *ApJ*, **280**, 825

Bondi, H. & Hoyle, F. 1944, *MNRAS*, **104**, 273

Bonnell, I. A., Bate, M. R., & Zinnecker, H. 1998, *MNRAS*, **298**, 93

Bonnell, I. A., Clarke, C. J., & Bate, M. R. 2006, *MNRAS*, **368**, 1296

Bonnor, W. B. 1956, *MNRAS*, **116**, 351

Bouchet, P., Lequeux, J., Maurice, E., *et al.* 1985, *A&A*, **149**, 330

Boutloukos, S. G. & Lamers, H. J. G. L. M. 2003, *A&A*, **338**, 717

Bowers, P. F., Johnston, K. J., & Spencer, J. H. 1983, *ApJ*, **274**, 733

Brandl, B., Brandner, W., Eisenhauer, F., *et al.* 1999, *A&A*, **352**, L69

Brandner, W., Grebel, E. K., Chu, Y.-H., & Weis, K. 1997, *ApJ*, **475**, L45

Bresolin, F., Gieren, W., Kudritzki, R.-P., *et al.* 2002, *ApJ*, **567**, 277

Bressan, A., Fagotto, F., Bertelli, G., & Chiosi C. 1993, *A&AS*, **100**, 647

Bromm V. & Larson, R. B. 2004, *ARA&A*, **42**, 79

Bromm, V., Coppi, P. S., & Larson, R. B. 1999, *ApJ*, **527**, L5

Bromm, V., Ferrera, A., Coppi, P. S., & Larson, R. B. 2001, *MNRAS*, **328**, 969

Brown, T. M., Heap, S. R., Hubeny, I., *et al.* 2002, *ApJ*, **579**, L75

Burbidge, E. M., Burbidge, G. R., Fowler, W. A., & Hoyle, F. 1957, *Rev. Mod. Phys.*, **29**, 547

Calzetti, D. 1997, *AJ*, **113**, 162

Calzetti, D. 2001, *PASP*, **113**, 1449

Calzetti, D., Kinney, A. L., & Storchi-Bergmann, T. 1994, *ApJ*, **429**, 582

Calzetti, D., Kennicutt, R. C., Engelbracht, C. W., *et al.* 2007, *ApJ*, **666**, 870

Cannon, R. C., Eggleton, P. P., Zytkow, A. N., & Podsiadlowski, P. 1992, *ApJ*, **386**, 206

Cappellaro, E., Evans, R., & Turatto, M. 1999, *A&A*, **351**, 459

Cardelli, J. A., Clayton, G. C., & Mathis, J. S. 1989, *ApJ*, **345**, 245

Carroll, J. A. 1933, *MNRAS*, **93**, 478

Cassinelli, J. P., Miller, N. A., Waldron, W. L., *et al.* 2001, *ApJ*, **554**, L55

Castor, J. I., Abbott, D. C., & Klein, R. I. 1975a, *ApJ*, **195**, 157

Castor, J. I., McCray, R., & Weaver, R. 1975b, *ApJ*, **200**, L107

Chandar, R., Leitherer, C., Tremonti, C., & Calzetti, D. 2003, *ApJ*, **586**, 939

Cherepashchuk, A. M., Eaton, J. A., & Khaliullin K. F., 1984, *ApJ*, **281**, 774

Chevalier, R. A. 1991, in *Massive Stars in Starbursts*, eds. C. Leitherer, N. R. Walborn, T. M. Heckman, & C. A. Norman (Cambridge: Cambridge University Press), 169

Chevalier, R. A., & Li, Z.-Y. 2000, *ApJ*, **536**, 195

Chlebowski, T., Harnden Jr., F. R., & Sciortino, S. 1989, *ApJ*, **341**, 427

Christlieb, N., Bessell, M. S., Beers, T. C., *et al.* 2002, *Nature*, **491**, 904

Chu, Y.-H. 2003, in *IAU Symp. 212, A Massive Star Odyssey: From Main Sequence to Supernova*, eds. K. A. van der Hucht, A. Herrero, & C. Esteban (San Francisco: Astronomical Society of the Pacific), 585

Ciardi, B., Ferrara, A., & White S. D. M. 2003, *MNRAS*, **344**, L7

Clark, J. S., Goodwin, S. P., Crowther, P. A., *et al.* 2002, *A&A*, **392**, 909

Clark, J. S., Negueruela, I., Crowther, P. A., & Goodwin, S. P. 2005, *A&A*, **434** 949

Clarke, C. J., & Bromm, V. 2003, *MNRAS*, **343**, 1224

Code, A. D., Davis, J., Bless, R. C., & Hanbury Brown, R. 1976, *ApJ*, **304**, 417

Condon, J. J. 1992, *ARA&A*, **30**, 575

Condon, J. J. & Yin, Q. F. 1990, *ApJ*, **357**, 97

Conti, P. S. 1976, *Mem. Soc. Roy. Sci. Liège*, **10**, 193

Conti, P. S. 1984, in *IAU Symp. 105, Observational Tests of the Stellar Evolution Theory*, eds. A. Maeder & A. Renzini (Kluwer: Dordrecht), 233

Conti, P. S. 1991, *ApJ*, **377**, 115

Conti, P. S. 1997, in *Luminous Blue Variables: Massive Stars in Transition*, eds. A. Nota & H. J. G. L. M. Lamers (San Francisco: Astronomical Society of the Pacific), 161

Conti, P. S. & Alschuler, W. R. 1971, *ApJ*, **170**, 325
Conti, P. S. & Blum, R. D. 2002, *ApJ*, **564**, 827
Conti, P. S. & Ebbets, D. 1977, *ApJ*, **213**, 438
Conti, P. S. & Frost, S. A. 1977, *ApJ*, **212**, 728
Conti, P. S. & Massey, P. 1989, *ApJ*, **337**, 251
Conti, P. S. & Vacca, W. D. 1994, *ApJ*, **423**, L97
Conti, P. S., Garmany, C. D., de Loore, C., & Vanbeveren, D. 1983, *ApJ*, **274**, 302
Conti, P. S., Garmany, C. D., & Massey, P. 1986, *AJ*, **92**, 48
Conti, P. S., Leitherer, C., & Vacca, W. D. 1996, *ApJ*, **461**, L87
Couchman, H. M. P. & Rees, M. J. 1986, *MNRAS*, **221**, 53
Cox, A. N. 1999, *Allen's Astrophysical Quantities* (New York: Springer-Verlag)
Crockett, R. M., Smartt S. J., Eldridge, J. J., *et al.* 2007, *MNRAS*, **381**, 835
Crowther, P. A. 1997, in *Luminous Blue Variables: Massive Stars in Transition*, eds. A. Nota &
    H. J. G. L. M. Lamers (San Francisco: Astronomical Society of the Pacific), 51
Crowther, P. A. 1998, in *IAU Symp. 189, Fundamental Stellar Properties*, eds. T. Bedding, A. Booth, & J. Davis
    (Dordrecht: Kluwer), 137
Crowther, P. A. 2005, in *IAU Symp. 227, Massive Star Birth: A Crossroads of Astrophysics*, eds. R. Cesaroni,
    M. Felli, E. Churchwell, & M. Walmsley (Cambridge: Cambridge University Press), 389
Crowther, P. A. 2007, *ARA&A*, **45**, 177
Crowther, P. A. & Hadfield, L. J. 2006, *A&A*, **449**, 711
Crowther, P. A., Smith, L. J., Hillier, D. J., & Schmutz, W. 1995, *A&A*, **293**, 427
Crowther, P. A., Dessart, L., Hillier, D. J., *et al.* 2002a, *A&A*, **392**, 653
Crowther, P. A., Hillier, D. J., Evans, C. J., *et al.* 2002b, *ApJ*, **579**, 774
Crowther, P. A., Lennon, D. J., & Walborn, N. R. 2006a, *A&A*, **446**, 279
Crowther, P. A., Morris, P. W., & Smith, J. D. 2006b, *ApJ*, **636**, 1033
Crowther, P. A., Prinja R. K., Pettini M., & Steidel, C. C. 2006c, *MNRAS*, **368**, 895
Crowther, P. A., Hadfield, L. J., Clark, J. S., *et al.* 2006d, *MNRAS*, **372**, 1407
Crowther, P. A., Carpano, S., Hadfield, L. J., & Pollock, A. M. T., 2007, *A&A*, **469**, L31
Davidson, K. & Humphreys, R. M. 1997, *ARA&A*, **35**, 1
Davis, M., Efstathiou, G., Frenk, C. S., & White, S. D. M. 1985, *MNRAS*, **292**, 371
De Marco, O., Schmutz, W., Crowther, P. A., *et al.* 2000, *A&A*, **358**, 187
De Pree, C. G., Mehringer, D. M., & Goss, W. M. 1997, *ApJ*, **482**, 307
Della Valle, M., Malesani, D., Benetti, S., *et al.* 2003, *A&A* , **406**, L33
Demoulin, M. H. & Burbidge, E. M. 1970, *ApJ*, **159**, 799
Denicoló, G., Terlevich, R., & Terlevich, E. 2002, *MNRAS*, **330**, 69
Dessart L. 2004, in *Evolution of Massive Stars, Mass Loss and Winds*, eds. M. Heydari-Malayeri, P. Stee, &
    J.-P. Zahn (Les Ulis: EDP Sciences), 251
Dessart L. & Chesneau, O. 2002, *A&A*, **395**, 209
Donati, J.-F., Babel, J., Harries, T. J., *et al.* 2002, *MNRAS*, **333**, 55
Dopita, M. A., Fischera, J., Sutherland, R. S., *et al.* 2006, *ApJS*, **167**, 177
Draine, B. T. 2003, *ARA&A*, **41**, 241
Dressler, A. & Gunn, J. E. 1983, *ApJ*, **270**, 7
Dyson, J. E. & Williams, D. A. 1997, *The Physics of the Interstellar Medium*, 2nd Edn. (Bristol: Institute of
    Physics)
Ebert, R. 1955, *Zeit. Astrophysik*, **37**, 217
Edmunds, M. G. & Pagel, B. E. J. 1978, *MNRAS*, **185**, 77P
Eggleton, P. 2006, *Evolutionary Processes in Binary and Multiple Stars*, Cambridge Astrophysics Series 40
    (Cambridge: Cambridge University Press)
Eldridge, J. J. & Vink, J. S. 2006, *A&A*, **452**, 295
Elson, R. A. W., Sigurdsson, S., Hurley, J., *et al.* 1998, *ApJ*, **499**, L53
Esteban, C., Peimbert, M., Torres-Peimbert, S., & Escalente, V. 1998, *MNRAS*, **295**, 401
Evans, C. J. & Howarth, I. D., 2003, *MNRAS*, **345**, 1223
Evans, C. J., Crowther, P. A., Fullerton, A. W., & Hillier, D. J. 2004a, *ApJ*, **610**, 1021
Evans, C. J., Lennon, D. J., Walborn, N. R., *et al.* 2004b, *PASP*, **116**, 909
Ewen, H. I. & Purcell, E. M. 1951, *Nature*, **168**, 356
Falk, S. W. & Arnett, W. D. 1977, *ApJS*, **33**, 515
Fall, S. M. & Rees, M. J. 1977, *MNRAS*, **181**, 37P

Fan, X., Hennawi, J. F., Richards, G. T., *et al.* 2004, *AJ*, **128**, 515

Fan, X., Carilli, C. L., & Veating, B. 2006, *ARA&A*, **44**, 415

Ferrarese, L. & Merritt, D. 2000, *ApJ*, **539**, L9

Figer, D. F. 2004, in *The Formation and Evolution of Massive Young Star Clusters*, eds. H. J. G. L. M. Lamers, L. J. Smith, & A. Nota (San Francisco: ASP), 49

Figer, D. F. 2005, *Nature*, **434**, 192

Figer, D. F., Najarro, F., Morris, M., *et al.* 1998, *ApJ*, **506**, 384

Figer, D. F., Kim, S. S., Morris, M., *et al.* 1999, *ApJ*, **525**, 750

Figer, D. F., MacKenty, J. W., Robberto, M., *et al.* 2006, *ApJ*, **643**, 1166

Firmani, C., Ghisellini, G., Avila-Reese, V., & Ghirlanda, G. 2006, *MNRAS*, **370**, 185

FitzGerald, M. P. 1970, *A&A*, **4**, 234

Fitzpatrick, E. L. 1985, *ApJS*, **59**, 77

Forrest, W. J., Shure, M. A., Pipher, J. L., & Woodward, C. E. 1987, in *The Galactic Center*, ed. D. C. Backer (New York: American Institute of Physics) 155, 153

Frail, D. A., Kulkarni, S. R., Sari, R., *et al.* 2001, *ApJ*, **562**, L55

Frebel, A., Aoki, W., Christlieb, N., *et al.* 2005, *Nature*, **434**, 871

Friend, D. B. & Abbott, D. C. 1986, *ApJ*, **311**, 701

Fruchter, A. S., Levan, A. J., Strolger, L., *et al.* 2006, *Nature*, **441**, 463

Fryer, C. L., Young, P. A., & Hungerford, A. L. 2006 *ApJ*, **650**, 1028

Fullerton, A. W., Massa, D. L., & Prinja, R. K. 2006, *ApJ*, **637**, 1025

Galama, T. J., Vreeswijk, P. M., van Paradijs, J., *et al.* 1998, *Nature*, **395**, 670

Gamov, G. 1943, *ApJ*, **98**, 500

Garnett, D. R. 2002, *ApJ*, **581**, 1019

Gayley, K. G., Owocki, S. P., & Cranmer, S. R. 1997, *ApJ*, **475**, 786

Gebhardt, K., Bender, R., Bower, G., *et al.* 2000, *ApJ*, **539**, L13

Gehrels, N., Sarazin, C. L., O'Brien, P. T., *et al.* 2005, *Nature*, **437**, 851

Gehrels, N., Norris, J. P., Barthelmy, S. D., *et al.* 2006, *Nature*, **444**, 1044

Genzel, R. & Cesarsky, C. J. 2000, *ARA&A*, **38**, 761

Genzel, R., Lutz, D., Sturm, E., *et al.* 1998, *ApJ*, **498**, 579

Ghirlanda, G., Ghiselini, G., & Lazzati, D. 2004, *ApJ*, **616**, 331

Ghosh, S. K., Iyengar, K. V. K., Rengarajan, T. N., *et al.* 1989, *ApJ*, **347**, 338

Giavalisco, M. 2002, *ARA&A*, **40**, 579

Gies, D. R. 1987, *ApJS*, **64**, 545

Gies, D. R. 2003, in *IAU Symp. 212, A Massive Star Odyssey: From Main Sequence to Supernova*, eds. K. A. van der Hucht, A. Herrero, & C. Esteban (San Francisco: Astronomical Society of the Pacific), 91

Gnedin, N. Y. & Ostriker, J. P. 1997, *ApJ*, **486**, 581

González Delgado, R. M., Heckman, T., Leitherer, C., *et al.* 1998, *ApJ*, **505**, 174

González Delgado, R. M., Leitherer, C., Stasińska, G., & Heckman, T. M. 2002, *ApJ*, **580**, 824

Goodwin, S. P. & Bastian, N., 2006, *MNRAS*, **373**, 752

Gräfener, G. & Hamann, W.-R. 2005, *A&A*, **432**, 633

Gray, D. F. 2005, *The Observation and Analysis of Stellar Photospheres*, 3rd Edn. (Cambridge: Cambridge University Press)

Grebel, E. K. 1999, in *IAU Symp. 192, The Stellar Content of Local Group Galaxies*, eds. P. Whitelock & R. Cannon (San Francisco: Astronomical Society of the Pacific), 17

Grebel, E. K. & Chu, Y. H. 2000, *AJ*, **119**, 787

Grosdidier, Y., Moffat, A. F. J., Joncas, G., & Acker, A. 1998, *ApJ*, **506**, L127

Gunn, J. E. & Peterson, N. A. 1965, *ApJ*, **142**, 1633

Gustaffson, B., Bell, R. A., Eriksson, K., & Nordlund, A. 1975, *A&A*, **42**, 407

Hadfield, L. J. & Crowther, P. A. 2007, *MNRAS*, **381**, 418

Hadfield, L. J., van Dyk, S. D., Morris, P. M., *et al.* 2007, *MNRAS*, **376**, 248

Hamann, W.-R., 1981, *A&A*, **93**, 353

Hamann, W.-R., Koesterke, L., & Wessolowski, U. 1993, *A&A*, **274**, 397

Hammer, F., Flores, H., Schaerer, D., *et al.* 2006, *A&A*, **454**, 103

Hanbury Brown, R., Davis, J., & Allen, L. R. 1974, *ApJ*, **167**, 121

Hanson, M. M., Howarth, I. D., & Conti, P. S. 1997, *ApJ*, **489**, 698

Hanson, M. M., Kudritzki, R.-P., Kenworthy, M. A., *et al.* 2005a, *ApJS*, **151**, 154

Hanson, M. M., Puls J., & Repolust, T. 2005b, in *IAU Symp. 227, Massive Star Birth: A Crossroads of Astrophysics*, eds. R. Cesaroni, M. Felli, E. Churchwell, & M. Walmsley (Cambridge: Cambridge University Press), 376

Harries, T. J., Hillier, D. J., & Howarth, I. D. 1998, *MNRAS*, **296**, 1072

Harris, G. J., Polyansky, O. L., & Tennyson J. 2002, *ApJ*, **578**, 657

Harris, W. E. 1991, *ARA&A*, **29**, 543

Heckman, T. M. 2003, in *Galaxy Evolution: Theory & Observations*, eds. V. Avila-Reese, C. Firmani, C. S. Frenk, & C. Allen, Rev. Mex. Astron. Astrofís. Ser. Conf., 17, 47

Heckman, T. M. & Leitherer, C. 1997, *AJ*, **114**, 69

Heckman, T. M., Armus, L., & Miley, G. K. 1990, *ApJS*, **74**, 833

Heckman, T. M., Robert, C., Leitherer, C., *et al.* 1998, *ApJ*, **503**, 646

Heckman, T. M., Sembach, K. R., Meurer, G. R., *et al.* 2001, *ApJ*, **554**, 1021

Heger, A., Langer, N., & Woosley, S. E. 2000, *ApJ*, **528**, 368

Heger, A., Fryer, C. L., Woosley, S. E., *et al.* 2003, *ApJ*, **591**, 288

Heger, A., Woosley, S. E., & Spruit, H. C. 2005, *ApJ*, **626**, 350

Henley, D. B., Stevens, I. R., & Pittard, J. M. 2003, *MNRAS*, **346**, 773

Herrero, A., Kudritzki, R.-P., Vílchez, J. M., *et al.* 1992, *A&A*, **261**, 209

Herrero, A., Puls, J., & Najarro, F. 2002, *A&A*, **396**, 949

van den Heuvel, E. P. J., & Heise, J. 1972, *Nature Phys. Sci.*, **239**, 67

Hilditch, R. W., Harries, T. J., & Bell, S. A. 1996, *A&A*, **314**, 165

Hillier, D. J. 1989, *ApJ*, **347**, 392

Hillier, D. J. 1991, *A&A*, **247**, 455

Hillier, D. J., Davidson, K., Ishibashi, K., & Gull, T. 2001, *ApJ*, **553**, 837

Hirata, K., Kajita, T., Koshiba, M., *et al.* 1987, *Phys. Rev. Lett.*, **58**, 1490

Hjorth, J., Sollerman, J., Møller, P., *et al.* 2003, *Nature*, **423**, 847

Hjorth, J., Watson, D., Fynbo, J. P. U., *et al.* 2005, *Nature*, **437**, 859

Ho, L. C. & Filippenko A. V. 1996, *ApJ*, **472**, 600

Hollenbach, D. & McKee, C. F. 1989, *ApJ*, **342**, 306

Howarth, I. D. 1983, *MNRAS*, **203**, 301

Howarth, I. D. 2004, in *IAU Symp. 215, Stellar Rotation*, eds. A. Maeder & P. Eenens (San Francisco: Astronomical Society of the Pacific), 33

Howarth, I. D. & Prinja, R. K. 1989, *ApJS*, **69**, 527

Howarth, I. D., Siebert, K. W., Hussain, G. A. J., & Prinja, R. K. 1997, *MNRAS*, **284**, 265

van der Hucht, K. A. 2001, *New Astron. Rev.*, **45**, 135

Hulse, R. A. & Taylor, J. H. 1975, *ApJ*, **195**, L51

Humphreys, R. M. 1978, *ApJS*, **38**, 309

Humphreys, R. M. 1979, *ApJ*, **231**, 384

Humphreys, R. M. & Davidson, K. 1979, *ApJ*, **232**, 409

Humphreys, R. M. & Davidson, K. 1994, *PASP*, **106**, 1025

Humphreys, R. M. & McElroy, D. B. 1984, *ApJ*, **284**, 565

Humphreys, R. M. & Sandage, A. 1980, *ApJS*, **44**, 319

Hunter, D. A., Shaya, E. J., Holtzman, J. A., *et al.* 1995, *ApJ*, **448**, 179

Hunter, D. A., Elmegreen, B. G., Dupuy, T. J., & Mortonson, M. 2003, *AJ*, **126**, 1836

Iben, I. Jr. 1967, *ARA&A*, **5**, 571

Iglesias, C. A. & Rogers, F. J. 1991, *ApJ*, **371**, 408

Indebetouw, R., Mathis, J. S., Babler, B. L., *et al.* 2005, *ApJ*, **619**, 931

Indebetouw, R., Whitney, B. A., Johnson, K. E., & Wood, K. 2006, *ApJ*, **636**, 362

Israelian, G., Rebolo, R., Basri, G., *et al.* 1999, *Nature*, **401**, 142

Ivanov, G. R., Freedman, W. L., & Madore, B. F. 1993, *ApJS*, **89**, 85

Izotov, Y. I. & Thuan, T. X. 2004, *ApJ*, **616**, 768

Izotov, Y. I., Thuan, T. X., & Guzeva, N. G. 2005, *ApJ*, **632**, 210

de Jager, C., Nieuwenhuijzen, H., & van der Hucht, K. A. 1988, *A&AS*, **72**, 259

Johnson, J. R. & Bromm, V. 2006, *MNRAS*, **366**, 247

Jungman, G., Kamionkowski, M., & Griest, K. 1996, *Phys. Rept.*, **267**, 195

Jura, M. & Kleinmann, S. G. 1990, *ApJS*, **73**, 769

Kaper, L., van Loon, J. Th., Augusteijn, T., *et al.* 1997, *ApJ*, **475**, L37

Kaper, L., van der Meer, A., & Tijani, A. H. 2004, in *IAU Coll. 191, The Environment and Evolution of Double and Multiple Stars*, eds. C. Allen & C. Scarfe, Rev. Mex. Astron. Astrofís. Ser. Conf., 21, 128

Käppeler, F., Wiescher, M., Giesen, U., *et al.* 1994, *ApJ*, **437**, 396

Kato, S. 1966, *PASJ*, **18**, 374

Kennicutt, R. C. Jr. 1984, *ApJ*, **287**, 116

Kennicutt, R. C. Jr. 1991, in *Massive Stars in Starbursts*, eds. C. Leitherer, N. R. Walborn, T. M. Heckman, & C. A. Norman (Cambridge: Cambridge University Press), 157

Kennicutt, R. C. Jr. 1998a, *ARA&A*, **36**, 189

Kennicutt, R. C. Jr. 1998b, *ApJ*, **498**, 541

van Kerkwijk, M. H., Geballe, T. R., King, D. L., *et al.* 1996, *A&A*, **314**, 521

Kerr, R. P., 1963, *Phys. Rev. Lett.*, **11**, 237

Kewley, L. & Kobulnicky, H. A. 2005, in *Starbursts: From 30 Doradus to Lyman Break Galaxies*, eds. R. de Grijs & R. M. González Delgado (Dordrecht: Springer), 307

Kim, S. S., Figer, D. F., Kudritzki, R. P., & Najarro, F. 2006, *ApJ*, **653**, L113

Kingsburgh, R. L., Barlow, M. J., & Storey, P. J. 1995, *A&A*, **295**, 75

Kinney, A. L., Bohlin, R. C., Calzetti, D., *et al.* 1993, *ApJS*, **86**, 5

Kinney, A. L., Calzetti, D., Bohlin, R. C., *et al.* 1996, *ApJ*, **467**, 38

Kippenhahn, R. & Weigert, A. 1967, *Zeitschrift für Astrophysik*, **65**, 251

Kiriakidis, M., Fricke, K. J., & Glatzel, W. 1993, *MNRAS*, **264**, 50

Koesterke, L. & Hamann, W.-R. 1995, *A&A*, **299**, 503

Koornneef, J. 1982, *A&A*, **107**, 247

Kouveliotou, C., Meegan, C. A., Fishman, *et al.* 1993, *ApJ*, **413**, L101

Kraus, S., Balega, Y. Y., Berger, J.-P., *et al.* 2007, *A&A*, **466**, 649

Kriss, G. A., Shull, J. M., Oegerle, W., *et al.* 2001, *Science*, **293**, 1112

Kroupa, P. 2001, *MNRAS*, **322**, 231

Kroupa, P., Aarseth, S., & Hurley, J. 2001, *MNRAS*, **321**, 699

Krumholz, M. R., McKee, C. F., & Klein, R. I. 2005, *ApJ*, **618**, L33

Krumholz, M. R., Klein, R. I., & McKee, C. F., 2007, *ApJ*, **665**, 478

Kudritzki, R.-P. & Puls, J. 2000, *ARA&A*, **38**, 613

Kudritzki, R.-P., Pauldrach, A., & Puls, J. 1987, *A&A*, **173**, 293

Kudritzki, R.-P., Puls, J., Lennon, D. J., *et al.* 1999, *A&A*, **350**, 970

Kudritzki, R.-P., Bresolin, F., & Przybilla., N. 2003, *ApJ*, **582**, L83

Kunz, R., Fey, M., Jaeger, M., *et al.* 2002, *ApJ*, **567**, 643

Kurt, C. M., Dufour, R. J., Garnett, D. R., *et al.* 1999, *ApJ*, **518**, 246

Kurtz, S. 2005, in *IAU Symp. 227, Massive Star Birth: A Crossroads of Astrophysics*, eds. R. Cesaroni, M. Felli, E. Churchwell, & M. Walmsley (Cambridge: Cambridge University Press), 111

Kurtz, S., Churchwell, E., & Wood, D. O. S. 1994, *ApJS*, **91**, 659

Kurucz, R. L. 1991, in *Precision Photometry: Astrophysics of the Galaxy*, eds. A. G. D. Philips, A. R. Upgen, & K. A. Janes (Schenectady: David Press), 27

Lamers, H. J. G. L. M. & Cassinelli, J. P. 1999, *Introduction to Stellar Winds* (Cambridge: Cambridge University Press)

Lamers, H. J. G. L. M. & Gieles, M. 2006, *A&A*, **455**, 117

Lamers, H. J. G. L. M. & Leitherer, C. 1993, *ApJ*, **412**, 771

Lamers, H. J. G. L. M., Cerruti-Sola, M., & Perinotto, M. 1987 *ApJ*, **314**, 726

Lamers, H. J. G. L. M., Snow, T. P., & Lindholm, D. M. 1995, *ApJ*, **455**, 269

Lamers, H. J. G. L. M., Zickgraf, F.-J., de Winter, D., *et al.* 1998, *A&A*, **340**, 117

Lamers, H. J. G. L. M., Nota, A., Panagia, N., *et al.* 2001, *ApJ*, **551**, 764

Langer, N. 1989, *A&A*, **220**, 135

Langer, N. 1998, *A&A*, **329**, 551

Langer, N. & Maeder, A. 1995, *A&A*, **295**, 685

Langer, N., Hamann, W.-R., Lennon, M., *et al.* 1994, *A&A*, **290**, 819

Lanz, T. & Hubeny, I. 2003, *ApJS*, **146**, 417

Larsen, S. S., Brodie, J. P., & Hunter, D. A. 2004, *AJ*, **128**, 2295

Larson, R. B. 2003, in *Galactic Star Formation Across the Stellar Mass Spectrum*, eds. J. M. De Buizer & N. S. van der Bliek (San Francisco: Astronomical Society of the Pacific), 65

Larson, R. B. & Tinsley, B. M. 1978, *ApJ*, **219**, 46

Lattimer, J. M. & Prakash, M. 2001, *ApJ*, **550**, 426

Leibundgut, B. & Suntzeff, N. B. 2003, in *Supernovae & Gamma-Ray Bursters*, ed. K. Weiler (New York: Springer), 77

Leitherer, C. 1998, in *Stellar Astrophysics for the Local Group: VIII Canary Islands Winter School of Astrophysics*, eds. A. Aparicio, A. Herrero, & F. Sánchez (Cambridge: Cambridge University Press), 527

Leitherer, C. 2000, in *Building Galaxies; from the Primordial Universe to the Present*, eds. F. Hammer, T. X. Thuan, V. Cayatte, B. Guiderdoni, & J. T. Thanh Van (Paris: Editions Frontières), 71

Leitherer, C. & Heckman, T. M. 1995, *ApJS*, **96**, 9

Leitherer, C., Chapman, J. M., & Koribalski, B. 1995, *ApJ*, **450**, 289

Leitherer, C., Vacca, W. D., Conti, P. S., *et al.* 1996, *ApJ*, **465**, 717

Leitherer, C., Schaerer, D., Goldader, J. D., *et al.* 1999, *ApJS*, **123**, 3

Leitherer, C., Leão, J. R. S., Heckman, T. M., *et al.* 2001, *ApJ*, **550**, 724

Lejeune, T. & Schaerer, D. 2001, *A&A*, **366**, 538

Lennon, D. J. 1997, *A&A*, **317**, 871

Lesh, J. R. 1968, *ApJS*, **17**, 371

Levesque, E. M., Massey, P., Olsen, K. A. G., *et al.* 2005, *ApJ*, **628**, 973

Levesque, E. M., Massey, P., Olsen, K. A. G., *et al.* 2006, *ApJ*, **645**, 1102

Livio, M. & Soker, N. 1988, *ApJ*, **329**, 764

van Loon, J. Th., Cioni, M.-R. L., Zijlstra, A. A., & Loup, C. 2005, *A&A*, **438**, 273

Lucy, L. B. & Abbott, D. C. 1993, *ApJ*, **405**, 738

Lucy, L. B. & Solomon, P. M. 1970, *ApJ*, **159**, 879

Lynds, C. R. & Sandage, A. R. 1963, *ApJ*, **137**, 1005

MacFadyen, A. I. & Woosley, S. E. 1999, *ApJ*, **524**, 262

Mackey, J., Bromm, V., & Hernquist, L. 2003, *ApJ*, **586**, 1

Madau, P. 1995, *ApJ*, **441**, 18

Madau, P. & Shull, J. M. 1996, *ApJ*, **457**, 551

Madau, P., Ferguson, H. C., Dickinson, M. E., *et al.* 1996, *MNRAS*, **283**, 1388

Maeder, A. 1987a, *A&A*, **173**, 247

Maeder, A. 1987b, *A&A*, **178**, 159

Maeder, A. & Meynet, G. 1987, *A&A*, **182**, 243

Maeder, A. & Meynet, G. 1994, *A&A*, **287**, 830

Maeder, A. & Meynet, G. 2000, *ARA&A*, **38**, 143

Maeder, A. & Meynet, G. 2005, *A&A*, **440**, 1041

Maíz-Apellániz, J., Walborn, N. R., Galue, H. A., & Wei, L. H. 2004, *ApJS*, **151**, 103

Maoz, D., Koratkar, A., Shields, J. C., *et al.* 1998, *AJ*, **116**, 55

Markarian, B. E. 1967, *Astrofizika*, **3**, 24

Martin, C. L. & Kennicutt, R. C., Jr. 2001, *ApJ*, **555**, 301

Martins, F. & Plez, B. 2006, *A&A*, **457**, 637

Martins, F., Schaerer D., & Hillier D. J. 2005, *A&A*, **436**, 1049

Massa, D., Fullerton, A. W., Nichols, J. S., *et al.* 1995, *ApJ*, **452**, L53

Massey, P. 1984, *ApJ*, **281**, 789

Massey, P. 2002, *ApJS*, **141**, 81

Massey, P. 2003, *ARA&A*, **41**, 15

Massey, P. & Conti, P. S., 1983, *ApJ*, **273**, 576

Massey, P. & Hunter, D. A. 1998, *ApJ*, **493**, 180

Massey, P. & Johnson, O. 1998, *ApJ*, **505**, 793

Massey, P. & Meyer, M. 2001, *Ency. Astr. Astrophys.* (Bristol: Institute of Physics)

Massey, P., Parker, J. Wm., & Garmany, C. D. 1989, *AJ*, **98**, 1305

Massey, P., Johnson, K. E., & Degioia-Eastwood, K. 1995a, *ApJ*, **454**, 151

Massey, P., Lang C. C., Degioia-Eastwood, K., & Garmany, C. D. 1995b, *ApJ*, **438**, 188

Massey, P., Penny, L. R., & Vukovich, J. 2002, *ApJ*, **565**, 982

Massey, P., Olson, K. A. G., Hodge, P. W., *et al.* 2006, *AJ*, **131**, 2478

Mazzali, P. A., Deng, J., Nomoto, K., *et al.* 2006, *Nature*, **442**, 1018

McGaugh, S. S. 1991, *ApJ*, **380**, 140

McGregor, P. J., Hyland, A. R., & Hillier, D. J. 1988, *ApJ*, **334**, 639, 1071

McKee, C. F. & Tan, J. C. 2003, *ApJ*, **585**, 850

Mehlert, D., Noll, S., Appenzeller, I., *et al.* 2002, *A&A*, **393**, 809

de Mello, D. F., Schaeres, D., Heldmann, J., & Leitherer, C. 1998, *ApJ*, **507**, 199

Melnick, J., Tenorio-Tagle, G., & Terlevich, R. 1999, *MNRAS*, **302**, 677

Mermilliod, J.-C. & García, B. 2001, in *IAU Symp. 200, The Formation of Binary Stars*, eds. H. Zinnecker & R. D. Mathieu (San Francisco: Astronomical Society of the Pacific), 191

Mészáros, P. 2002, *ARA&A*, **40**, 137

Meurer, G. R. 2000, in *Massive Stellar Clusters*, eds. A. Lançon, & C. Boily (San Francisco: Astronomical Society of the Pacific), 81

Meurer, G. R., Freeman, K. C., Dopita, M. A., & Cacciari, C. 1992, *AJ*, **103**, 60

Meurer, G. R., Heckman, T. M., Leitherer, C., *et al.* 1995, *AJ*, **110**, 2665

Meynet, G. 2008, *European Physical Journal*, Sp. Topic, **156**, 257

Meynet, G. & Maeder, A. 1997, *A&A*, **321**, 465

Meynet, G. & Maeder, A. 2003, *A&A*, **404**, 975

Meynet, G. & Maeder, A. 2005, *A&A*, **429**, 581

Meynet, G., Maeder, A., Schaller, G., *et al.* 1994, *A&AS*, **103**, 97

Mezger, P. G. & Henderson, A. P. 1967, *ApJ*, **167**, 471

Mihalas, D. 1978, *Stellar Atmospheres*, 2nd Edn. (San Francisco: Freeman)

Mihalas, D., Hummer, D. G., & Conti, P. S. 1972, *ApJ*, **175**, L99

Millour, F., Petrov, R. G., Chesneau, O., *et al.* 2007, *A&A*, **464**, 107

Milne, E. A. 1926, *MNRAS*, **86**, 459

Modjaz, M., Kewley, L., Kirschner, R. P., *et al.* 2008, *AJ*, **135**, 1136

Moffat, A. F. J. & Shara, M. M., 1983, *ApJ*, **273**, 554

Moffat, A. F. J., Drissen, L., Lamontagne, R., & Robert, C. 1988, *ApJ*, **334**, 1038

Mokiem, M. R., de Koter, A., Evans, C. J., *et al.* 2006, *A&A*, **456**, 1131

Mokiem, M. R., de Koter, A., Vink, J. S., *et al.* 2007, *A&A*, **473**, 603

Moore, B. D., Hester, J. J., & Scowen, P. A. 2000, *AJ*, **119**, 2991

Moore, B. D., Walter, D. K., Hester, J. J., 2002, *AJ*, **124**, 3313

Morgan, W. W., Keenan, P. C., & Kellman, E. 1943, *An Atlas of Stellar Spectra* (Chicago: University of Chicago Press)

Morisset, C., Schaerer, D., Martín-Hernández, N. L., *et al.* 2002, *A&A*, **386**, 558

Morris, P. W., van der Hucht, K. A., Crowther, P. A., *et al.* 2000, *A&A*, **353**, 624

Morton, D. C. 1967, *ApJ*, **147**, 1017

Muench, A. A., Lada, E. A., & Lada, C. J. 2000, *ApJ*, **533**, 358

Muno, M. P., Clark, J. S., Crowther, P. A., *et al.* 2006, *ApJ*, **636**, L41

Murakami, H., Baba, H., & Barthel, P. 2007, *PASJ*, **59**, 369

Nelan, E. P., Walborn, N. R., Wallace, D. J., *et al.* 2004, *AJ*, **128**, 323

Nomoto, K., Sugimoto, D., Sparks, W. M., *et al.* 1982, *Nature*, **299**, 803

Nomoto, K., Maeda, K., Nakamura, T., *et al.* 2000, in *Gamma-ray Bursts*, eds. R. M. Kippen, R. S. Mallozzi, & C. J. Fishman (New York: American Institute of Physics), 622

Nota, A. & Clampin, M. 1997, in *Luminous Blue Variables: Massive Stars in Transition*, eds. A. Nota & H. J. G. L. M. Lamers (San Francisco: Astronomical Society of the Pacific), 303

Nota, A., Livio, M., Clampin, M., & Schulte-Ladbeck, R. 1995, *ApJ*, **448**, 788

Nugis, T., Crowther, P. A., & Willis, A. J. 1998, *A&A*, **333**, 956

Nugis, T. & Lamers, H. J. G. L. M. 2000, *A&A*, **360**, 227

O'Connell, R. W., Gallagher, J. S., & Hunter, D. A. 1994, *ApJ*, **433**, 65

Okamoto, Y. K., Kataza, H., Yamashita, T., *et al.* 2003, *ApJ*, **584**, 368

Oliva, E., Origlia, L., Kotilainen, J. K., & Moorwood, A. F. M. 1995, *A&A*, **301**, 55

Oort, J. H. 1958, *Ricerche Astron.*, **5**, 507

Oppenheimer, J. R. & Volkoff, G. 1939, *Phys Rev.*, **55**, 374

Origlia, L., Leitherer, C., Aloisi, A., *et al.* 2001, *AJ*, **122**, 815

Osterbrock, D. E. & Ferland, G. J. 2006, *Astrophysics of Gaseous Nebulae and Active Galactic Nuclei*, 2nd Edn. (Sausalito: University Science Books)

Owocki, S. R. 2001, in *Encyclopedia of Astronomy and Astrophysics*, ed. P. Murdin (London: Taylor & Francis), eaa.crcpress.com

Owocki, S. R. 2003, in *IAU Symp. 212, A Massive Star Odyssey: From Main Sequence to Supernova*, eds. K. A. van der Hucht, A. Herrero, & C. Esteban (San Francisco: Astronomical Society of the Pacific), 281

Owocki, S. R. 2004, in *Evolution of Massive Stars, Mass Loss and Winds*, eds. M. Heydari-Malayeri, P. Stee, & J-P. Zahn (Les Ulis: Edition Diffusion Press), 163

Owocki, S. R., Castor, J. I., & Rybicki, G. B. 1988, *ApJ*, **355**, 914

Owocki, S. R., Cranmer, S. R., & Gayley, K. G. 1996, *ApJ*, **472**, 115

Owocki, S. R., Gayley, K. G., & Shaviv, N. J. 2004, *ApJ*, **616**, 525

Paciesas, W. S., Meegan, C. A., Pendleton, G. N., *et al.* 1999, *ApJS*, **122**, 465

Paczynski, B. 1967, *Acta Astron*, **17**, 355

Paczynski, B. 1971, *ARA&A*, **9**, 183

Paczynski, B. 1976, in *IAU Symp 73, Structure & Evolution of Close Binary Systems*, eds. P. Eggleton, S. Mitton & J. Whelan (Dordrecht: Reidel), 75

Pagel, B. E. J. & Edmunds, M. G. 1981, *ARA&A*, **19**, 77

Palmer, D. M., Barthelmy, S., Gehrels, N., *et al.* 2005, *Natures*, **434**, 1107

Panagia, N., 1973, *AJ*, **78**, 929

Panagia, N. & Felli, M. 1975, *A&A*, **39**, 1

van Paradijs, J., Groot, P. J., Galama, T., *et al.* 1997, *Nature*, **386**, 868

Parker, J. Wm. 1993, *AJ*, **106**, 560

Pauldrach, A. W. A. & Puls, J. 1990, *A&A*, **237**, 409

Pauldrach, A. W. A, Puls, J., & Kudritzki, R.-P. 1986, *A&A*, **184**, 86

Pauldrach, A. W. A., Hoffmann, T. L., & Lennon, M. 2001, *A&A*, **375**, 161

Perlmutter, S., Aldering, G., Goldhaber, G., *et al.* 1999, *ApJ*, **517**, 565

Perrin, G., Ridgway, S. T., Coude de Foresto, V., *et al.* 2004, *A&A*, **418**, 675

Peterson, B. M. 1997, *An Introduction to Active Galactic Nuclei* (Cambridge: Cambridge University Press)

Petrenz, P. & Puls, J. 2000, *A&A*, **358**, 956

Pettini, M. 2004, in *Cosmochemistry, XIII Canary Islands Winter School*, eds. C. Esteban, R. J. García López, A. Herrero, & F. Sánchez (Cambridge: Cambridge University Press), 257

Pettini, M. 2006, in *Stellar Evolution at Low Metallicity: Mass Loss, Explosions, Cosmology*, eds. H. J. G. L. M. Lamers, N. Langer, T. Nugis & K. Annuk (San Francisco: ASP), 363

Pettini, M. & Pagel, B. E. J. 2004, *MNRAS*, **348**, L59

Pettini, M., Steidel, C. C., Adelberger, K. L., *et al.* 2000, *ApJ*, **528**, 96

Pettini, M., Shapley, A. E., Steidel, C. C., *et al.* 2001, *ApJ*, **554**, 981

Pettini, M., Rix, S. A., Steidel, C. C., *et al.* 2002, *ApJ*, **569**, 742

Phillips, M. M., 1993, *ApJ*, **413**, 105

Pian, E., Mazzali, P. A., Masetti, N., *et al.* 2006, *Nature*, **442**, 1011

Podsiadlowski, P., Joss, P. C., & Hsu, J. J. L. 1992, *ApJ*, **391**, 246

Podsiadlowski, P., Langer, N., Poelarends, A. J. T., *et al.* 2004, *ApJ*, **612**, 1044

Pogson, N. 1856, *MNRAS*, **17**, 12

Portegies-Zwart, S. F., Makino, J., McMillan, S. L. W., & Hut, P. 2001, *ApJ*, **546**, L101

Portegies-Zwart, S. F., Baumgardt, H., Hut, P., *et al.* 2004, *Nature*, **428**, 724

Poveda, A., Ruiz, J., & Allen, C. 1967, *Bol. Obs. Tonantzintla & Tacubaya*, **4**, 86

Prestwich, A. H., Kilgard, R., Crowther, P. A., *et al.* 2007, *ApJ*, **669**, L21

Prinja, R. K., Barlow, M. J., & Howarth, I. D. 1990, *ApJ*, **361**, 607

Puls, J., Kudritzki, R.-P., Herrero, A., *et al.* 1996, *A&A*, **305**, 171

Puls, J., Springmann, U., & Lennon, M., 2000, *A&AS*, **141**, 23

Puls, J., Markova, N., Scuderi, S., *et al.* 2006, *A&A*, **454**, 625

Puxley, P. J., Doyon, R., & Ward, M. J. 1997, *ApJ*, **476**, 120

Rauw, G. 2004, in *Evolution of Massive Stars, Mass Loss and Winds*, eds. M. Heydari-Malayeri, Ph. Stee, & J.-P. Zahn (Les Ulis: Edition Diffusion Press Sciences), series 13, 293

Rauw, G., De Becker, M., Naze, Y., *et al.* 2004, *A&A*, **420**, L9

Rieke, G. H., Lebofsky, M. J., Thompson, R. I., *et al.* 1980, *ApJ*, **238**, 24

Repolust, T., Puls, J., & Herrero A. 2004, *A&A*, **415**, 349

Rhoads, J. E. & Malhotra, S. 2001, *ApJ*, **563**, L5

Rieke, G. H. & Lebofsky, M. J. 1985, *ApJ*, **288**, 618

Rix, S. A., Pettini, M., Leitherer, C., *et al.* 2004, *ApJ*, **615**, 98

Rousseau, J., Martin, N., Prévot, L., *et al.* 1978, *A&AS*, **31**, 243

Salpeter, E. E. 1955, *ApJ*, **121**, 161

Salzer, J. J., Gronwall, C., Lipovetsky, V. A., *et al.* 2000, *AJ*, **120**, 80

Salzer, J. J., Jangren, A., Gronwall, C., *et al.* 2005a, *AJ*, **130**, 2584

Salzer, J. J., Lee, J. C., Melbourne, J., *et al.* 2005b, *ApJ*, **624**, 661

Sanders, D. B. 1997, in *Starburst Activity in Galaxies*, eds. J. Franco, R. Terlevich, & A. Serrano, Rev. Mex. Astron. Astrofís. Ser. Conf., **6**, 42

Sanders, D. B. & Mirabel, I. F. 1996, *ARA&A*, **34**, 749

Sargent, W. L. W. & Searle, L. 1970, *ApJ*, **162**, L155

Scalo, J. M. 1998, in *The Initial Mass Function*, eds. G. Gilmore & D. Howell (San Francisco: Astronomical Society of the Pacific), 201

Schaerer, D. 1999, in *IAU Symp. 193, Wolf–Rayet Phenomena in Massive Stars and Starburst Galaxies*, eds. K. A. van der Hucht, G. Koenigsberger, & P. R. J. Eenens (San Francisco: Astronomical Society of the Pacific), 539

Schaerer, D. 2002, *A&A*, **382**, 28

Schaerer, D. & Maeder, A. 1992, *A&A*, **263**, 129

Schaerer, D. & Schmutz, W. 1994, *A&A*, **288**, 231

Schaerer, D. & Stasińska, G. 1999, *A&A*, **345**, L17

Schaerer, D. & Vacca, W. D. 1998, *ApJ*, **497**, 618

Schaerer, D., Meynet, G., Maeder, A., & Schaller, G. 1993, *A&AS*, **98**, 523

Schaller, G., Schaerer, D., Meynet, G., & Maeder, A. 1992, *A&AS*, **96**, 269

Scharmer, G. B. 1981, *ApJ*, **249**, 720

Schechter, P. 1976, *ApJ*, **203**, 297

Schmidt, M. 1959, *ApJ*, **129**, 243

Schmutz, W. 1997, *A&A*, **321**, 268

Schmutz, W., Hamann, W.-R., & Wessolowski, U. 1989, *A&A*, **210**, 236

Schmutz, W., Leitherer, C., & Gruenwald, R. 1992, *PASP*, **104**, 1164

Searle, L. & Sargent, W. L. W. 1972, *ApJ*, **173**, 25

Searle, L., Sargent, W. L. W., & Bagnuolo, W. G. 1973, *ApJ*, **179**, 427

Seaton, M. J. 1979, *MNRAS*, **187**, P73

Seaton, M. J., Yan, Y., Mihalas, D., & Pradhan, A. K. 1994, *MNRAS*, **266**, 805

Seitz, S., Saglia, R. P., Bender, R., *et al.* 1998, *MNRAS*, **298**, 945

Seyfert, C. K. 1943, *ApJ*, **97**, 28

Shapley, A. E., Steidel, C. C., Pettini, M., & Adelberger, K. L. 2003, *ApJ*, **588**, 65

Shapley, A. E., Erb, D. K., Pettini, M., *et al.* 2004, *ApJ*, **612**, 108

Shapley, A. E., Steidel, C. C., Pettini, M., *et al.* 2006, *ApJ*, **651**, 688

Shields, J. C. 1992, *ApJ*, **399**, L27

Shu, F. H., Adams, F. C., & Lizano, S. 1987, *ARA&A*, **25**, 23

Shull, J. M. 1993, in *Massive Stars: Their Lives in the Interstellar Medium*, eds. J. P. Cassinelli & E. B. Churchwell (San Francisco: Astronomical Society of the Pacific), 327

Shull, J. M. & Saken, J. M. 1995, *ApJ*, **444**, 663

Slettebak, A., Collins, G. W., Parkinson, T. D., *et al.* 1975, *ApJS*, **29**, 137

Smartt, S. J., Maund, J. R., Hendry, M. A., *et al.* 2004, *Science*, **303**, 499

Smith, L. F. 1968a, *MNRAS*, **138**, 109

Smith, L. F. 1968b, *MNRAS*, **140**, 409

Smith, L. F. 1973, in *IAU Symp. 49, Wolf–Rayet and High-Temperature Stars.*, eds. M. K. V. Bappu & J. Sahade (Dordrecht: Reidel), 15

Smith, L. F. & Hummer, D. G., 1988, *MNRAS*, **230**, 511

Smith, L. F., Shara, M. M., & Moffat, A. F. J. 1996, *MNRAS*, **281**, 163

Smith, L. J. & Gallagher, J. S. 2001, *MNRAS*, **326**, 1027

Smith, L. J., Crowther, P. A., & Prinja, R. K. 1994, *A&A*, **281**, 833

Smith, L. J., Norris, R. P. F., & Crowther, P. A. 2002, *MNRAS*, **337**, 1309

Smith, N. 2008, in *Massive Stars: From Pop III and GRBs to the Milky Way*, ed. M. Livio (Cambridge: Cambridge University Press), in press (astro-ph/0607457)

Smith, N. & Conti P. S., 2008, *ApJ* **679**, 1467

Smith, N. & Owocki, S. P. 2006, *ApJ*, **146**, L45

Smith, N., Davidson, K., Gull, T., & Ishibashi, K. 2003, *ApJ*, **586**, 432

Smith, N., Li, W., Foley, R. J., *et al.* 2007, *ApJ*, **666**, 1116

Sobolev, V. V. 1960, *Moving Envelopes of Stars* (Cambridge: Cambridge University Press)

Soderberg, A. M., Berger, E., Page, K. L., *et al.* 2008, *Nature*, **453**, 469

Sokasian, A., Yoshida, N., Abel, T., Hernquist, L., & Springel, V. 2004, *MNRAS*, **350**, 47

Somerville, R. S., Primack, J. R., & Faber, S. M. 2001, *MNRAS*, **320**, 504

Sonneborn, G., Altner, B., & Kirshner, R. P. 1987, *ApJ*, **323**, L35

Spergel, D. N., Verde, L., Peiris, H. V., *et al.* 2003, *ApJS*, **148**, 175

Spergel, D. N., Bean, R., Doré, O., *et al.* 2007 *ApJS*, **170**, 377

Spitzer, L. J. 1958, *ApJ*, **127**, 17

Springmann, U. 1994, *A&A*, **289**, 505

Spruit, H. C. 2002, *A&A*, **381**, 923

St-Louis, N., Moffat, A. F. J., Lapointe, L., *et al.* 1993, *ApJ*, **410**, 342

St-Louis, N., Chené, A.-N., de la Chevrotiere, A., & Moffat A. F. J. 2008, in *Mass Loss from Stars and the Evolution of Stellar Clusters*, eds. A. de Koter, L. J. Smith, & R. Waters (San Francisco: ASP) Conf. Ser., in press

Stahl, O., Jankovics, I., Kovács, J., *et al.* 2001, *A&A*, **375**, 54

Stahler, S. W. & Palla, F. 2004, *The Formation of Stars* (Weinheim: Wiley-VCH)

Stasińska, G. & Leitherer, C. 1996, *ApJS*, **107**, 661

Steidel C. C. & Hamilton, D. 1993, *AJ*, **105**, 2017

Steidel, C. C., Giavalisco, M., Pettini, M., *et al.* 1996, *ApJ*, **462**, L17

Steidel, C. C., Pettini, M., & Adelberger, K. L. 2001, *ApJ*, **546**, 665

Stevens, I. R., Blondin, J. M., & Pollock, A. M. T. 1992, *ApJ*, **386**, 265

Stolte, A., Brandner, W., Brandl, B., & Zinnecker, H. 2006, *AJ*, **132**, 253

Stothers, R. B. & Chin, C.-W. 1993, *ApJ*, **408**, L85

Sturm, E., Lutz, D., Tran, D., *et al.* 2000, *A&A*, **358**, 481

Sugerman B. E. K., Ercolano, B., Barlow, M. J., *et al.* 2006, *Science*, **313**, 196

Swings, J. P. 1976, in *IAU Symp. 70, Be and Shell Stars*, ed. A. Sletteback (Dordrecht: Reidel), 219

Taniguchi, Y., Shioya, Y., & Murayama, T. 2000, *AJ*, **120**, 1265

Tayler, R. J. 1973, *MNRAS*, **161**, 365

Telles, E., Melnick, J., & Terlevich, R. 1997, *MNRAS*, **288**, 78

Terlevich, R. 1994, in *Violent Star Formation – from 30 Doradus to QSOS*, ed. G. Tenivio-Tagle (Cambridge: Cambridge University Press), 329

Terlevich, R. 1997, in *Starburst Activity in Galaxies*, eds. J. Franco, R. Terlevich, & A. Serrano, Rev. Mex. Astron. Astrofís. Ser. Conf., **6**, 1

Terlevich, R. & Melnick, J. 1985, *MNRAS*, **213**, 841

Testi, L. & Sargent, A. I. 1998, *ApJ*, **508**, L91

Thielemann, F.-K., Nomoto, K., & Hashimoto, M. 1996, *ApJ*, **460**, 408

Thompson, D., Djorgovski, S., & Trauger, J. 1995, *AJ*, **110**, 963

Thorne, K. S. & Żytkov, A. N. 1977, *ApJ*, **212**, 832

Tielens, A. G. G. M. 2005, *The Physics and Chemistry of the Interstellar Medium* (Cambridge: Cambridge University Press)

Tinsley, B. M. 1968, *ApJ*, **151**, 547

Todini, P. & Ferrara, A. 2001, *MNRAS*, **325**, 726

Townsend, R. H. D., Owocki, S. P., & Howarth, I. D. 2004, *MNRAS*, **350**, 189

Townsley, L. K., Broos, P. S., Feigelson, E. D., *et al.* 2006a, *AJ*, **131**, 2140

Townsley, L. K., Broos, P. S., Feigelson, E. D., *et al.* 2006b, *AJ*, **131**, 2164

Tremonti, C. A., Calzetti, D., Leitherer, C., & Heckman, T. M. 2001, *ApJ*, **555**, 322

Trundle, C. & Lennon, D. J. 2005, *A&A*, **434**, 677

Tumlinson, J. 2006, *ApJ*, **641**, 1

Tumlinson, J. & Shull, J. M. 2000, *ApJ*, **528**, L65

Turatto, M. 2003, in *Supernovae & Gamma-Ray Bursters*, ed. K. Weiler (New York: Springer), 21

Tuthill, P. G., Monnier, J. D., & Danchi, W. C. 1999, *Nature*, **398**, 487

Ulrich, M.-H. 1978, *ApJ*, **219**, 424

Underhill, A. B. 1949, *MNRAS*, **109**, 562

Underhill, A. B., Divan, L., Prévot-Burnichon, M.-L., & Doazan V., 1979, *MNRAS*, **189**, 601

Vacca, W. D., Robert, C., Leitherer, C., & Conti, P. S. 1995, *ApJ*, **444**, 647

Vacca, W. D., Garmany, C. D., & Shull, J. M. 1996, *ApJ*, **460**, 914

Vanbeveren, D., De Loore, C., & Van Rensbergen, W. 1998, *A&A Rev.*, **9**, 63

Vázquez, G. A., Leitherer, C., Heckman, T. M., *et al.* 2004, *ApJ*, **600**, 162

Veilleux, S. 2001, in *Starburst Galaxies: Near and Far*, eds. L. Tacconi & D. Lutz (Heidelberg: Springer), 88

Venn, K. A. 1995a, *ApJS*, **99**, 659

Venn, K. A. 1995b, *ApJ*, **449**, 839

Venn, K. A. 1999, *ApJ*, **518**, 405

Vidal, C. R., Cooper, J., & Smith, E. W. 1973, *ApJS*, **25**, 37

Vila-Costas, M. B. & Edmunds, M. G. 1992, *MNRAS*, **259**, 121

Vink, J. S. & de Koter, A. 2005, *A&A*, **442**, 587

Vink, J., de Koter, A., & Lamers, H. J. G. L. M. 2000, *A&A*, **362**, 295

Vink, J., de Koter, A., & Lamers, H. J. G. L. M. 2001, *A&A*, **369**, 574

Walborn, N. R. 1971a, *ApJS*, **23**, 257

Walborn, N. R. 1971b, *ApJ*, **167**, L31

Walborn, N. R. 1973, *ApJ*, **182**, L21

Walborn, N. R. 1976, *ApJ*, **205**, 419

Walborn, N. R. 1982, *ApJ*, **254**, L15

Walborn, N. R. 1991, in *Massive Stars in Starbursts*, eds.
    C. Leitherer, N. R. Walborn, T. M. Heckman, & C. A. Norman (Cambridge: Cambridge University Press), 145

Walborn, N. R. & Fitzpatrick, E. L. 1990, *PASP*, **102**, 379

Walborn, N. R. & Fitzpatrick, E. L. 2000, *PASP*, **112**, 50

Walborn, N. R., Nichols-Bohlin, J., & Panek, R. J. 1985, *NASA Ref. Publ.*, 1155

Walborn, N. R., Barbá, R. H., Brandner, W., *et al.* 1999, *AJ*, **117**, 225

Walborn, N. R., Howarth, I. D., Lennon, D. J., *et al.* 2002, *AJ*, **123**, 2754

Walter, F. M. & Lattimer, J. M., 2002 *ApJ*, **576**, L145

Walter, F. M., Wolk, S. J., & Neuhauser, R. 1996, *Nature*, **379**, 233

Wampler, E. J., Wang, L., Baade, D., *et al.* 1990, *ApJ*, **362**, L13

Wang, Q. D. 1999, *ApJ*, **510**, L139

Weaver, R., McCray, R., Castor, J., *et al.* 1977, *ApJ*, **218**, 377

Weedman, D. 1991, in *Massive Stars in Starbursts*, eds.
    C. Leitherer, N. R. Walborn, T. M. Heckman, & C. A. Norman (Cambridge: Cambridge University Press), 317

Weedman, D. W., Feldman, F. R., Balzano, V. A., *et al.* 1981, *ApJ*, **248**, 105

Weidner, C. & Kroupa, P. 2006, *MNRAS*, **365**, 1333

Welch, W. J., Dreher, J. W., Jackson, J. M., Terebey, S., & Vogel, S. N. 1987, *Science*, **238**, 1550

Wellstein, S. & Langer, N. 1999, *A&A*, **350**, 148

White, S. D. M. & Rees, M. J. 1978, *MNRAS*, **183**, 341

Whitmore, B. C. 2003, in *A Decade of Hubble Space Telescope Science*, eds. M. Livio, K. Noll, & M. Stiavelli
    (Cambridge: Cambridge University Press), 153

Whitmore, B. C., Chander, R., & Fall, S. M., 2007, *AJ*, **133**, 1067

Whitmore, B. C., Zhang, Q., Leitherer, C., *et al.* 1999, *AJ*, **118**, 1551

Williams, P. M., van der Hucht, K. A., & Thé, P. S. 1987, *A&A*, **182**, 91

Williams, R. E., Blacker, B., Dickinson, M., *et al.* 1996, *AJ*, **112**, 1335

Williams, P. M., Dougherty, S. M., Davis, R. J., *et al.* 1997, *MNRAS*, **289**, 10

Willis, A. J., Schild, H., & Stevens, I. R. 1995, *A&A*, **298**, 549

de Wit, W. J., Testi, L., Palla, F., & Zinnecker, H. 2005, *A&A*, **437**, 247

Wolf, C. J. E. & Rayet, G. 1867, *Comptes Rendus*, **65**, 292

Wolfire, M. G. & Cassinelli, J. P. 1987, *ApJ*, **319**, 850

Woltjer, L. 1964, *ApJ*, **140**, 1309

Woltjer, L. 1972, *ARA&A*, **10**, 129

Wood, D. O. S. & Churchwell, E. D. 1989, *ApJS*, **69**, 831

Woods, P. M. & Thompson, C. 2006, in *Compact
    Stellar X-ray Sources*, eds. W. Lewin & M. van der Klis (Cambridge: Cambridge University Press), 547

Woosley, S. E. 1993, *ApJ*, **405**, 273

Woosley, S. E. & Bloom, J. S. 2006, *ARA&A*, **44**, 507

Woosley, S. E. & Janka, H.-T. 2005, *Nat. Phys.*, **1**, 147

Woosley, S. E. & Weaver, T. A. 1995, *ApJS*, **101**, 181

Woosley, S. E., Langer, N., & Weaver, T. A. 1993, *ApJ*, **411**, 823

Woosley, S. E., Wilson, J. R., Mathews, G. J., *et al.* 1994, *ApJ*, **433**, 229

Wright, A. E. & Barlow, M. J. 1975, *MNRAS*, **170**, 41

Yee, H. K. C., Ellingson, E., Bechtold, J., *et al.* 1996, *AJ*, **111**, 1783

Yoon, S.-C. & Langer, N. 2005, *A&A*, **443**, 643

Yorke, H. W. & Krügel, E. 1977, *A&A*, **54**, 183

Yorke, H. W. & Sonnhalter, C. 2002, *ApJ*, **569**, 846

Zahn, J. P. 1992, *A&A*, **265**, 115

von Zeipel, H. 1924, *MNRAS*, **84**, 665

Zhang, W., Woosley, S. E., & MacFadyen, A. I. 2003, *ApJ*, **586**, 356

Zickgraf, F.-J., Wolf, B., Stahl, O., *et al.* 1985, *A&A*, **143**, 421

Zinnecker H. & Yorke, H. W., 2007, *ARA&A*, **45**, 481

Zwicky, F. & Zwicky, M. A. 1971, *Catalogue of Selected Compact Galaxies and of Post-Eruptive
    Galaxies* (Guemligen: Zwicky)

Acronyms

AGB, Asymptotic Giant Branch, 30
AGN, active galactic nuclei, 14
ATCA, Australia Telescope Compact Array, 80
AXP, anomalous X-ray pulsar, 125

BATSE, Burst and Transient Source Experiment, 287
BC, bolometric correction, 18
BSG, blue supergiant, 24

CAK, Castor, Abbott, & Klein, 71
CCD, charge-coupled device, 17
CGRO, Compton Gamma Ray Observatory, 287
CIR, co-rotating interaction regions, 90
CMB, Cosmic Microwave Background, 15
COBE, Cosmic Background Explorer, 267

DAC, discrete absorption component, 90
DLA, damped Lyman alpha (systems), 282
DM, distance modulus, 18

EUV, extreme ultraviolet, 159

FUV, far ultraviolet, 161
FUSE, Far Ultraviolet Spectroscopic Explorer, 40
FWHM, full-width at half maximum, 226

GALEX, Galaxy Evolution Explorer, 222
GMC, giant molecular cloud, 10
GOODS, Great Observatories Origins Deep Survey, 283
GRB, gamma ray burst, 8

HCHII, hypercompact HII region, 154
HD, Henry Draper (catalog), 35
H-D, Humphreys–Davidson (limit), 4
HMXB, high mass X-ray binary, 140
H-R, Hertzsprung–Russell (diagram), 1
HST, Hubble Space Telescope, 1
HUT, Hopkins Ultraviolet Telescope, 40

IGM, intergalactic medium, 15
IMBH, intermediate mass black hole, 177
IMF, initial mass function, 11

IRAS, Infrared Astronomical Satellite, 161
ISM, interstellar medium, 26
ISO, Infrared Space Observatory, 158
IUE, International Ultraviolet Explorer, 39

JWST, James Webb Space Telescope, 272

LBG, Lyman break galaxies, 16
LBV, luminous blue variable, 6
LMXB, low mass X-ray binary, 140
LTE, local thermodynamic equilibrium, 5

M, Messier (catalog), 184
MJD, modified Julian date, 27
MK, Morgan Keenan (classification), 35
MS, Main Sequence, 1
MSX, Midcourse Space Experiment, 161
MYSO, massive young stellar object, 154

NGC, New General Catalog, 21

PAH, polycyclic aromatic hydrocarbons, 159
PDMF, present day mass function, 232
PPK, Pauldrach, Puls, & Kudritzki, 71

QSO, quasi-stellar object, 229

RLOF, Roche lobe overflow, 131
RSG, red supergiant, 4

S-AGB, Super-Asymptotic Giant Branch, 107
SDSS, Sloan Digital Sky Survey, 267
SED, spectral energy distribution, 13
SEI, Sobolov with exact integration, 89
SFE, star formation efficiency, 178
SFR, star formation rate, 215
SGR, Soft Gamma Ray Repeater, 125
SN, supernova, 8
SPH, smooth particle hydrodynamics, 168
SSC, super-star cluster, 176

Symbols

α, CAK force multiplier, 71
α, IMF power law exponent, 11
α, overshooting parameter, 104
$\alpha_\lambda^{\mathrm{eff}}$, effective recombination coefficient, 186
α_B, case B recombination coefficient, 186
α_{CE}, efficiency factor for common envelope ejection, 136

β, velocity law exponent, 71
β, ultraviolet extinction attenution slope, 181
β, cluster power law exponent, 166

γ, IMF power law exponent, 164
Γ, IMF power law exponent, 164
Γ_e, Eddington parameter, 68

δ, CAK force multiplier, 72

ε, emission measure, 156
ε_λ, emissivity law, 152

ζ, fraction of ionizing photons absorbed by gas, 160
$\zeta(M)$, cluster mass function, 166

η, parameter relating cluster radius and velocity dispersion to dynamical mass, 177
$\eta_0(t)$, ratio of equivalent to actual O star population at time t, 186

κ, opacity, 69

λ, wavelength, 12

μ, mean molecular weight, 165

ν, frequency, 18
ν_t, turnover frequency, 156
ν_{SN}, core-collapse supernova rate, 218

ξ, macroturbulence, 25
ξ, efficiency of pulsational instabilities, 110
$\xi(M)$, logarithmic mass distribution, 164

ρ_0, Holmberg radius, 22
$\rho(r)$, density at radius r, 67

ρ_e, equatorial density, 96
ρ_p, polar density, 96

σ, Stefan–Boltzmann constant, 2
σ, velocity dispersion of star cluster/galaxy, 177/200
σ_e, Thompson (electron) scattering cross section, 68

Σ_{SFR}, star formation rate surface density, 241
Σ_{Gas}, gas surface density, 241

τ, optical depth, 69
τ_ν, optical depth to thermal Bremsstrahlung, 156
τ_e, optical depth to electron scattering, 269
τ_{MS}, main-sequence lifetime, 102
τ_{SF}, duration of star formation event, 231
τ_{Ross}, Rosseland optical depth, 55

Υ, ratio of radiative acceleration from lines to electron scattering, 72

$\phi(M)$, linear mass distribution, 11, 163

χ, ionization energy, 160

$\psi(t)$, stellar mass formed per unit time, 163

ω, rotational frequency, 146

Ω, ratio of rotation to critical velocity, 95
Ω_Λ, vacuum energy density, 273
Ω_m, matter density, 268
Ω_b, baryon density, 268

a, semi-major axis of orbit, 130
A_λ, extinction at wavelength λ, 181
A_{V}, visual extinction, 18

b, Doppler line-broadening parameter, 277
$(B-V)_0$, intrinsic color, 20

c, velocity of light, 67
c_0, isothermal sound speed, 251
$c(\mathrm{H}\beta)$, extinction coefficient, 183

d, distance, 18
D_{mom}, reduced wind momentum, 73

308

Object index

Subject index

abundances, 5, 62–64, 283–289
active galactic nuclei, 257–268
atmospheres
  LTE, 49, 50
  non-LTE, 50–59

B[e] supergiants, 28, 29
Be stars, 28, 38, 47, 98, 141
binaries
  high mass X-ray, 140–144
  interacting, 134, 135, 137–140
  statistics, 130, 131
blue supergiants, 21, 24, 25, 35–38, 40
bolometric correction, 18, 19, 61
bubbles (wind blown), 188–190, 192

convection, 105, 106

distances, 19–21
dust, extinction, 181–185
dust formation
  supernovae, 122, 123
  W-R stars, 150–154

Eddington limit, 58, 69, 88, 96

galaxies
  HII, 221–229
  Lyman-break, 15, 16, 275–283
  starburst, 231–233, 239, 243–251, 275, 279, 281
  ultraluminous infrared, 268
gamma ray bursts, 16, 128, 146, 147, 289–297
gravity
  flux weighted, 62
  surface, 61, 62, 69

HII regions, 157, 185–187, 201, 210
  giant, 199–210
  ultracompact, 156–164
H-R diagram, 1, 2

initial mass function, 11, 164–166, 233, 234, 236–239, 241–243

luminosity
  stellar, 18, 53, 54, 57–59, 74
luminous blue variables, 6, 25–28

magnetar, 143
magnetic fields, 47, 48, 116, 127
magnitudes, absolute, 17, 18, 20, 74, 119
mass
  stellar, 62, 131–133, 164, 166
mass-loss rates, 58, 74, 81, 84–86
  clumping, 93–96
  IR/radio, 81, 83
  optical, 82–84
  UV, 89, 90, 93

Population III stars, 15, 242, 269, 270, 272–275
pulsar, 125, 127, 141–143, 146
pulsation, 26, 90, 92, 98, 106

radiation pressure, 2, 68
red supergiants, 29, 30, 58, 59, 61, 90
runaways, 140

star clusters, 11, 166–168, 171–180, 209, 235
superbubbles, 252
supernovae, 117–121, 192, 220, 281, 282, 290, 291, 294, 297
  electron capture, 108
  pair-production, 124, 125, 128, 274
  progenitors, 123, 124
superwinds, 252–255, 257

temperature, effective, 2, 18, 54, 56, 57, 61, 69, 86, 87

velocity
  escape, 72, 76
  rotation, 45, 96–99, 112, 294
  terminal wind, 38, 72, 74–77, 81

314